The Water Environment of Cities

Lawrence A. Baker
Editor

The Water Environment of Cities

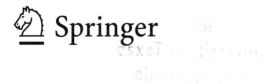

Editor
Lawrence A. Baker
University of Minnesota and WaterThink, LLC
Water Resources Center
1985 Buford Ave.
Cpy Paul MN 55108
173 MacNeal Hall
St. Paul, MN
USA
baker127@umn.edu

ISBN 978-0-387-84890-7   e-ISBN 978-0-387-84891-4
DOI 10.1007/978-0-387-84891-4

Library of Congress Control Number: 2008932847

© Springer Science+Business Media, LLC 2009
All rights reserved. This work may not be translated or copied in whole or in part without the written permission of the publisher (Springer Science+Business Media, LLC, 233 Spring Street, New York, NY 10013, USA), except for brief excerpts in connection with reviews or scholarly analysis. Use in connection with any form of information storage and retrieval, electronic adaptation, computer software, or by similar or dissimilar methodology now known or hereafter developed is forbidden.
The use in this publication of trade names, trademarks, service marks, and similar terms, even if they are not identified as such, is not to be taken as an expression of opinion as to whether or not they are subject to proprietary rights.

Printed on acid-free paper

springer.com

Preface

The concept for the *Water Environment of Cities* arose from a workshop "Green Cities, Blue Waters" workshop held in 2006.[1] The workshop assembled experts from engineering, planning, economics, law, hydrology, aquatic ecology, geomorphology, and other disciplines to present research findings and identify key new ideas on the urban water environment. At a lunch discussion near the end of the workshop, several of us came to the recognition that despite having considerable expertise in a narrow discipline, none of us had a vision of the "urban water environment" as a whole. We were, as in the parable, blind men at opposite ends of the elephant, knowing a great deal about the parts, but not understanding the whole. We quickly recognized the need to develop a book that would integrate this knowledge to create this vision. The goal was to develop a book that could be used to teach a complete, multidisciplinary course, "The Urban Water Environment", but could also be used as a supplemental text for courses on urban ecosystems, urban design, landscape architecture, water policy, water quality management and watershed management. The book is also valuable as a reference source for water professionals stepping outside their arena of disciplinary expertise.

The Water Environment of Cities is the first book to use a holistic, interdisciplinary approach to examine the urban water environment. We have attempted to portray a holistic vision built around the concept of water as a core element of cities. Water has multiple roles: municipal water supply, aquatic habitat, landscape aesthetics, and recreation. Increasingly, urban water is reused, serving multiple purposes. In this vision, humans are not merely inhabitants of cities, but an integral part of the urban water environment. Humans alter the urban hydrologic cycle and the chemical and physical integrity of urban water systems and are recipients of these alterations. Some of those changes are beneficial, like being able to enjoy a well-planned park with water features whereas others are harmful, like exacerbated flooding caused by poorly planned development upstream. These changes alter the sustainability and resilience of cities in ways that can reasonably be predicted, or at least, anticipated.

---

[1] Novotny, V. and P. Brown, 2007. Cities of the Future: Towards Integrated Sustainable Water and Landscape Management. Proceedings of an international workshop held July 12–14, 2006 at the Wingspread Conference Center, Racine, WI. IWA Publishing, London.

To reach a multidisciplinary audience, we have written the book for a scientifically literate audience – a reader with a B.S. degree but who would not necessarily have specialized education in hydrology, engineering, law, or other topics. We used several techniques to achieve this goal. First, we explored the same six cross-cutting themes in each chapter – water scarcity, multiple uses of water, water management institutions, integration of new knowledge, sustainability, and resilience. Key paradigms from our specialties, which both guide and limit us, are explained to build context for each chapter. Third, we tried to limit specialized jargon to the extent possible. When specialized terms are needed to achieve precision of meaning, they are defined and included in a glossary. Chapters were cross-reviewed by chapter authors from other disciplines to assure that chapters are readily understood by readers from other disciplines. Finally, last chapter is a synthesis, developed in a workshop held in January 2006 at the Riverwood Inn in Otsego, Minnesota, after authors had written their core chapters.

Minnesota, USA   Lawrence A. Baker

Acknowledgments

We would like to thank Vladimir Novotny for organizing the Green Cities, Blue Waters Workshop, a project that has catalyzed thinking about urban water and has led to several ongoing, interrelated projects. I would like to thank several people for making the synthesis workshop a success. First, I thank two discussants who aptly guided us in our search for synthesis: Lance Neckar, from the University of Minnesota's Department of Landscape Architecture, and Joan Nassauer, a landscape architect in the School of Natural Resources and the Environment at the University of Michigan. I would also like to thank Jana Caywood, a graduate student at the University of Minnesota, who did an extraordinary job organizing the logistics of the workshop, as well as contributing perceptions of a sociologist to the synthesis discussion. Finally, I would like to acknowledge support from the National Science Foundation for supporting the synthesis workshop and related activities (award CBET 0739952 to the University of Minnesota).

Contents

1 **Introduction** 1
   Lawrence A. Baker

2 **The Urban Water Budget** 17
   Claire Welty

3 **Groundwater in the Urban Environment** 29
   Peter Shanahan

4 **Urban Infrastructure and Use of Mass Balance Models
   for Water and Salt** 49
   Paul Westerhoff and John Crittenden

5 **New Concepts for Managing Urban Pollution** 69
   Lawrence A. Baker

6 **Streams and Urbanization** 93
   Derek B. Booth and Brian P. Bledsoe

7 **Urban Water Recreation: Experiences, Place Meanings,
   and Future Issues** 125
   Ingrid E. Schneider

8 **Urban Design and Urban Water Ecosystems** 141
   Kristina Hill

9 **Legal Framework for the Urban Water Environment** 171
   Robert W. Adler

10 **Institutions Affecting the Urban Water Environment** 195
   Robert W. Adler

11 **Institutional Structures for Water Management in the Eastern United States** 217
    Cliff Aichinger

12 **Adaptive Water Quantity Management: Designing for Sustainability and Resiliency in Water Scarce Regions** 235
    Jim Holway

13 **Demand Management, Privatization, Water Markets, and Efficient Water Allocation in Our Cities** 259
    K. William Easter

14 **Principles for Managing the Urban Water Environment in the 21st Century** 275
    Lawrence A. Baker, Peter Shanahan, and Jim Holway

Glossary 291

Index 301

Contributors

Robert W. Adler   University of Utah, S.J. Quinney College of Law, Salt Lake City, Utah, adlerr@law.utah.edu

Cliff Aichinger   Ramsey-Washington Metro Watershed District, St. Paul, Minnesota, MN, USA, cliff@rwmwd.org

Lawrence A. Baker   Minnesota Water Resources Center, University of Minnesota, and WaterThink, LLC, St. Paul, Minnesota, USA, baker127@unm.edu

Brian P. Bledsoe   Colorado State University, Fort Collins, Colorado, USA, bbledsoe@goku.engr.colostate.edu

Derek B. Booth   Quaternary Research Center, University of Washington, Seattle WA 98195; Stillwater Sciences Inc., 2855 Telegraph Avenue, Berkeley, CA 94705, dbooth@stillwatersci.com

John Crittenden   Arizona State University, Tempe, AZ, USA, jcritt@asu.edu

K. William Easter   University of Minnesota, Minneapolis and Saint Paul, MN, USA, kweaster@umn.edu

Kristina Hill   University of Virginia, Charlottesville, VA, USA, keh3u@virginia.edu

Jim Holway   Arizona State University, Tempe, AZ, USA, jim.holway@asu.edu

Peter Shanahan   Department of Civil and Environmental Engineering at Massachusetts Institute of Technology, Cambridge, MA, USA, peteshan@mit.edu

Ingrid E. Schneider   University of Minnesota, Minneapolis and Saint Paul, MN, USA, ingridss@umn.edu

Claire Welty   University of Maryland, Baltimore County, MD, USA, weltyc@umbc.edu

Paul Westerhoff   Arizona State University, Tempe, AZ, USA, p.westerhoff@asu.edu

Author Biographies

Robert W. Adler, J.D., is Associate Dean for Academic Affairs and James I. Farr Chair and Professor at the University of Utah, S. J. Quinney College of Law. His writings include "The Clean Water Act: Twenty Years Later" (Island Press 1993, with Landman and Cameron), "Restoring Colorado River Ecosystems: A Troubled Sense of Immensity" (Island Press, forthcoming 2007), "Environmental Law: A Conceptual and Functional Approach" (Aspen Publishers forthcoming 2007, with Driesen), and numerous book chapters and scholarly articles about water pollution, water law, and other aspects of environmental law and policy. He was a Senior Attorney and Clean Water Program Director at the Natural Resources Defense Council, Executive Director of Trustees for Alaska, and an Assistant Attorney General with the Pennsylvania Department of Environmental Quality. He has a J.D. *cum laude* from the Georgetown University Law Center and a B.A. in ecology from Johns Hopkins University.

Cliff Aichinger is currently the Administrator for the Ramsey-Washington Metro Watershed District, a watershed management organization in St. Paul, Minnesota. Under his leadership, the District has completed three Watershed Management Plans, lake studies, TMDL studies, development rules, and more than $30 million in capital improvements. He has had over 35 years experience in water management and environmental planning in Minnesota and has held positions at local, regional and state levels of government. Cliff has also been involved in a number of state and national committees and studies evaluating water management science and institutional structure. He was a member of the National Research Council team that authored "New Strategies for America's Watersheds". He has a B.S. from the University of Minnesota.

Lawrence A. Baker, Ph.D. (Editor), is a Senior Fellow in the Minnesota Water Resources Center and owner of WaterThink, LLC. His research examines human ecosystems, at scales from households to urban regions, with the goal of developing novel approaches for reducing pollution that are more effective, cheaper and fairer than conventional approaches. Recent research has included development of phosphorus balances for watersheds of recreational lakes, examination of drivers change in water quality, source reduction for storm water pollution, the role of human choice on generation of pollutants through households, and nutrient flows through cities and farms. He has published more than 100 technical publications and one book, "Environmental Chemistry of Lakes and Reservoirs". He was a primary technical contributor to the *Integrated Assessment*, the final report of the National Acidic Precipitation Assessment Program to the U.S. Congress, and chaired the Human Health Committee of the Arizona Comparative Environmental Risk Project, which produced a "blueprint" for future environmental initiatives in Arizona. He writes articles and columns for the popular press, including the *Minneapolis Star and Tribune* and the *Minnesota Journal*.

Brian P. Bledsoe, Ph.D., P.E., has 20 years of experience as an engineer and environmental scientist in the private and public sectors. He is currently an Associate Professor in the Department of Civil and Environmental Engineering at Colorado State University. His research and teaching interests are focused on the interface between engineering and ecology with emphasis on multi-scale linkages between land use, hydrologic processes, sedimentation, channel stability, and water quality. Prior to entering academia, he served as Non-point Source Program Coordinator for the State of North Carolina. He has authored over fifty publications related to stream and watershed processes, restoration and water quality, and is a licensed professional engineer in NC and CO.

Derek B. Booth, Ph.D., P.E., P.G., is President and Senior Geologist at Stillwater Sciences, Inc.; he is also Adjunct Professor of Civil Engineering and Earth and Space Sciences at the University of Washington, where he is senior editor of the international journal Quaternary Research. Previously, he was director of the Center for Urban Water Resources Management (and its successor, the Center for Water and Watershed Studies) at the University. He maintains active research into the causes of stream-channel degradation, the effectiveness of storm water mitigation strategies, and the physical effects of urban development on aquatic systems, with several dozen publications and a wide range of national and international invited presentations on the topic.

John Crittenden, Ph.D., P.E., is the Richard Snell Professor of Civil and Environmental Engineering at Arizona State University. His areas of expertise include sustainability, pollution prevention, physical–chemical treatment processes in water, nanotechnology synthesis and environmental applications and implications, and catalysis. Dr. Crittenden has more than 120 publications in refereed journals and 135 reports, or contributions to proceedings, 4 copyrighted software products and three patents in the areas of pollution prevention, stripping, ion exchange, advanced oxidation/catalysis, adsorption and groundwater transport. In 2005, Dr. Crittenden coauthored the book, "Water Treatment Principles and Design" (1984 pages and 22 chapters), which has been recognized as the most authoritative text on theory and practice of water treatment. He has acted as a consultant to over 100 utilities, companies, and universities, worldwide on air and water treatment, sustainability science and engineering, and pollution prevention since 1975. For his contributions to water treatment theory and practice, Dr. Crittenden was elected to the National Academy of Engineering in 2002. Dr. Crittenden is an Associate Editor of "Environmental Science & Technology".

K. William Easter, Ph.D., has been a faculty in the Department of Applied Economics at the University of Minnesota since 1970 and was Director of the Center for International Food and Agricultural Policy from July 1999 to June 2003. His research interests include resource economics, economic development and environmental economics, with a special focus on water and land problems and resource pricing issues. He has worked extensively on international agricultural and water issues on projects in India and throughout Asia for the Ford Foundation, US-AID, the East-West Center, the World Bank, the International Water Management Institute, and other groups on projects in Thailand, Egypt, Pakistan, Nepal, Bangladesh, and Chile. Dr. Easter has published widely in professional journals and has co-authored, edited, or co-edited 12 books dealing with a range of natural resources and environmental economics issues, but with a focus on water resources. His most recent effort is an edited volume, "The Economics of Water Quality".

Kristina Hill is a Professor of Landscape Architecture and Urban Design at the University of Virginia. She received her doctoral training in design and ecology at Harvard University, and taught urban design at the Massachusetts Institute of Technology and the University of Washington in Seattle before joining the UVA faculty. Kristina has served as a professional consultant on urban design, infrastructure and water for states and municipalities in the U.S. and Germany, as well as for private sector development clients. She edited the book "Design and Ecology" (Johnson and Hill, Island Press, 2002), and is about to release a book on case studies of urban ecological design and water with the University of Washington Press. Her current focus is on the relationship between urban design and the health of near shore marine ecosystems.

Jim Holway, Ph.D., is the Associate Director of the Global Institute of Sustainability at Arizona State University where he directs the Sustainability Partnership (SP). He focuses on building partnerships between policymakers, practitioners, researchers and educators to connect science with practice. He is also a Professor of Practice in Civil and Environmental Engineering and the School of Sustainability, and the ASU Coordinator for the Arizona Water Institute. Prior to joining ASU in 2005, Dr. Holway served as Assistant Director of the Arizona Department of Water Resources. His responsibilities included overseeing the state's Active Management Area, conservation, assured water supply, recharge, well permitting, and groundwater and surface water rights programs. Holway's research interests focus on the connections between water, growth and land use and he holds a Ph.D and Masters in Regional Planning from the University of North Carolina.

Ingrid E. Schneider, Ph.D., is Professor in the Department of Forest Resources at the University of Minnesota as well as Director of the University of Minnesota's

Tourism Center. She teaches and does research on recreation resource visitors and nature-based tourists as well as the communities and organizations that host them. With a primary focus on visitor/community behavior and attitudes, Dr. Schneider works with various national, state, and local organizations to understand and plan for enhanced visitor management. Schneider has published numerous refereed journal articles, technical reports, and made presentations in a variety of academic and other professional settings in the U.S. and beyond. Schneider is also co-editor of the text on Diversity and the Recreation Profession. A South Dakota native, Ingrid received her B.S. in Technical Communication and M.S. in Forest Resources from the University of Minnesota and her doctorate in Parks, Recreation, and Tourism Management from Clemson University.

Peter Shanahan, Ph.D., is a Senior Lecturer in the Department of Civil and Environmental Engineering at MIT. At MIT, Dr. Shanahan concentrates on teaching engineering practice as well as applied research in water quality and hydrology. Prior to his academic career, he was an engineering practitioner for thirty years, specializing in hydrology, water quality, and hazardous waste site engineering with an emphasis on the application of mathematical modeling. He holds a Ph.D. in environmental engineering from MIT, an MS in environmental earth sciences Stanford University, and BS degrees in civil engineering and earth sciences from MIT. He is a registered professional engineer. Dr. Shanahan is the author of numerous publications on water quality and computer modeling of surface- and ground-water systems.

Claire Welty, Ph.D., is the Director of the Center for Urban Environmental Research and Education and Professor of Civil and Environmental Engineering at University of Maryland, Baltimore County. Dr. Welty's research focuses on watershed-scale urban hydrology, particularly on urban groundwater. At UMBC she serves as the PI on a NSF IGERT grant, "Water in the Urban Environment". Dr. Welty is Chair of National Research Council's Water Science and Technology Board and is the past Chair of the Consortium of Universities for the Advancement of Hydrologic Science Inc. She received her Ph.D degree in Civil and Environmental Engineering from MIT.

Paul Westerhoff, Ph.D., P.E., is a Professor and Chair of the Department of Civil and Environmental Engineering at Arizona State University (ASU, Tempe, AZ). His research is supported by water and wastewater research foundations (AWWRF and WERF), U.S. Environmental Protection Agency, NSF and local municipalities. Dr. Westerhoff's research focuses on water and wastewater treatment, specializing in organic matter and emerging pollutants. He is a member of the AWWRF Expert Panel on Endocrine Disruptors and AWWARFs Public Council. He graduated from University of Colorado, Boulder in 1995.

Chapter 1
Introduction

Lawrence A. Baker

1.1 The Water Environment of Cities

Few of us, even among professionals who think about specific aspects of water every day, have ever thought about the "water environment of cities". What does this mean, and why is it important? Some aspects may be familiar, whereas others are out of sight, and others are conceptual constructs. Parts of the urban water environment are obvious. Water features are often the heart and soul of many cities. Chicago's Lake Michigan shoreline, the Chain of Lakes park system and the Mississippi riverfront in Minneapolis (Fig. 1.1), and Boston's "Emerald Necklace" define these cities and make them uniquely livable. These water features renew the soul.

Water also contributes to the economic lifeblood of a city. Most coastal cities are located at the mouths of large rivers and have deep harbors. Early industrial cities were often located on rivers, which were used to provide both hydro-based energy and transportation. In an earlier era, most freight was moved by water. Even today, ports that transport three-fourths of our international trade (on a tonnage basis) dominate the shorelines of coastal cities. As we will see, the decline of inland waterway transportation has led to a major transformation of the urban waterfronts, now dominated by parks and housing. Water was also the dominant form of energy for early industrialization, spurring the growth of hundreds of small cities on high-gradient rivers prior to the advent of economies based on fossil fuels.

A less obvious part of the urban water environment is the municipal water supply system, which brings water from outside a city's boundaries, treats it, and distributes it throughout the city via a subterranean network of pipes. Much of this water becomes wastewater, which flushes human and industrial wastes out of the urban core via an extensive network of underground sewers. These sewers once emptied directly to rivers but now nearly always discharge wastes to sophisticated wastewater treatment plants, often capable of digesting 95% of the organic waste before discharging relatively pure water into rivers.

L.A. Baker (✉)
Minnesota Water Resources Center, University of Minnesota, and WaterThink, LLC, St. Paul, Minnesota, USA
e-mail: baker127@umn.edu

Fig. 1.1 An urban riparian environment: the Mississippi River as it flows through Minneapolis. Source: National Park Service - Mississippi National River and Recreation Area

Cities have entirely different hydrologic environments than the natural environments they replaced. Precipitation that once infiltrated into forest or agriculture soils becomes surface runoff when the impervious surfaces of cities (roads, driveways, parking lots, and rooftops) replace pervious, vegetated landscapes. Storm sewers in urban *watersheds* drain impervious surfaces, transporting water very quickly to urban streams, lakes, and rivers. The hydrology, morphology, and aquatic biota of streams are severely altered by urbanization, transforming them into clearly distinguishable "urban" streams when approximately 10%–20% of the watershed area becomes covered with impervious surface (Booth et al. 2002).

Even further below the surface than storm sewers (usually!), there may be large underground groundwater *aquifers*. These are often used for water supply, and sometimes for waste disposal, and sometimes for both. Overuse of aquifers not only depletes them, but can cause land to subside and fissure, damaging urban infrastructure. In coastal areas, overdraft of groundwater results in seawater intrusion.

Cities also include important aquatic ecosystems. Fifty years ago, rivers downstream from cities were often grossly polluted with untreated sewage, often creating "dead zones" of severely oxygen-depleted waters. Many of these have been restored to be *fishable and swimmable*, sometimes providing excellent angling within view of skyscrapers. Some types of wetland ecosystems are so valuable to urban dwellers that mere proximity to them increases residential property values (Boyer and Polaksy 2004). The ecosystems of urban lakes and reservoirs used to store municipal source water are particularly important, because eutrophication caused by encroaching urbanization can greatly impair the quality of drinking water, especially through the production of taste and odor compounds by blue-green algae.

Most importantly, the urban water environment includes humans! Water features in the urban landscape – whether natural streams and lakes or constructed fountains – restore body and mind and help create our "sense of place" (*place identity*) in the world. In studies of landscape preference, subjects nearly always indicate preference for landscapes with water features (Ulrich 1993). Desirable landscapes not only are preferred but may have actual restorative properties, including improvement of higher order cognition. One study suggests that a view of open water might speed up the recovery from open heart surgery (Ulrich 1993).

Finally, we cannot conceptualize the water environment of cities without consideration of the legal and institutional systems that shape our biophysical world and provide connections to larger political systems. In fact, as we will see, many urban regional institutions developed partly, and some exclusively, to manage various aspects of the urban water environment.

The water environment of cities includes all of these things, but more importantly, it is the whole of these things – a vast, interconnected system of human nature. Yet, even after 200 years of industrial urbanization, water management tends to focus on individual parts of the urban water environment, not the whole. For example, we know that new development which adds impervious surface increases flooding downstream, but flood policy mainly focuses on amelioration of flood effects in downstream communities. Stormwater pollution management focuses mainly on treating stormwater after it enters a storm sewer, rather than prevention of pollution in the first place. Recycling wastewater, a well-intended water conservation effort, can accelerate accumulation of salts in desert cities, with poorly understood consequences. Our policies to manage water and pollution are often fragmented, dealing with one part of the picture. Policies have major gaps, and are sometimes even antagonistic, working at cross-purpose with other policies.

The overarching goal of this book is to develop a holistic view of the urban water environment, in order to manage it more effectively. There are two key reasons for doing this now, one a problem and the other an opportunity. The problem is that we have major urban water problems that cannot be solved using conventional, compartmentalized thinking. These problems are becoming more severe as urban populations swell, high-quality source water becomes scarcer, demands for environmental quality increase, and climate change brings new uncertainties into play. New thinking is needed to yield solutions that are cheaper, more effective and fairer. Second, we are at a moment in history with unparalleled opportunity: our emerging information technologies. Our ability to acquire, store, and process data is accelerating exponentially, enabling entirely new ways of creating and using knowledge to improve management of water resources. These new ways of thinking and new technologies can bring the concept of "design with nature", envisioned nearly 40 years ago (McHarg 1971), to fruition.

The next section is a brief history of the water environment of the modern city, from which we can learn several key themes that will help us to look into the future. We then identify six cross-cutting themes which are developed throughout the book.

1.2 A Brief History of the Urban Water Environment

1.2.1 Advent of the Industrial City

The pessimistic reader might be suspicious of our goals from the outset. Aren't governments bound to be incompetent? Doesn't the Katrina disaster illustrate how poorly we have prepared for water-related disasters? Isn't Atlanta a case study in the failure of urban water supply management? Aren't Phoenix and Las Vegas poised for catastrophe, as water demand outstrips water supply? What new knowledge and new concepts have evolved that would allow us to manage our urban water environments in a fundamentally sounder fashion than we have in the past?

The history of the water environment of the modern, industrialized city reveals more complex, nuanced view of human progress. Although the cities with a million or more people existed more than a thousand years ago in warm climates (Baghdad was the first with over one million, in 800 AD; Chandler 1987), the industrialized city of the North Temperate Zone is a relatively modern institution in human history. London was the first of these, attaining the one million mark in 1810 (Fig. 1.2). Modern urbanization was made possible by industrialization, which in turn was driven by two parallel, symbiotic developments: coal mining and the development of the steam engine, which in turn enabled more intensive mining. By 1800, London desperately needed a new source of energy to augment dwindling forests and found it in the mines of Newcastle-on-Tyne, readily accessible by ocean transport (Freese 2003). During the 19th century, London's population grew nearly 8-fold, propelled by a 15-fold expansion in energy use, supplied mainly by coal. Growth in cities on the eastern coast of the United States lagged that of London, but then took off with a vengeance: Philadelphia grew 20-fold during this period and New York grew nearly 70-fold since the early 19th century (Fig. 1.2). From 1800 to 1975, the cumulative population of the world's ten largest cities increased 20-fold.

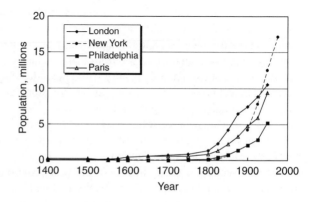

Fig. 1.2 Population growth of key cities during industrialization

1.2.2 Evolution of Modern Water and Sewage Works: The London Experience

Halliday (2001) reviewed the early development of London's modern water and sewage systems. In London, water was brought into the city since Roman times, augmented by many public wells. The average person hauled water from distribution points, whereas wealthier citizens employed the service of water carriers. London and other cities had developed extensive drainage systems by the early 19th century, often built to follow natural streambeds. But these were designed originally to convey only rainwater and urban drainage, not household sewage. At the beginning of the 19th century, the standard practice for urban sanitation was to dump excrement into latrines and cesspools – basically pits for storing human feces. London had about 200,000 cesspools in 1810. Human manure was considered an important agricultural resource, collected by "nightsoil men", who hauled human manure to farms, a practice that was continued in Beijing, China until 2000 (Browne 2000). The Romans had well-developed sanitary sewer systems and even limited household sewage disposal (only for the very wealthy, of course), but the idea of conveying sewage through underground pipes fell out of the public mind until the mid-19th century.

Widespread adoption of the water closet – the precursor to the modern flush toilet – changed that. As one might expect, water closets were very popular, and by 1850 there were about 250,000 water closets in London. Cesspools of the time were not designed to handle the higher flows; hence they overflowed, creating a stinking mess. One can only imagine a London gentleman, the proud owner of his new water closet, stepping off his porch into the stinking mire of an overflowing cesspool, first in shock (&%$#@!), and then having a hydrologic Zen moment. To compound the problem, the value of human manure declined for several reasons. First, as London grew, it became more expensive to haul manure to more distant agricultural fields. Simultaneously, England farmers began importing newly discovered South American guano for fertilizer, and they became more proficient in cultivation of legumes, which replenished nitrogen to soils (Smil 2001).

By the mid-1800s London's sewage situation had become dire. Four cholera epidemics occurred between 1800 and 1860. The prevailing wisdom, which maintained that cholera was spread by foul air, a theory known as the "miasmatic" theory, prevented Londoners from taking straightforward action until the mid-19th century, when Dr. John Snow and other epidemiologists gradually accumulated evidence that cholera was spread by contaminated water, not air. This cleared the way for Parliament to pass the "Cholera Bill" in 1845, which mandated that all new and existing buildings be connected to the existing storm sewers. Parliament also formed the Metropolitan Commission of Sewers in 1848, which compiled a series of reports and plans for dealing with the growing sewage problem and the Metropolitan Board of Works in 1955 – the precursor to modern regional urban sewage authorities.

As is often the case, the solution to one problem (filth in the streets) often creates another problem. In this case, the Thames River, which received most of London's new sewage, became so polluted from the new discharges of sewage that the stench forced Parliament, located on the banks of the Thames, to adjourn! The "Great Stink", as the 1858 event became known, motivated Parliament to provide the funding for an extensive renovation of its sewer system, which eventually led to the disappearance of cholera (Halliday 2001).

1.2.3 Urbanization and Water in the Eastern United States

The pattern of development of urban water supply systems in the United States reflected a strong sense of individualism, which delayed the construction of public waterworks in many cities until the 1870s (Ogle 1999). Many households relied on rain cisterns, local wells, and small water companies for water supply. Early public water systems were often "segmented", supplying water only to neighborhoods that could pay for them, meaning relatively the wealthy ones. Philadelphia constructed the first municipal water system in the United States in 1802, but even as late as 1880 there were only 598 public water systems in the United States (Tarr 1996). As public water systems and indoor plumbing became widespread, water use increased dramatically, exacerbated the sewage problem as it had done in London. A curious transformation of U.S. water systems occurred after the mid-19th century, with nearly all major cities eventually adopting public water supply systems to virtually all households, at minimal cost. Even today, nearly all major cities in the United States have publicly owned, and mostly publicly operated, water supply systems which provide universal service.

On the downstream side, nearly all larger U.S. cities (except New Orleans) decided to discharge sewage into existing storm sewers rather than create separate "sanitary" sewers (Tarr 1996). No U.S. city had human sewage disposal systems by 1850, though most developed them between 1850 and 1900. Baltimore was the last U.S. city to build a sewer system, only after a fire destroyed much of the city in 1904, catalyzing the need for new infrastructure (Boone 2003).

The provision of sewage treatment occurred even later. Rudimentary sewage treatment by land application was the main sewage treatment technology of the late 19th century. Sewage treatment was particularly important in the United States because many cities were located downstream from other cities; hence their water intakes were subject to contamination by sewage produced upstream. The engineering community of the era developed a consensus in the late 1800s that sewage treatment was not economically justified because dilution and natural purification would be adequate (Tarr 1996). Downstream *water treatment plants*, it was reasoned, could then further purify the water using sand filtration, first used in London in 1827. Data compiled by Tarr (1996) suggested that adoption of sewage collection, with no provision for treatment, may have increased typhoid mortality rates in several cities, presumably by diffusing the typhoid bacterium downstream.

The first century of urbanization did not go well from a public health perspective. By 1890 there was a substantial "urban penalty" for urban living: Mortality rates in U.S. cities were 30% higher than in rural areas (Culter and Miller 2005). The displacement of the miasmatic theory by the germ theory of disease following breakthroughs by Koch and Pasteur in the late 1800s provided a solid theoretical basis for new practices in clean water technology. Chlorination became a key water treatment process, first started in Jersey City in 1908 and becoming nearly universal for major U.S. cities within a decade (Culter and Miller 2005). Chlorination also became widely used to treat sewage, which together with primary treatment of sewage (simple sedimentation), greatly reduced the discharge of human pathogens to rivers and other waterways.

Treatment of drinking water in water treatment plants by filtration and chlorination, cessation of discharges of wastewater near water treatment plant intakes, and exploitation of new and cleaner source waters greatly improved the quality of urban life in the early 20th century. Typhoid mortality dropped precipitously, to near zero levels by the 1940s. Life expectancy increased from 43 years to 63 years; clean water (treatment of water and wastewater) accounted for nearly half (43%) of the improvement (Culter and Miller 2005). When the *British Medical Journal* polled its readers to determine what they thought were the most important medical advances since 1840, the winner was "sanitation" (clean water and sewage disposal; BMJ 2007).

Cities continued to discharge untreated, or minimally treated, sewage for many more years. Cities in the United States were largely sewered by 1940, but only 57% of sewered areas had sewage treatment (Tarr 1996). The discharge of raw sewage grossly polluted rivers, often causing oxygen depletion and fish kills, and little had been done to curb industrial pollution. The wake-up call occurred in 1969, when *Time Magazine* showed the Cuyahoga River burning (Fig. 1.3). In reporting the story, *Time Magazine* described the river as "chocolate-brown, oily, bubbling with subsurface gases, it oozes rather than flows".

The Cuyahoga River fire was a catalyst for passage of the Clean Water Act in 1972. The Clean Water Act had a major impact on urban water, mandated specific water quality standards for rivers, establishing treatment standards for both

Fig. 1.3 The photo of the Cuyahoga River on fire which appeared in the August 1, 1969 issue of *Time Magazine*. Reprinted with permission from AP Photos

industrial and municipal wastewater treatment, requiring the pre-treatment of industrial wastes discharged to sewers, and providing cost-sharing for the construction of municipal *wastewater treatment plants* throughout the country. By the late 1990s, the organic loading of municipal wastewater had declined by 45% even while the U.S. urban population increased by nearly 50% (USEPA 2000). This decreased the number of oxygen-depleted dead zones below cities and allowed the resurgence of fish populations. Since 2000, the new thrust in reducing urban water pollution has been improved management of urban runoff.

1.2.4 Energy and Water Transportation

Early industrialization, water was a major source of energy and route of transportation for growing cities. Most large cities were coastal ports and many smaller industrial towns were located on high-gradient rivers which provided hydraulic energy. Heavy freight was moved on water whenever possible (Smil 1994). For example, the steel industry of Pittsburgh relied on the Ohio, Monongahela, and Allegheny Rivers to ship coal and ore to the city and finished steel from the city. As railroads expanded during the last half of the 19th century, some cities served as connecting points for rail and water transportation. Chicago is perhaps the best example. Railways connected Chicago to the east and west and waterways connected Chicago to the north (Lake Michigan and the St. Lawrence Seaway) and south (the Chicago Canal and the Mississippi River) (Cronan 1991). In the 20th century, railways became the dominant movers of freight in the United States. By 2000, railways dominated U.S. domestic transportation, accounting for 42% of ton-miles. Trucks accounted for another 28% of ton-miles, whereas water transportation accounted for only 13%. However, water is still the dominant form of international freight transport, carrying about three-fourths of the tonnage, and port cities on ocean fronts still maintain their water transportation function.

Urban waterfronts, once the hub of commercial life of many cities, declined through the first half of the 20th century. Even where shipping remained important, older port areas were supplanted by larger, modernized ports that could handle larger ships and larger cargoes, generally located downstream of the original ports (Marshall 2001). The trend toward decline was reversed about 40 years ago with urban renewal that converted decaying ports into thriving commercial and residential centers. One key factor in this resurgence was the environmental cleanup of rivers, driven by improved municipal and industrial wastewater and by the cleanup of old industrial "brownfields". Urban waterfronts exemplify the concept of succession in urban systems, an analog to succession in natural plant communities (Fig. 1.4).

The changing role of water in energy and transportation also affected smaller, upstream industrial cities. During early industrialization, prior to the widespread use of fossil fuels, hydropower was a major source of energy for small factories and mills. Because this power was often provided by small, low-head dams, many

Fig. 1.4 Transitions of an urban waterfront. The St. Paul riverfront, seen from the Wabasha Bridge, in 1930 (*left*), and in 2008. The earlier photo shows multiple rail lines along the shore, and buildings on the bluff, with no visible landscape amenities. A single rail line remains in 2008, along with a tree-lined boulevard, but a pedestrian walkway overlooks the river, and a park has been built on the bluff. The 1930-era photo with permission from the Minnesota Historical Society; the 2008 photo was taken by the author

small industrial cities located on high-gradient rivers far upstream from the mouths of rivers flourished and grew. As coal (and later other fossil fuels) became available, hydropower fell out of vogue. Many of these cities declined, at least temporarily, and hundreds of small dams that were once critical to small cities fell into disuse. Many of these focal points of local cultures are now being torn down to avoid the cost of structural repairs and to restore the natural course of rivers. In Wisconsin alone, more than 100 small dams have been removed, many of which were once central to the economy of the towns in which they were located. Perhaps in the coming years of energy scarcity some of these cities will restore or rebuild some of these dams to again provide hydropower.

1.2.5 Water and Urbanization in the Arid Southwest

Most cities in the temperate regions of the United States were built without severe constraints on water supply. The situation was very different in the arid southwestern United States as cities started to develop in the mid-20th century: Water was a critical issue from the very start. The difference can be represented by comparing water footprints – the size of watershed needed to provide water for a single person. For New York City, where precipitation is about 40 inches per year, the water footprint is about one-third of an acre. By contrast, residents of Phoenix, Arizona use 2.5 times more water per person (the higher use is mostly due to landscape irrigation) and local rainfall is only seven inches. The resulting water footprint for a Phoenix resident is four acres – 13 times larger!

Southwestern cities like Phoenix, Los Angeles, Denver, and Las Vegas rely on spring runoff that originates from snowmelt at high elevations and is stored in large reservoirs. They may also utilize water conveyed across watershed boundaries. For example, San Diego gets much of its water from the 240-mile Colorado River Aqueduct, which transports water from the Colorado River. Flows in the Colorado River above the point where water is diverted to the Colorado River Aqueduct are maintained by a series of major dams upstream on the Colorado River and numerous smaller dams located on tributaries.

Building the huge dams that supply water to southwestern cities depended on massive federal subsidies, guided by a once-massive federal agency, the Bureau of Reclamation, in the early 1900s. The initial motivation for most of these dams was not urban development, but agricultural development, part of a broad policy to encourage farmers to settle in the American West (Reisner 1993). From 1908 onward through the 1970s, the bureau eventually built 345 dams. Several factors affecting rapid urbanization of southwestern cities were (1) expansion of railroads and highway as key movers of people and freight, (2) the invention of air conditioning, (3) availability of "water rights" that accompanied farmlands sold for urban development, and (4) constraints on the selling price of water imposed by the original water utility charters that kept water prices far below market value. To augment the highly subsidized surface water, most cities also withdrew freely from *groundwater* aquifers, often dropping the level of the aquifer by hundreds of feet. Moreover, these cities have often literally sucked rivers dry, resulting in severe damage to downstream aquatic ecosystems.

1.2.6 Flooding

Urbanization drastically alters the hydrologic cycle of a watershed, as we will see in Chapter 2. One of the most important aspects of altered hydrology is increased flooding. Cities both cause floods and are impacted by flooding, often being located on the banks of waterways to take advantage of waterborne transportation. A city can cause downstream flooding by increasing the percentage of impervious surface – surfaces such as roads, rooftops, and parking lots, by filling wetlands, resulting in loss of water storage, and by increasing the drainage density through the installation of storm sewers. In many urban regions, increased flow from upper parts of the watershed has increased flooding in downstream parts of the watershed (Konrad 2003).

Throughout the first half of the 20th century, the main responses to urban flooding were the construction of flood control reservoirs upstream, increased drainage to route water downstream as quickly as possible, and the construction of levees to keep flood flows within riverbanks. Still, many cities along major river basins in the eastern United States periodically flooded. I happened to grow up in one of these towns, New Martinsville, West Virginia, on the banks of the Ohio River. The downtown area flooded so regularly that there was a local law requiring that motorboats cruising on Main Street not exceed 15 miles per hour – a law intended to minimize breakage of shop windows!

Since 1968, the Federal Flood Insurance Act has limited the types of buildings that can be insured on flood plains, allowing only those that can withstand periodic flooding without damage. Since then, historical commercial and residential districts located in frequently flooded areas near rivers have gradually been moved uphill and have been replaced with parks and other land uses that can withstand flooding with minimal damage. A more recent development has been to "soften" urban landscapes, reducing the amount of impervious surface and building stormwater ponds, wetlands, swales, and other *best management practices* (BMPs) to limit the deleterious impact of urbanization. Many municipalities and other local units of government now require that new developments retain a specified amount of precipitation "on site", storing it in ponds or infiltrating it to groundwater in infiltration basins. Some planners and hydrologists envision creating low-impact development designs that nearly mimic natural hydrologic conditions.

1.3 Summary

This brief history offers several insights regarding the urban water environment.

1. *Urbanization often occurs very quickly*: Rapid urbanization often overwhelm the ability of local municipal governments to evolve new water management systems quickly, leading to water crises. London was unable to respond adequately to its urban water crisis until the Metropolitan Water Board was formed. The formation of water management institutions is often needed to avert crisis during urbanization.
2. *Progress often has unintended, unpleasant consequences*: The widespread adoption of water closets in London without provision for managing the excess water entering cesspools is a classic example. Downstream flooding caused by upstream development, industrial pollution, and depletion of urban aquifers are other common, unintended consequences of urbanization.
3. *Crisis drives innovation*: Epidemics of cholera and typhoid spurred scientific advances in epidemiology and microbiology during the 19th century, which led to improved water sanitation, one of the most important scientific advances of modern civilization. Some of this innovation is positive, irreversible cultural evolution – gains in wisdom that will be transmitted through generations.
4. *Management of the urban water environment has lacked holism*: Most cities manage discrete parts of their water environments, such as wastewater or water supply, but rarely connect the parts. Very few cities are managed using complete hydrologic balances. We are starting to tie pieces together – like reducing downstream flooding through changes in upstream land use practices – but we are not very far along in understanding and managing this holism.
5. *The urban water environment will continue to evolve*: This evolution reflects broader economic, technological, and social change.

1.4 Looking Forward

The central premise of this book is that we can improve the quality of urban life by thinking more holistically about the urban water environment, incorporating ideas from hydrology, engineering, planning, law, and ecology to develop a view of the "urban water environment" as a central organizing concept for cities.

1.4.1 The Magnitude of the Problem

Urban regions throughout the world are gaining population, placing increasing pressure on water resources. Metropolitan areas in the United States have been growing at a rate of about 1% per year since 1960. Continued growth at this rate would result in 65 million more urban dwellers by 2030, a 30% gain. Throughout the world, half of the world's population (some 3.2 billion people) are now living in cities, and the urban population is expected to swell to nearly 5 billion by 2030 (Table 1.1). Nearly all of the world's population growth will be in cities in less developed regions of the world, where poverty and corruption often stand in the way of developing appropriate water infrastructure.

As urban populations increase, and particularly where this expansion is also accompanied by gains in average wealth, water supplies often become strained, especially in arid lands, where urban needs and agricultural needs compete. Many cities in the southwestern United States already have severe problems with long-term water supply, as do some cities in wetter regions, such as Atlanta and Miami. Many U.S. cities will also have to replace water infrastructure within the next few decades, requiring increases in utility fees to a level that may strain budgets of low-income families (CBO 2002).

Downstream impacts of urban drainage remain a serious urban water problem 35 years after passage of the Clean Water Act. Urban stormwater is badly polluted with sediments, nutrients, metals, and salts. In some cases, these pollutants accumulate within urban systems, contaminating groundwater and soils. Moreover, the hydrology urban streams have often been severely disrupted, aggravating downstream flooding and damaging aquatic habitats.

Finally, water-related policies are often fragmented, outdated, and ineffective. "End-of-pipe" solutions that worked well to treat municipal sewage don't

Table 1. World's projected urban population growth, 2005–2030, in billions

	2005	2030	Percentage of change
World	3.15	4.91	56
More developed regions	0.90	1.01	13
Less developed regions	2.25	3.90	73
Less developed, as %	71	79	–

Source: U.N. (2006)

necessarily work to reduce pollution in urban stormwater or to improve hydrologic conditions. Although some institutions have emerged that match watershed boundaries (surface drainages or groundwater basins) and have broad mandates, urban water environments are generally managed by a mish-mash of agencies and governmental units, each with narrow agendas.

1.4.2 Cause for Hope

In a sense, the glass is half full. The history of the urban water environment has shown that out of crisis comes innovation and renewal. At present, an obvious innovation is the enormous advances in information technologies – the 1000-fold increase in computing speed over the past 10 years, with parallel improvement in the quality of satellite imagery. These advances allow great technological advances. For example, we are now starting to use sophisticated mathematic models rather than statistical analysis of historical records to forecast flooding, and we are developing whole new ways of communicating hydrologic knowledge to policy makers and even ordinary citizens. These advances in information technology are comparable with the invention of the steam engine that spearheaded the Industrial Age or the discovery of germ theory that led to innovations such as water treatment and vaccination that rapidly increased lifespan in the early 20th century. We are therefore hopeful that this book does not merely lead to incremental improvements in managing urban water environments, but serves as a prolegomenon for a new paradigm of environmental management, to be fully developed by our readers in the next few decades..

1.4.3 Cross-Cutting Themes

To build coherence across a variety of topics, we developed each chapter on the same six cross-cutting themes. The first is *water scarcity*. All other policies regarding urban water are increasingly being linked to the issue of scarcity. To a great extent, we are able to see that urban water scarcity can occur even in regions with ample rainfall. The second theme is *multiple uses of water*. In most locations, urban water supply no longer means simply acquiring source water, using it, and flushing it downstream. In today's urban water environment, urban stormwater might be collected in an infiltration basin and used to recharge depleted aquifers. Suburban streams, once considered little more than conduits for drainage, are now being restored for trout fishing. In coastal areas, management of urban flows and reduction of pollutants is becoming increasingly necessary to maintain economically important finfish and shellfish industries and beach recreation.

The third theme is *water management institutions*. No urban regions have institutions specifically designed to address the entire spectrum of water issues. Management is often based around utilities (regional water supply or wastewater treatment authorities); a few urban regions have strong institutions for management of either surface watersheds or groundwater systems.

As scientists, we implicitly believe that new knowledge will benefit humanity, but our ability as a society to find and incorporate new knowledge into urban design and management is often sluggish. Therefore, our fourth theme is *integration of new knowledge*. New design prototypes can allow development to adapt successfully to new regulatory demands, using new knowledge, and change the ultimate performance of cities as water-using and water-producing systems. The combination of new needs and new opportunities may drive us toward new management models – for example, greater use of *participatory research* and *adaptive management*. There is also a pressing need to integrate the human dimension into transdisciplinary knowledge of human ecosystems. Hydrology alone will not solve our urban water problems.

The final two themes are sustainability and resilience. Using the definition from the Brundtland Commission (U.N. 1987), *sustainable development* "meets the needs of the present without compromising the ability of future generations to meet their own needs". In the context of the urban water environment, this includes economic and social environment, as well as the biophysical condition of the environment. The related concept, *resilience*, is the capacity of a system to withstand perturbations without major system changes. In the context of the water environment of cities, these perturbations include droughts and flooding. One specific concern is climate change, which may alter the entire hydrology of a city, creating a whole new set of problems. In addition to "natural" perturbations, there may be perturbations to the built water environment, such as dam collapse or major failures in the water delivery system. Gaining a better understanding of how we can increase urban resilience to these extreme events can help avoid loss of basic services and minimize the damage done by decisions made in crises that ignore medium and long-term consequences.

1.5 Chapter Topics

The first section of this book examines the flow of water and materials through urban systems. In Chapter 2, Claire Welty examines the water budget of cities, developing concepts that will be revisited in several subsequent chapters. Chapter 3, by Peter Shananan, examines a part of the urban water environment that few of us have seen – the groundwater systems that can provide sustenance or wreak havoc, depending on how we manage them. Paul Westerhoff and John Crittenden look at the engineered water environment – our modern water supply, sewage, and stormwater systems – in Chapter 4. Chapter 5 explores the movement of nutrients and materials through cities. Section II focuses on broader uses of the urban water environment. Derek Booth and Brian Bledsoe (Chapter 6) examine the physical and biological structure of urban streams, with an eye towards restoration of urban aquatic habitats. In Chapter 7, Ingrid Schneider looks at the importance of urban water recreation and implications with regard to broader water management. Finally, planner Kristina

Hill integrates many of these ideas into a modern perspective of water in urban design (Chapter 8).

The third section examines legal and institutional aspects of the urban water environment. Robert Adler develops the legal framework that guides urban water management (Chapter 9). The two chapters that follow describe very different types of water management institutions that are used in the eastern United States (Chapter 11, with emphasis on watershed management) and the western United States (Chapter 12, with an emphasis on groundwater basins), written by experienced practitioners who have headed these types of institutions (Cliff Aichinger, Director of the Ramsey-Washington Watershed District in the Twin Cities of Minnesota and James Holway, formerly Assistant Director of the Arizona Department of Water Resources). Finally, William Easter's chapter on economics (Chapter 13) focuses on water pricing and privatization.

After writing the core topical chapters, chapter authors came together to develop a synthesis chapter to integrate their ideas. The last chapter, "Principles for managing the urban water environment in the 21st century", outlines five core principles for water management in post-industrial cities of the 21st century.

Acknowledgments Development of this chapter was supported by NSF Grant 0739952.

References

Boone, C. G. 2003. Obstacles to infrastructure provision: the struggle to build comprehensive sewer works in Baltimore. Historical Geography **31**:151–168.

Booth, D., D. Hartley, and R. Jackson. 2002. Forest cover, impervious-surface area, and the mitigation of stormwater impacts. J. Am. Water Resour. Assoc. **38**:835–845.

Boyer, T., and S. Polasky. 2004. Valuing urban wetlands: a review of non-market economic studies. Wetlands **24**:744–755.

Browne, A. 2000. Beijing scoops the last ladles of nightsoil. Reuters News Service, December 23, 2000.

CBO. 2002. Future Investment in Drinking Water and Wastewater Infrastructure. Congressional Budget Office, Washington, DC.

Chandler, T. 1987. Four Thousand Years of Urban Growth: An Historical Census. Edwin Mellen.

Cronan, W. 1991. Nature's Metropolis: Chicago and the Great West. W.W. Norton, New York.

Culter, D., and G. Miller. 2005. The role of public health improvements in health advances: the twentieth century United States. Demography **42**:1–22.

Freese, B. 2003. Coal: A Human History. Penguin Books, New York.

Halliday. 2001. The Great Stink of Paris and the Nineteenth-Century Struggle Against Filth and Germs. Stroud Publisher, Sutton, England.

Konrad, C. P. 2003. Effects of Urban Development on Floods. U.S. Geological Survey Fact Sheet 076-03. U.S. Geological Survey, Washington, DC.

Marshall, R. 2001. Waterfronts in Post-Industrial Cities. Spoon Press, New York.

McHarg, I. 1971. Design with Nature. The American Museum of Natural History, New York.

Ogle, M. 1999. Water supply, waste disposal and culture of privatism in the mid-nineteenth century. J. Urban History **25**:321–347.

Reisner, M. 1993. Cadillac Desert. Viking Penguin, New York.

Smil, V. 1994. Energy in World History. Westview Press, Boulder, CO.

Smil, V. 2001. Enriching the Earth: Fritz Haber, Carl Bosch and the Transformation of World Food Production. MIT Press, Boston.

Tarr, J. A. 1996. The Search for the Ultimate Sink: Urban Pollution in Historical Perspective. University of Akron Press, Akron, OH.

U.N. 1987. Our Common Future: Report of the World Commission on Environment and Development. Annex to document A/42/427 – Development and International Co-operation: Environment, United Nations World Commission on Environment and Development, New York.

U.N. 2006. World Urbanization Prospects: The 2005 Revision. United Nations, Department of Economic and Social Affairs, Population Division, New York.

Ulrich, R. S. 1993. Biophilia, biophobia and natural landscapes. In S. E. Kellert and E. O. Wilson, editors. Biophilia, Biophobia, and Natural Landscapes. Island Press, Washington, DC.

USEPA. 2000. Progress in Water Quality: An Evaluation of the National Investment in Municipal Wastewater Treatment. U.S. Environmental Protection Agency, Washington, DC.

Chapter 2
The Urban Water Budget

Claire Welty

2.1 Basic Concepts

In this chapter our goal is to highlight some of the many ways that urbanization affects the water budget and water cycle. A *water budget* describes the stores or volumes of water in the surface, subsurface and atmospheric compartments of the environment over a chosen increment of time. The *water cycle* has to do with characterizing the flow paths and flow rates of water from one store to another. Understanding how urbanization affects the water budget and water cycle first requires an appreciation of how conditions work in a natural system.

The sun drives the hydrologic cycle, whereby water is evaporated by solar radiation from oceans, inland water bodies and soil, condenses and falls on land as precipitation, and returns to receiving water bodies by either surface runoff or groundwater discharge (Fig. 2.1). There are many critical sub cycles within the overall hydrologic cycle. For example, a portion of precipitation is returned to the atmosphere by evaporation before it reaches the ground. A portion of precipitation that is stored on vegetation (interception storage), on the land surface in puddles (depression storage), or in shallow soil pores, also evaporates rather than moving downward to groundwater or running off to surface water channels. Precipitation infiltrating the soil that is not lost to evaporation can flow downward to *recharge* groundwater, contributing to a rise in the *water table*, or flow shallowly in a lateral direction and discharge to streams. Flow in streams that is not due to surface or shallow subsurface runoff from the land is termed *base flow*; base flow in natural systems arises from deep and shallow groundwater discharging to streams during both storm and non-storm periods.

A water budget or water mass balance can be calculated for any time increment for a chosen *control volume*, where

$$\text{Inflows} - \text{Outflows} = \Delta \text{Storage} \qquad (2.1)$$

C. Welty (✉)
University of Maryland, Baltimore County, MD, USA
e-mail: weltyc@umbc.edu

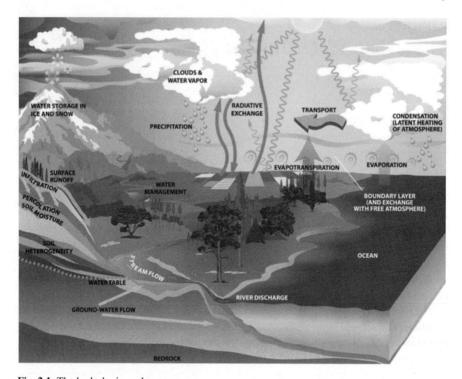

Fig. 2.1 The hydrologic cycle.
Source: US Global Change Research Program, http://www.usgcrp.gov/usgcrp/ProgramElements/water.htm

For natural systems, a control volume is often defined laterally by watershed boundaries (topographic highs) and vertically from the top of vegetation to the bottom extent of water-bearing subsurface sediments or fractured rock. Inflows are precipitation and groundwater flowing into the control volume; outflows are evaporation from surface water, vegetation, and soils, transpiration from plants (together called *evapotranspiration*), streamflow (runoff) and groundwater flow exiting the domain. A mass balance for a watershed-based control volume can therefore be expressed as

$$\text{Precipitation} - \text{Runoff} - \text{Net Groundwater Outflow} - \text{Evapotranspiration} = \Delta \text{Storage} \qquad (2.2)$$

where the net groundwater outflow is groundwater inflow minus outflow, and all terms are measured in volumes over the time period of interest. The change in storage includes changes in both the amount of water stored in groundwater (or *aquifer* storage) as well as in surface reservoirs. Measurement records of precipitation, streamflow, groundwater levels, and surface reservoir levels are available from governmental agencies; evapotranspiration can be calculated because it is the only unknown. A word of caution should accompany this kind of calculation: Although records are widely available, there are sampling and measurement errors associated with the data, and therefore evapotranspiration approximated this way reflects these uncertainties.

The urban water budget can be significantly affected by infrastructure moving water across natural flow boundaries. Piped water to and from a watershed (potable water supply, stormwater, and wastewater) can be an important part of the water balance. For urban areas, the water budget for a watershed-based control volume therefore may be altered to reflect these additional considerations and expressed as

$$\text{Precipitation} - \text{Runoff} - \text{Net Groundwater Outflow} + \text{Net Potable Water Imported} \\ + \text{Net Wastewater Imported} + \text{Net Stormwater Imported} - \text{Evapotranspiration} \\ = \Delta \text{Storage}$$

(2.3)

where net water imported is water imported minus exported by piped systems (potable water, wastewater, stormwater). A developed watershed may have any combination of water, wastewater, and stormwater imports and exports, which can lead to a net loss or gain to the system compared to the natural system. For example, water supply may be imported, and a portion may be used for lawn watering and disposal to a septic system, which would result in a net gain. Other portions of imported water may be disposed to a wastewater collection system transporting water out of the watershed, which would offset the gain in a partial loss. Box 2.1 shows calculations of a water budget for Valley Creek watershed in suburban Philadelphia, Pennsylvania where groundwater withdrawals from wells are a significant export. Calculation of a water budget for any urbanized area requires site-specific understanding in terms of the sources and disposition of all water budget components.

Box 2.1 Example Water Budget Calculation

This example illustrates calculation of the 1985 annual water budget for Valley Creek watershed, an urbanizing area in suburban Philadelphia, Pennsylvania. The area of interest is 20.8 mi^2 draining past a USGS gage recording the streamflow. (See http://waterdata.usgs.gov/pa/nwis/uv?site_no=01473169). The example is based on the work of Sloto (1990).

The relevant annual water budget for this case is given as:

$$\text{Precipitation} - \text{Runoff} - \text{Groundwater Withdrawals} - \text{Evapotranspiration} \\ = \Delta \text{Storage}$$

where groundwater withdrawals are the amount pumped from wells and exported from the basin for water use elsewhere. In this case the net natural groundwater outflow from across the basin divides is considered to be zero. It is also assumed that all wastewater generated within the basin is treated/disposed within the basin and therefore there is no net flow of wastewater across watershed divides to account for.

For 1985, the total precipitation measured at a raingage in the region was 42.71 inches, the average annual streamflow as measured at the USGS streamgage was reported as 23.83 ft^3/sec, groundwater withdrawal from wells, obtained from utility records, was 1.89 mgd (million gallons per day), and the change in groundwater storage was estimated as 0.10 inches (change in depth of water in a representative well from one year to the next multiplied by specific yield.) (Specific yield is an aquifer characteristic of the volume of water that is released by drainage from a unit volume of aquifer material.) The unknown in this water balance is therefore evapotranspiration:

Evapotranspiration = Precipitation − Runoff − Groundwater Withdrawals
  − Δ Storage

Putting all terms into like units,

Precipitation = 42.71 in/yr

Runoff = [23.83 ft^3/sec × (12 in/ft) × 86,400 sec/d
  × 365 d/yr]/[20.8 mi^2 × (5280 ft/mi)^2]
  = 15.5 in/yr

Groundwater withdrawals = 1.89 × 10^6 gal/d × (365 d/yr)
  × (12 in/ft)/[7.48 gal/ft^3
  × [20.8 mi^2 × (5280 ft/mi)^2] = 1.91 in/yr

Δ Storage = Δ groundwater storage = 0.42 in/yr (observed change in well
  water level × specific yield)

Note that in carrying out the calculations, the volume of runoff and the volume of well water withdrawn are spread over the area of the watershed to obtain equivalent depths, so that the units are comparable to those of precipitation.

Substituting for all terms in the water budget equation,

Evapotranspiration = (42.71 − 15.5 − 1.91 − 0.42) in/yr = 24.8 in/yr

This calculation shows that evapotranspiration is about 58% of the precipitation input.

In ultra-urban areas, the area drained by the dense underground system of pipes criss-crossing what remains of natural watershed boundaries may be so significantly different from the area drained by natural boundaries, that it may be desirable to delineate *sewersheds* drained by wastewater and stormwater pipes separately from the natural watershed, for purposes of water management (Mitchell et al., 2001). Delineation of sewersheds is carried out by spatial analysis of the system of pipe layouts to determine the area drained or serviced by the pipe system.

2.2 Impacts of Urbanization on the Water Cycle

Figure 2.2 is a map of an urbanized 14.3-km^2 watershed in Baltimore, Maryland, which shows how drastically the density and connectivity of the flow channels have been altered due to construction of buildings and roads. In some locations on this map it is difficult to discern the direction of surface water flow, owing to the high degree of landscape alteration. The vast network of pipes and utility conduits underlying urban areas can also significantly affect the cycling of water, although this effect is much more difficult to quantify because the systems are hidden.

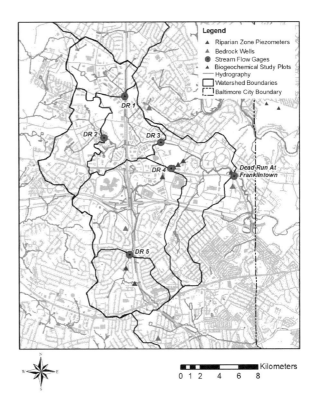

Fig. 2.2 Evidence of altered surface drainage in Dead Run watershed in the Baltimore, MD area. Pale blue lines are surface drainage; gray indicates impervious surfaces. (Courtesy of Michael P. McGuire, Geospatial Data Analysis Laboratory, Center for Urban Environmental Research and Education, UMBC)

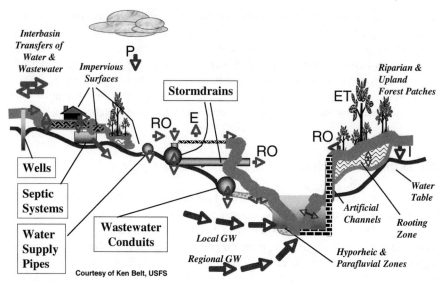

Fig. 2.3 The urban water cycle (Courtesy of Kenneth Belt, USDA Forest Service, Baltimore, Maryland)

Figure 2.3 is a conceptual cross-section through the surface and subsurface, where blue arrows depict a portion of the natural system flow paths, and red arrows show how the built environment alters the route that water takes. The effects of urbanization on the water cycle can include short-circuiting or path-lengthening compared to the natural system. Leaking pressurized water distribution systems can contribute to groundwater recharge. Cracks in parking lots and other sealed surfaces can act as focused recharge points to the subsurface (Sharp et al., 2006). Joints and cracks in sanitary and storm sewers can serve as drains for groundwater; conversely these pipes can also leak to groundwater, causing water quality impairment. Utility conduits and tunnels themselves can act as preferential flow channels for groundwater (Sharp et al., 2003), drastically altering the effective permeability of the subsurface. Treated wastewater discharged into a stream supplements base flow and in some cases accounts for the greatest portion of streamflow.

As pointed out in the introductory chapter, humans enhance the conditions for flood production in cities by hardening the land surface. The urbanized watershed of Fig. 2.2 shows impervious areas (parking lots, roads, sidewalks, roofs) in pale gray. In this particular watershed, impervious surfaces constitute 41% of the watershed area. Impervious area and compacted soils cause stormwater to run off instead of seeping into the ground; the highly engineered storm drain system carries water quickly away from properties.

Figure 2.4 depicts quantitatively the modeled effect of urbanization on storm runoff. This graph shows storm runoff from application of a 6-hour duration storm with a *return period* of 2 years to a 6-acre site in Valley Creek watershed, PA under conditions of (1) predevelopment, (2) post-development with no controls, and (3) post-development with stormwater detention (Emerson, 2003). (Plots of flow rate versus time are termed *hydrographs*.) The peak flow rate of the post-development hydrograph (with no detention) is more than three times greater than the hydrograph of the pre-development case; and the first arrival of surface runoff of the post-development case with no detention is within several minutes as compared to a delay of about 90 min for runoff from the pre-development case. Clearly the area under the curve, the total stormwater volume, is much greater for the post-development case with no detention control compared to the pre-development case.

Stormwater management regulations often require reduction of peak flow rates by routing flows to a *detention basin* designed for this purpose; a photograph of a detention basin is provided in Fig. 2.5. A detention basin or pond is the most commonly used *best management practice* (BMP) used to control runoff and nonpoint source pollution. Figure 2.4 shows that routing runoff through a detention pond reduces the hydrograph peak to a rate below that of pre-development flow rate, and delays the first arrival time of runoff compared to the post-development flow without detention. However, the area under the curve, which represents the runoff volume, remains much greater for the post-development case with detention compared to the pre-development case. In addition, the post-development case with detention control still has an earlier first arrival of flow initiation than pre-development flow in the example shown.

Storm detention systems designed to reduce higher flows typically pass lower flows through unattenuated; these flows plus volumes of larger detained flows all discharging into a watershed main channel in an additive fashion can result in

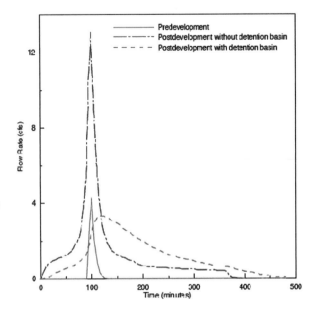

Fig. 2.4 Model results for pre-development surface water flow (*solid line*), post-development flow (*dash-dot line*), and post-development flow with detention (*dashed line*), for a storm having a duration of 6 hours and a return period of 2 years, applied to a 6-acre site from Valley Creek watershed, Pennsylvania (© Emerson, 2003; reprinted with permission)

Fig. 2.5 Stormwater detention basin in suburban Baltimore, MD. Photo by Claire Welty

exacerbated flooding in downstream areas (Emerson et al., 2005). A net result is that storm flow in downstream area has a greater flow rate and total volume than storm flow in downstream area before urbanization as a result of decreased travel time and increased water volume being directed toward the stream. Resolving this problem is one of the most challenging issues in stormwater management of the present day. In addition to actually posing life-threatening situations with high water velocities, greater streamflows contribute to increased bank erosion, often resulting in streams becoming wider and shallower in urban areas. This topic is discussed in further detail in Chapter 6. Down cutting of stream beds due to erosion can suppress water table elevations and reduce soil moisture available to *riparian zones* (Groffman et al., 2003).

A companion effect of increased runoff from impervious surfaces is reduced infiltration and groundwater recharge, which in turn can result in reduced groundwater discharge to streams and lower stream base flows. However, these reductions may partly offset by enhanced groundwater recharge in areas served by septic systems (Burns et al., 2005), as well as summertime irrigation in residential areas, especially during low base-flow months when such effects would be most pronounced (Meyer, 2005). Even in cases where groundwater is not used for water supply, groundwater in urban areas needs to be considered as a critical component of the hydrologic cycle owing to its important contribution to base flow and hence maintenance of instream flows for aquatic life. A more detailed discussion on the importance of groundwater in urban systems is provided in Chapter 3.

Evapotranspiration is often assumed to be negligible in highly developed areas owing to lack of trees. While this may be true for downtown or industrial areas, suburban residential areas often have a higher density of trees than rural farmed areas, and can have significant evapotranspiration rates. Grimmond and Oke (1991) found that evapotranspiration was 40% of the annual and 80% of the summer water budget of Vancouver, British Columbia. Using eddy covariance equipment to measure water vapor flux and to calculate evapotranspiration rates, they also found that evapotranspiration was often sustained in residential areas in summer by use of the potable water supply for garden irrigation and that the evapotranspiration rate could actually exceed the precipitation rate (Grimmond and Oke, 1999).

2.3 Effects of New Approaches to Water Management on the Water Cycle

In recent years there has been an emphasis on so-called "low-impact" development (LID) technologies for managing stormwater, including use of green roofs, cisterns, rain gardens, and on-site infiltration ponds. These practices can have a significant impact on the urban water cycle by increasing on-site recharge and reducing runoff rates. In evaluating the hydrologic response to three types of landscapes – undeveloped, developed, and LID – Holman-Dodds et al. (2003) were able to show that LID was most effective for small, frequent storms and that further stormwater protection (e.g., detention basins) was needed for larger events. The optimal kinds of sites where stormwater infiltration can be promoted are in upland areas where the soils are better suited for infiltration. Williams and Wise (2006) reached a similar conclusion in modeling the hydrologic response to four types of land development patterns, concluding that land preservation and infiltration could be effective at reducing stormwater flows, although detention ponds were needed to control peak storm flows.

2.4 Future Directions

The basis for sound water management starts with reliable water budget calculations (Healy et al., 2007). In other words, one needs to understand the urban water system before making recommendations to modify it. Example types of sustainable water management scenarios that would be dependent on water budget calculations are listed in Table 2.1. The concept of "water use regime" has been recently advanced (Weiskel et al., 2007) as a system to quantify the sustainable use of water by humans compared to water cycling that would be expected in a natural system. Four bounding conditions are defined: Natural-flow-dominated (undeveloped), human-flow-dominated ("churned"), withdrawal-dominated ("depleted"), and

Table 2.1 Sustainable water management strategies in urban systems (after Eiswirth, 2002)

Component of the Urban Water System	Example of Sustainable Development
Wastewater drainage	Automated leak detection system
	Scheduled infrastructure replacement and repair
Wastewater treatment	Reuse of treated wastewater for irrigation or groundwater recharge
	Advanced wastewater treatment
	Direct wastewater to potable reuse
Stormwater drainage	Separate storm sewer system
	Promotion of infiltration where feasible
	Reuse of stormwater on-site
	Disconnection of impervious surfaces
	Management on a watershed scale
	Managing small, frequent storms differently than large, infrequent events
Water supply	Leak detection and leak reduction
	Scheduled infrastructure replacement and repair
All	Manage the water system as a single resource
	Set up data collection, management, and archiving system

return-flow-dominated ("surcharged"), against which alternative and historic scenarios can be compared. This classification system depends strongly on the availability of water budget information. There is currently also a desire to assess the impact of climate change on urban water systems. A recent study involving water balance calculations in Connecticut showed that climate change had a far more significant impact on the water budget than land cover change brought about by urbanization (Claessens et al., 2006).

There is a very real need for coordinated data collection of all components in the urban water budget, and sometimes at small time increments (as small as a day for some applications) and small watershed scales (on the order of square kilometers), to be able to calibrate numerical hydrologic models to current conditions or to make reliable predictions. A consistent methodology must be established for data collection, processing, and archiving, in a consistent set of units, such that the data can easily be assimilated into models. Such an approach has been promoted by the Consortium of Universities for the Advancement of Hydrologic Science, Inc. (http://www.cuahsi.org). In addition to the "natural" components of streamflow, groundwater levels, precipitation, and evapotranspiration, it is of paramount importance to quantify imported and exported water to the basin of interest, so that highly managed systems can be accurately assessed. Once a coherent coordinated data collection system is in place, fully coupled water-cycle models can be calibrated, enabling predictions of various scenarios involving growth and climate change. There are many challenges in modeling of urban systems, especially how to represent the fine granularity of the landscape in a total water cycle approach. New approaches to modeling the effects of pipe infrastructure over large scales are needed, as well as greater utilization of remote sensing (satellite, aircraft) data for variables such as land surface temperature and soil moisture.

Finally, holistic water management is likely to be thwarted unless management agencies (water supply, wastewater treatment, stormwater management) operate under either one umbrella or in a more coordinated fashion so that water is managed as a single resource. Institutional issues are discussed in Chapters 11 and 12.

References

Burns, D., Vitvar, A. T., McDonnell, J., Hassett, J., Duncan, J., and Kendall, C. 2005. Effects of suburban development on runoff generation in the Croton River Basin, New York, USA. Journal of Hydrology, 311, 266–281.

Claessens, L., Hopkinson, C., Rastetter, E., and J. Vallino, 2006. Effect of historical changes in land use and climate on the water budget of an urbanizing watershed. Water Resources Research, 42, W03426, doi:10.1029/2005WR004131.

Eiswirth, M., 2002. Hydrogeological factors for sustainable urban water systems. In Howard, K. W. F. and Israfilov, R. G., eds. Current Problems of Hydrogeology in Urban Areas, Urban Agglomerates and Industrial Centres, NATO Science Series, IV. Earth and Environmental Science, Vol. 8, Kluwer Academic Publishers, 159–183.

Emerson, C. H., 2003. Evaluation of the Additive Effects of Stormwater Detention Basins at the Watershed Scale. MS thesis, Department of Civil, Architectural, and Environmental Engineering, Drexel University, Philadelphia, PA.

Emerson, C. H., Welty, C., and Traver R. G., 2005. A watershed-scale evaluation of a system of stormwater detention basins. Journal of Hydrologic Engineering, 10(3), 237–242.

Grimmond, C. S. B., and Oke, T. R. 1991. An evapotranspiration-interception model for urban areas. Water Resources Research, 27, 1739–1755.

Grimmond, C. S. B., and Oke, T. R. 1999. Evapotranspiration rates in Urban Areas. Impacts of Urban Growth on Surface Water and Groundwater Quality, 259, 235–243.

Groffman, P. M., Bain, D. J., Band, L. E., Belt, K. T., Brush, G. S., Grove, J. M., Pouyat, R. V., Yesilonis, I. C., and Zipperer, W. C., 2003. Down by the riverside: Urban riparian ecology. Frontiers in Ecology and Environment, 1(6), 315–321.

Healy, R. W., Winter, T. C., LaBaugh, J. W., and Franke, O. L., 2007, Water budgets: Foundations for effective water-resources and environmental management: U.S. Geological Survey Circular 1308, p. 90.

Holman-Dodds, J. K., et al., 2003. Evaluation of hydrologic benefits of infiltration based urban storm water management. Journal of the American Water Resources Association, 39, 205–215.

Meyer, S.C., 2005. Analysis of base flow trends in urban streams, northeastern Illinois, USA Hydrogeology Journal, 13(5–6), 871–885.

Mitchell, V.G., et al., 2001. Modeling the urban water cycle. Environmental Modeling and Software, 16, 615–629.

Sharp, J. M., Christian, L. N., Garcia-Fresca, B., Pierce, S. A., and Wiles, T. J., 2006, Changing Recharge and Hydrogeology in an Urbanizing Area – Example of Austin, Texas, USA. Philadelphia Annual Meeting (22–25 October 2006), Geological Society of America, Abstracts with Programs, 38(7), 289.

Sharp, J. M., Krothe, J. N., Mather, J. D., Garcia-Fresca, B., and Stewart, C. A., 2003. Effects of urbanization on groundwater systems. In: Earth Science in the City: A Reader, 257–278. American Geophysical Union, Washington, DC.

Sloto, R. A., 1990. Geohydrology and Simulation of Groundwater Flow in the Carbonate Rocks of the Valley Creek Basin, Eastern Chester County, Pennsylvania, USGS Water-Resources Investigations Report 89-4169.

Weiskel, P. K., Vogel, R. M., Steeves, P. A., Zarriello, P. J., DeSimone, L. A., and Ries III, K. G., 2007. Water use regimes: Characterizing direct human interaction with hydrologic systems. Water Resources Research, 43, W00402, doi:10.1029/2006WR005062.

Williams, E. S., and Wise, W. R., 2006. Hydrologic impacts of alternative approaches to storm water management and land development. Journal of the American Water Resources Association, 42, 443–455.

Chapter 3
Groundwater in the Urban Environment

Peter Shanahan

Groundwater is a critical resource for many of the world's cities. While a few cities (for example, New York) rely upon protected surface-water reservoirs for their supply, many more depend on groundwater. Conservation, protection, and management of groundwater are thus necessities for most cities. This chapter reviews the basics of groundwater hydrology, supply, and water quality, and then goes on to examine groundwater in the specific context of the urban environment.

3.1 Introduction to Groundwater

Groundwater is but one component of the hydrologic cycle that moves water through the environment. Precipitation (rain and snow) carries moisture from the atmosphere to the land surface. Other than in arid areas, the majority of precipitation runs off the land surface, concentrating in streams and then rivers and eventually reaches the oceans. Evaporation from the oceans carries water to the atmosphere once again, restarting the cycle. The portion of precipitation that does not run off the land surface is available to infiltrate into the ground. Some of that water evaporates or is taken up by plants which transpire water to the atmosphere. This combined loss of water to the atmosphere is known as evapotranspiration. The portion of the water that remains in the ground becomes groundwater and is known as recharge.

Groundwater lies beneath all landscapes in the empty space that exists between soil particles or in cracks in bedrock. Often, the soil near the surface is unsaturated, meaning that the space between soil particles is not completely filled with water. This water slowly seeps downward under the influence of gravity. Eventually, it reaches the water table, which is the top of the zone in which the soil is saturated

---

P. Shanahan (✉)
Department of Civil and Environmental Engineering at Massachusetts Institute of Technology, Cambridge, MA, USA
e-mail: peteshan@mit.edu

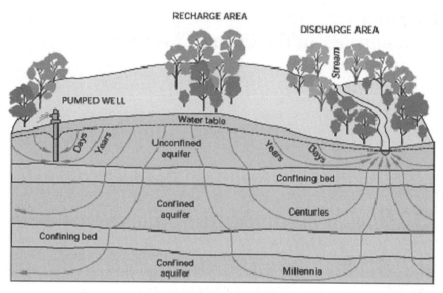

Fig. 3.1 Illustration of confined and unconfined aquifers and associated travel paths (Winter et al., 1998)

with water (i.e., all space between soil particles is filled with water). Conditions in which there is a water table are called *unconfined*, to distinguish them from confined, or artesian, conditions. In confined conditions, groundwater is kept under pressure by a capping layer of rock or soil that is too tight to allow water to flow through it easily (a confining bed). Figure 3.1 illustrates the concepts of confined and unconfined aquifers and illustrates that multiple aquifers may lie beneath the land surface.

Groundwater exists in the subsurface in geologic formations of different character. Often, it is useful to distinguish between bedrock formations and unconsolidated formations (also called overburden). Bedrock is the basement of rock that lies at depth beneath the land surface. Some types of rock, such as sandstone, are made up of cemented particles that have openings between them. Others, such as granite, have essentially no openings except where there are cracks, known as fractures, in the otherwise solid rock. Either of these types of rock can store and transmit groundwater. Although bedrock may be at the land surface it is usually deeper and covered by unconsolidated soil and geologic deposits in which the particles are not cemented together. For example, the sand and gravel that forms the bed of a river would be such an unconsolidated formation. Flowing water, glaciers, blowing wind, and landslides on mountain slopes are among the processes that can leave unconsolidated deposits. Landfilling, digging, and earthmoving by man can also create unconsolidated deposits. As with bedrock formations, unconsolidated deposits can store and transmit groundwater.

3.1.1 Groundwater Flow

Groundwater in the saturated zone flows from areas of high potential energy (known as potential, potentiometric head, or simply head) to areas of low potential energy. On a water table, this would be the same as flowing downhill from high water-table elevations to low elevations. Under confined conditions, the concept is similar but the confining bed keeps the groundwater from forming an actual water surface with highs and lows. Instead, pressure builds up under the confining bed, with higher potential in some areas than in others. The potential can be mapped with highs and lows of pressure on what is known as a potentiometric surface map much as actual elevations can be mapped on a water-table map. The water-table map is in fact a type of potentiometric surface map. Figure 3.2 shows an example potentiometric surface map for historical conditions in the confined Chalk aquifer beneath London in the UK.

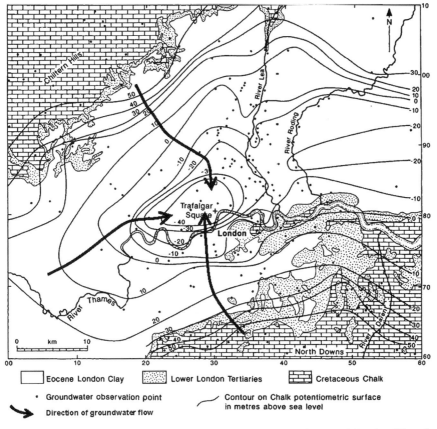

Fig. 3.2 Map of the potentiometric surface in the confined Chalk aquifer beneath London (Hiscock, 2005). Potentiometric surface elevations are given in meters above mean sea level

Groundwater moves from any area of high potential to areas of low potential, whether horizontally or vertically. Thus, if potential decreases with depth underground, the groundwater will tend to move downward. Groundwater will similarly move horizontally to areas of low potential. Often, the water table is a muted reflection of the land surface and groundwater typically flows horizontally from areas of high ground to areas of low ground, eventually exiting the subsurface to flow into streams and rivers. This natural pattern can be disrupted locally where wells withdraw large quantities of groundwater. Because of the relationship between head and groundwater flow, a potentiometric surface map can be used to infer the direction of flow. For example, in Fig. 3.2, the numbered lines on the map are lines of equal potential head with the number indicating the elevation in meters relative to mean sea level. The map can be read much like a topographic map of the land surface: as indicated by the arrows on the map, water flows from the higher heads at the periphery of the city into the city center where substantial pumping from wells lowers the potentiometric surface.

Groundwater movement is restrained by frictional resistance to flow in the tight spaces between soil and rock particles. The ability of a soil to allow water movement is called its hydraulic conductivity or sometimes permeability. For example, water has a much more difficult time moving through tight clay than it does through much more open gravel and coarse sands. Sand and gravel is much more permeable (has a much higher hydraulic conductivity) than clay. Hydraulic conductivity is in units of distance over time (for example, centimeters per second) and represents the amount of groundwater that will flow through a unit area (for example, a 1 cm by 1 cm area through which the water would flow) under a unit potential gradient (for example, a water table with a slope of 1 cm vertical over 1 cm horizontal). A hydraulic conductivity of 1 cm/sec does not mean groundwater moves at a speed of 1 cm/sec; in fact, the speed of groundwater movement is almost always many, many times less than the hydraulic conductivity because hydraulic gradients are typically very small. We discuss groundwater velocities further below.

The relationship between hydraulic conductivity and the rate at which groundwater moves from areas of high potential to areas of low potential is given by Darcy's Law. The law was discovered by a French engineer, Henri Darcy, when he undertook laboratory tests in 1856 while working on the water supply for the city of Dijon, France. This simple but elegant equation is the fundamental law of groundwater hydrology and therefore we include it in this otherwise non-mathematical textbook. Darcy's Law states that the flow of water in a porous medium is given by:

$$q = \frac{Q}{A} = -K\frac{\Delta h}{\Delta L}$$

where,

q is known as the specific discharge, the amount of water flowing per unit time through a unit cross-sectional area of the porous medium, and is given in units of length per time;

Q is the discharge, the total amount of water flowing, given in units of volume per time;

A is the total cross-sectional area through which the water is flowing, given in units of length squared;

K is the hydraulic conductivity, in units of length per time; and,

$\Delta h/\Delta L$ is the hydraulic gradient, which is the change in potentiometric head (Δh) over distance (ΔL). Since Δh and ΔL are both in units of length, $\Delta h/\Delta L$ is dimensionless.

Darcy's Law indicates that the amount of water that will flow through a porous medium is a function of the energy that drives the flow (as represented by the potential gradient, $\Delta h/\Delta L$) and the degree to which the medium allows movement of water (captured by hydraulic conductivity, K). Hydraulic conductivity varies widely for different kinds of geologic materials. Going back to the examples given above, a tight, compact clay resists the movement of water and therefore has a low value of K—about 0.00009 m/day. In contrast, water readily moves through coarse sand and gravel which has a high value of K—about 300 m/day. Table 3.1 shows typical hydraulic property values for a variety of geologic materials.

These last paragraphs have emphasized a key concept in groundwater—the importance of hydraulic conductivity in dictating the rate of groundwater flow. In most of the subsurface there is present a variety of different geologic materials or soil types, often with widely varying hydraulic conductivity. Given two pathways for flow, groundwater will tend to follow the pathway of least resistance—the pathway of highest hydraulic conductivity. Thus, if there is a seam of sand and gravel in an otherwise clayey soil, almost all of the groundwater will flow in the sand and gravel rather than the clay. These zones of high groundwater hydraulic conductivity are also the logical places to install groundwater supply wells.

In many instances, we are interested in the speed at which groundwater flows. Particularly, if groundwater is contaminated by a chemical substance, we often want to know how fast the contaminated water will travel to a point where it can affect

Table 3.1 Typical hydraulic properties of geologic materials (based on Mercer et al., 1982)

Material	Hydraulic conductivity (cm/sec)	Porosity	Groundwater velocity*(m/day)	Groundwater velocity (m/year)
Gravel	4×10^{-1}	0.34	5	1900
Coarse sand	5×10^{-2}	0.39	0.6	200
Fine sand	3×10^{-3}	0.43	0.03	11
Silt	3×10^{-5}	0.46	3×10^{-4}	0.1
Clay	1×10^{-7}	0.42	1×10^{-6}	0.004
Weathered granite	2×10^{-3}	0.45	0.02	7
Basalt	1×10^{-5}	0.17	3×10^{-4}	0.09
Sandstone	5×10^{-4}	0.34	0.006	2

*Assuming hydraulic gradient of 0.005.

people. The speed of groundwater movement is affected by the fact that water and chemical contaminants can only flow within the spaces between the rock and soil particles. The space between the rock and soil particles is known as the pore space and the fraction of the aquifer taken up by pore space is known as the porosity. The speed at which groundwater flows through these pores, v, is the specific discharge divided by the porosity or $v = q/n$, where q is the specific discharge that is computed using Darcy's Law and n is the porosity. Porosity is a fraction between zero and one, but is typically around 0.3 to 0.4 for sand. It can be much less in fractured bedrock, where only a small part of the rock is open space. Table 3.1 includes typical porosity values for porous media.

It is instructive to consider the speed at which groundwater moves in the subsurface. Typically, the water table is gently sloped, falling perhaps 5 m in 1000 m. Thus the hydraulic gradient $\Delta h/\Delta L$ is 0.005. Assume the soil is a fine sand, with a hydraulic conductivity of $K = 0.003$ cm/sec $= 3$ m/day. A typical porosity for fine sand is $n = 0.43$. With these values, we can use Darcy's Law to compute the speed of groundwater flow as $v = K/n \, (\Delta h/\Delta L) = 0.03$ m/day, only about 11 m/year. Groundwater moves slowly! Typical groundwater speeds for a range of geologic materials are shown in Table 3.1.

3.1.2 Groundwater Supply

Water under the ground represents a resource that can be tapped to provide water supply for individual homes, small villages, or large towns or industries. According to the United States Geological Survey, approximately one quarter of the fresh water used in the United States in the year 2000 originated from groundwater (Hutson et al., 2004). Groundwater is also an ecological resource in that groundwater typically flows into streams, providing the baseflow that sustains aquatic ecosystems through dry spells. Finally, groundwater serves as a structural element in the subsurface, contributing to the geologic integrity of the land. Removal of groundwater can destroy this integrity and lead to subsidence and creation of sinkholes.

Groundwater supplies come from aquifers. Water-bearing zones in the subsurface are called aquifers if they are sufficiently permeable to transmit readily useful quantities of water to wells, springs, or streams under ordinary hydrologic conditions. Aquifers thus have economic value—they can provide a water supply to a home, a village, or a city. However, an aquifer that can provide adequate water for a single residence might not be able to supply a municipality. Thus, it is common to consider the yield of an aquifer, which is the long-term sustainable quantity of water that may be withdrawn from it. Aquifers with large areal extent and high hydraulic conductivity have high yield.

Groundwater may naturally discharge to the land surface at springs or into streams, however the most common means to extract groundwater is via a well. A water well is simply a hole dug or drilled into the earth and used to obtain water. In a pumped well, an electric or mechanical pump mechanism is installed below the water surface in the well and used to withdraw water. The removal of water causes a localized lowering of the water table which, due to Darcy's Law, causes an inward flow of groundwater from the surrounding aquifer (Fig. 3.2). Around the well, the water-table surface forms a characteristic shape, known as a cone of depression, which features a progressive lowering of the water table and steepening of the hydraulic gradient inward to the well.

When multiple wells are located near each other, their cones of depression can overlap and cause a general lowering of the water table. Figure 3.2 shows a contour map of the potentiometric surface of the confined Chalk aquifer beneath London in 1994. Prior to pumping, potentiometric elevations in the aquifer were above sea level and wells in the Chalk aquifer flowed under artesian pressure. However, pumping over time lowered the potential in the aquifer and by 1994 potential head was at least 10 m below sea level through much of the city and over 40 m below mean sea level in the city center.

Just as removing groundwater from wells can lower potential in the aquifer, adding water into wells can raise the potential. There is now an extensive aquifer recharge system around London in which excess surface water is injected into the aquifer. Such artificial recharge schemes can be particularly useful in areas with large seasonal variations. Excess surface water can be recharged to aquifers during wet seasons and stored so it is available for withdrawal during the dry season. Similarly, treated wastewater is now being recharged to aquifers in water-short areas.

3.1.3 Groundwater Quality

The value of the groundwater resource can be completely or partially lost when groundwater is contaminated. While a spill of chemicals to a river can cause great damage, the immediate threat is short lived: the flowing stream can flush out the chemical in a matter of days. In contrast, we have seen that groundwater moves slowly. Thus, a spill to a groundwater aquifer usually persists for many years as the chemical slowly spreads in the aquifer. Further, water can be rendered unfit for drinking by only small amounts of chemicals. For example, benzene, a component of gasoline, is limited to a concentration of 5 μg/L in drinking water by U.S. Environmental Protection Agency regulations. This is equivalent to 5 parts of benzene in a billion parts of water—about the same as 1 drop of benzene in a railroad tank car of water. A very small spill can thus contaminate a very large quantity of water.

Water quality can be compromised by more than just an isolated spill. Systemic practices may also cause problems. Some examples that have affected cities are widespread discharge of wastewater to subsurface disposal systems and groundwater overdraft in coastal cities leading to salt-water intrusion. Areas transitioning from agricultural to urban land use may have legacies of excessive pesticides or nitrogen-based fertilizers in the groundwater, and salt accumulation from past irrigation practices. Finally, it is important to recognize that some groundwater is naturally contaminated by high mineral concentrations.

Compromises to groundwater quality can often be remedied by treatment. Indeed, groundwater used for municipal drinking water is always treated by disinfection and additional treatment is common. Other uses of groundwater may or may not require treatment—agricultural use is not nearly as demanding as drinking-water use while some specialized industrial uses require more extensive treatment than drinking water. Whatever the use, treatment should be recognized as a viable option to enabling use of groundwater.

3.1.4 Additional Reading

There are many good references that provide additional information on groundwater. A very accessible and comprehensive reader is a U.S. Geological Survey report by Heath (1982), available on-line. Advanced texts on groundwater hydrology and hydrogeology include Freeze and Cherry (1979), Todd and Mays (2005), Fetter (2001), and Domenico and Schwartz (1998). A very practical field-oriented reference is Driscoll (1986) and an excellent and very accessible text on contaminants in groundwater is Fetter (1999).

3.2 Groundwater in Cities

Up to this point in the chapter, we have been discussing groundwater only in general and not focusing on its place in the urban environment. That place is subtle and usually hidden, but can be significant to the overall life of the city. Natural hydrologic processes are greatly modified in the urban environment, with consequent effects on groundwater. The city disrupts the natural cycle in which precipitation recharges the aquifer and groundwater flows through the aquifer from recharge areas to discharge areas along streams and rivers. In the city, roads, buildings, and other impervious structures cover the land surface and reduce recharge. But, the city's water infrastructure (water-supply and sewer pipes) may leak water in some locales, providing localized sources of recharge (some potentially contaminated). Irrigated parks and gardens may also act as pockets of concentrated recharge. On the discharge side of the cycle, withdrawals from wells that supply the city may capture the groundwater that would otherwise flow to streams and rivers. Subsurface utilities and structures may act as drains that draw water from the ground. These various disruptions of the recharge and discharge components of the natural system must all be accounted for when considering the interaction between groundwater and the city. Figure 3.3 illustrates these interactions in a schematic chart of flows to and from the urban aquifer. Still another potential disruption is to the character of the subsurface. Filled land may include bricks and other debris, with large openings and much higher hydraulic conductivity than the natural soils, thus creating artificial pathways for preferential flow of groundwater.

3.2.1 Phases in the Relationship Between a City and Its Groundwater

Before considering how urban infrastructure interacts with the groundwater system, we must recognize that the city is not static—it grows and changes with time, and

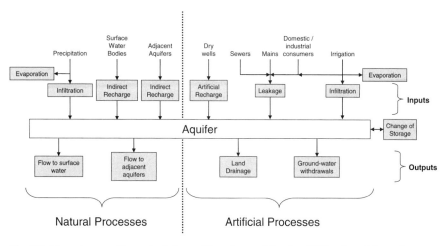

Fig. 3.3 The urban aquifer water balance (Shanahan and Jacobs, 2007, adapted from Simpson, 1994)

its effects upon the groundwater system change with time. This section considers the different phases in how a city interacts with the underlying groundwater, using London, England as an example.

The pre-industrial city—using local groundwater: The nascent city is small and decentralized. Even after people have started to gather in a single locale, the services and infrastructure that we typically associate with mature cities take time to develop. Thus, residents in the young city typically draw upon springs, streams, and local shallow aquifers to self-supply water. Simultaneously, they develop cesspools and septic systems to dispose of wastewater on a residence-by-residence basis. Most of today's great cities had such humble beginnings. London, for example, was established by the Romans around 50 AD, deriving its water from the Thames River and several small streams and springs within the town. This local supply sufficed for about twelve hundred years (Richards and Payne, 1899).

While local water supply and local wastewater disposal typify the first phase of urban development, these methods are unsustainable for growing cities and invariably fail to keep up with population growth. In London by the early thirteenth century, the original supplies were no longer adequate and conduits began to be constructed to carry water from outlying districts to public cisterns in the city proper. Further improvements commenced in 1582 when Peter Morrys constructed beneath London Bridge a water wheel that pumped Thames River water into a small network of pipes servicing a part of the city. Other systems followed, including in the first two decades of the 1600s construction of the New River, a 68-kilometer-long artificial canal to bring water to the city from groundwater springs to the north (Ward, 2003). Despite these beginnings of London's modern water-supply infrastructure, some city residents still depended upon wells within the city through most of the 1800s.

The experience of London in the Victorian era—around the mid-1800s—points out another force that moved cities away from local groundwater supplies: groundwater pollution. In 1854, Dr. John Snow showed the link between a cholera epidemic in the Golden Square neighborhood of London and consumption of water from the Broad Street well, an 8.5 m-deep well tapping the shallow aquifer. The Broad Street well was immediately adjacent to and in fact hydraulically connected to a cesspool, which proved to be the source of the bacterium that caused the cholera epidemic (Prescott and Horwood, 1935; Johnson, 2006). Snow's findings eventually led to the germ theory of disease and modern approaches to public health, including protection of water supplies from pollution. Groundwater pollution can be extensive where on-site disposal is the predominant means of handling domestic wastewater as well as near industrial facilities. Lawrence et al. (2000) cite a modern-day example in the developing world: wastewater discharges from the city of Hat Yai, Thailand have so altered the quality and chemistry of the underlying aquifer as to dissolve natural arsenic and further contaminate the aquifer.

The industrial city—falling groundwater levels: The initial phase of urban effects on groundwater thus gives way to a second phase in which local supplies are inadequate or too contaminated to provide water from within the city. Often, this phase coincides with industrial development and the increase in water demands that industry brings. The second phase of urban effect is thus one in which the water infrastructure is more or less fully developed. A water distribution system is in place along with wastewater collection. Both depend upon high-volume centralized facilities for water supply and wastewater treatment. In London, the second half of the nineteenth century led to the development of a modern infrastructure for water supply and wastewater collection and disposal. Water supply was developed first by private companies that initially competed fiercely for business, later achieved a truce by which each company serviced a particular area within London, and finally were consolidated into a single public entity (Ward, 2003). By 1866, under the farsighted leadership of engineer Joseph Bazalgette, the city constructed a comprehensive sewer system for collection and disposal of the city's wastewater (Halliday, 1999). Storm sewers also replaced natural streams; the Fleet River that once flowed through the center of the old city of London was now hidden, buried in an underground brick tunnel as much as 8 m below the surface (Barton, 1992, p. 26).

For the urban aquifer, the second phase of urban development is often a period during which the water table is lowered as regional (usually deep) groundwater aquifers are tapped. The regional Chalk aquifer underlies southwest England including London. It is about 180 to 240 m thick and is in most places confined by the 60 m-thick London Clay formation (Walters, 1936). While some wells drew from the Chalk aquifer as early as 1800 (Clow, 1989), beginning in 1846, a number of deep wells were constructed to tap the aquifer north of London (Ward, 2003, p. 181). Walters (1936) reports that by the mid-1930s there were 750 wells in the London metropolitan area drawing over ten thousand m^3/day. This large and extensive withdrawal resulted in a general lowering of the water level in the aquifer by 15 m or more through most of the metropolitan area and over 45 m at a pumping center north of the city (Fig. 3.4).

The falling groundwater levels beneath the industrial city have many effects. One particular problem is land subsidence. In London, the layer of London Clay has shrunk as it has become dewatered, leading to a lowering of the land surface in

3 Groundwater in the Urban Environment

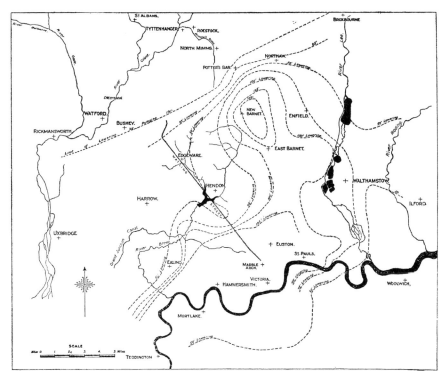

Fig. 3.4 Decline in water level (in feet; 1 ft = 0.3 m) in the London Basin Chalk aquifer (Walters, 1936)

Central London by 20 to 25 cm since 1865 (Clow, 1989) with localized maximum settlement of about 50 cm (Downing, 1994). Other cities have seen much more dramatic changes—Mexico City has subsided by meters—but even a fraction of a meter is enough to create potential structural problems for buildings that experience uneven settling. There are other problems as well: supply wells must be deepened as water levels fall; pumping costs increase because groundwater must be lifted more; salt water may intrude into previously fresh aquifers in coastal areas; and the bearing strength of foundation soils may change as soils become unsaturated (Downing, 1994). A peculiarly insidious problem is the dewatering of the wood piles that support structures in some areas of older cites (Lambrechts, 2000). As long as wood piles remain saturated by groundwater, they resist rot and can last centuries. If the water table is lowered below the pile tops, however, wood rot is rapid and building support may collapse in a matter of years.

Overall, the lowering of water tables implies a loss in water-supply resilience: local supplies are no longer adequate or perhaps even available to supply the populace during times of drought or water-supply emergency. The city becomes dependant, even hostage, to distant sources and the infrastructure needed to bring that faraway water to the city. Chapter 12 discusses this type of situation in Arizona in the arid southwestern United States.

The post-industrial city—rising groundwater levels: The mature city often passes through a third phase in its relationship with its underlying groundwater. This phase can be triggered by one or more factors including deliberate restoration of past groundwater levels, the transition to a post-industrial economy, the loss of groundwater supplies due to industrial contamination or overdraft, or the replacement of nearby groundwater supplies with more distant safer supplies. The result is reduced use of local groundwater and a slow, steady rise in groundwater elevation. Often this is a desired outcome, but it can lead to problems. Such is the case in London where a rising water table is now intruding on subsurface structures built during the time of lower water table. Figure 3.5 shows the water level measured in an observation well at Trafalgar Square in central London. Water levels reached their lowest levels in the 1950s and early 1960s, but have since recovered nearly 50 m due to reduced groundwater pumping.

As with falling groundwater levels, rising groundwater levels bring a host of problems (Johnson, 1994; Brassington, 1990). Rising water levels affect the structural properties of soils and can reduce bearing capacity, cause swelling, and create hydrostatic uplift pressures. As it approaches the land surface, a rising water level can intrude on structures, cause surface ponding, jeopardize on-site wastewater

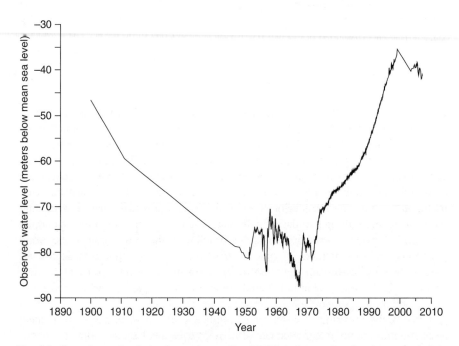

Fig. 3.5 Groundwater levels in observation well at Trafalgar Square, London, in meters below mean sea level (data provided by DEFRA, 2007)

systems, infiltrate into sewers, and attack masonry and concrete. In karstic areas where soluble limestone formations are near the land surface, rising water levels can dissolve the rock and create sinkholes. Most problematic of these are the interferences with subsurface infrastructure, to the point that in London the major organizations responsible for infrastructure have banded together to form the General Aquifer Research, Development and Investigation Team (GARDIT) to manage groundwater levels beneath the central city (Cherry, 2007). Since the year 2000, pumping in London has been increased in order to bring water levels back down and prevent groundwater intrusion into subsurface structures.

The future city under climate change and population growth: Two trends will have great influence on groundwater in the city of the future: climate change and population growth in the world's largest cities. Climate change will influence groundwater primarily by changing the quantity and distribution of groundwater recharge. IPCC (2007) indicates that changes in climate include more frequent intense rainfall, more frequent drought, less rainfall at subtropical latitudes, and more rainfall at tropical and high latitudes. The geographical variation of these trends implies different responses in different areas, a phenomenon addressed by numerous region-specific studies in the technical literature (see Hiscock, 2005, pp. 302–316, for a summary of some of these studies). As the result of climate change, many regions will experience less groundwater recharge. Even areas that see more rainfall overall may have less recharge if there is less frequent, more intense rain that runs off the land surface rather than soaks into the soil. Climate change may also increase salinity intrusion in coastal aquifers. One cause of this is a rise in sea level (IPCC, 2007). An even greater influence would be reduction in recharge: lighter freshwater floats on heavier salt water in coastal areas and a drop in the surface elevation of the freshwater (as might be caused by a reduction in recharge) would have a roughly forty-fold greater effect than an equivalent rise in the salt water level. Overall, disruption of groundwater recharge patterns has the potential to reduce the availability of the groundwater resource in both coastal and inland areas.

Another phenomenon of great significance to cities in the future is the growth of cities generally and the megacities of 10 million people or more particularly. Cities of this size only appeared late in the twentieth century but there were 16 new megacities by the year 2000 (WWAP, 2003). Most of these megacities lie in water-short areas, often in developing countries. Moreover, they reflect a worldwide trend of increasing population in cities of all sizes—54% of the population in the year 2015 is forecasted to be urban versus only 38% in the year 1975. Considering also the overall increase in population (7.2 billion in 2015 from 4.1 billion in 1975), the urban population will have more than doubled by 2015 (from 1.5 billion to 3.9 billion). Clearly, the world's cities will need to provide much more water and deal with much more wastewater and pollution potential as these trends continue. Within this context, groundwater supplies will be vulnerable to both greater pollution and overexploitation.

Summary: We can define three phases in the relationship between a maturing city and its underlying groundwater (as well as foresee a possible fourth phase associated with climate change). The young city draws upon shallow groundwater as both its source of drinking water and its disposal medium for wastewater. With population growth and industrialization, this arrangement becomes unsustainable and the city moves to a second phase, drawing groundwater in quantity from distant or deep aquifers, often lowering groundwater levels dramatically. Finally, groundwater withdrawals decrease dramatically in the post-industrial city, causing groundwater levels to start rising, affecting subsurface infrastructure created in times of lower water tables.

3.2.2 Cities and Groundwater Quality

A large number of studies have shown that urban and suburban development is associated with a degradation of groundwater quality (Shanahan and Jacobs, 2007). A variety of potential contamination sources exist in urban areas including chemical handling at industrial and commercial facilities, storage of petroleum fuels, on-site disposal of wastewater, infiltration of contaminated runoff, and application of deicing chemicals to roadways. Thus, there is in general a concentration of potential sources in urban areas that contribute to overall higher pollutant loads. The character of the resulting problems is not unique to urban settings and their investigation and remediation is addressed in textbooks such as Fetter (1999) and Sharma and Reddy (2004).

A water-quality issue unique to cities is the potential for leaking sewers to contaminate the groundwater. Sanitary sewers are known to leak and the released sewage would clearly carry contaminants. There is scant evidence that leaking municipal sewers have caused actual observed instances of groundwater pollution (Lerner et al., 1994). Barrett et al. (1999) sampled for specific chemicals associated with sewage in an urban setting and failed to find a strong indication of leakage. An exception from these general findings is the author's experience in the case of industrial facilities that historically discharged toxic organic chemicals—leaking sewers at industrial facilities have been found to be sources of groundwater contamination in investigations of contaminated sites. Overall, leaking sewers may be sources of groundwater contamination in some instances, but seem not to be a problematic general source based on the limited information available.

3.2.3 Analysis of Urban Groundwater

Review of the previously cited textbooks on groundwater hydrology and hydrogeology will show that analysis of groundwater is a well established discipline with carefully developed tools and methods for analysis of groundwater systems. For example, there are known equations for predicting how the water table will draw down (that is, become lower) after a pumping well starts to withdraw water from an aquifer. Unfortunately, these techniques require special care and attention when applied to the particular circumstances of the city.

The most unique aspect of urban hydrogeology is the degree to which the subsurface has been modified by man. Most cities have been excavated and filled, are crisscrossed by pipes and conduits of various kinds, may be pockmarked by old unused water-supply wells, and are drained at tunnels and basements—in short, have been extensively modified in ways that affect the groundwater system. Consider construction of a sanitary sewer. First, a trench is cut into the earth. Next, a gravel bedding layer is placed in the bottom of the trench. Often, a drain line is then installed within the gravel (and beneath the eventual sewer line) to drain away groundwater during construction. Next, the sewer line itself is laid. At last, the trench is backfilled to restore the original land-surface grade. Throughout this process, materials of high hydraulic conductivity have been placed in contiguous deposits creating preferential pathways for flow of groundwater—mini-aquifers in essence. These mini-aquifers typically affect a local area only, but sometimes with dramatic effect. In Boston, Massachusetts, a sewer line along St. James Avenue lowered the water table beneath the nearby public library and Trinity Church—two classic buildings supported on wood piles. The lowered water table (Fig. 3.6) exposed the tops of the wood piles, caused them to rot and jeopardized the underpinnings of both buildings (Snow, 1936).

The existence of preferential pathways along pipelines, conduits, and old wells also has implications for contaminant transport in groundwater. It is not uncom-

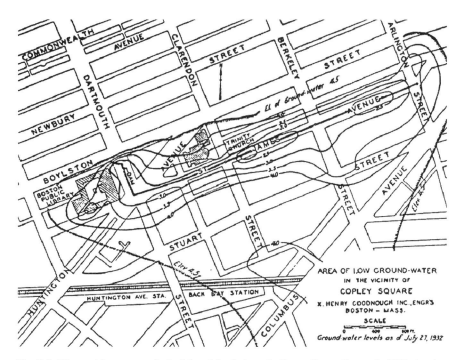

Fig. 3.6 Water-table contours (in ft; 1 ft = 0.3 m) along St. James Street, Boston in 1932 showing impact of St. James Avenue sewer (Aldrich and Lambrechts, 1986)

mon for underground gasoline storage tanks to leak, releasing gasoline to the subsurface. In an urban setting, the leaked product can follow a zigzagging pathway from one street to the next as it travels along sewer trenches (Lundy and Gogel, 1988). In an undisturbed geologic setting, we would expect entirely different behavior, with a tongue-shaped plume heading off in a single direction aligned with the regional groundwater flow. In contrast, the disparate fill material used in many cities can create preferential pathways for groundwater flow and contaminant migration. Similarly, old wells can create preferential pathways for vertical flow, sometimes spreading contaminants from a contaminated shallow aquifer to deep uncontaminated aquifers.

These examples illustrate a special challenge of urban groundwater analysis: subsurface infrastructure can have significant but local effect on groundwater flow and contaminant transport and must be taken into account. Unfortunately, the tools for analyzing groundwater systems—computer models—typically look at larger spatial scales and are cumbersome for the street-by-street analysis needed for many urban groundwater problems. For many problems, the urban groundwater analyst must investigate and consider finer spatial details in order to understand the groundwater system. Thus there is an intrinsic knowledge gap in urban groundwater analysis: rarely is there sufficient detailed information to understand completely the urban groundwater system.

The urban hydrologist must also account for the idiosyncrasies of particular climates and locales. For example, several researchers have reported on the paradox of groundwater flooding in desert cities (Shanahan and Jacobs, 2007). Over time, low-hydraulic-conductivity hard pans develop in desert soils from the residuals left by evaporation. When a city is created atop these hard pans, the water added by leakage from water and sewer lines and irrigation of parks and gardens can pond on the hard-pan layer and cause surface flooding and water intrusion into buildings. Kuwait City, for example, has seen water tables rise as much as 5 m (Al-Rashed and Sherif, 2001). Local conditions have caused unique responses to altered groundwater hydrology in other locations as well. Harking back to prior examples in this chapter, the dramatic subsidence in Mexico City stems from its location over thick clay beds while the contamination by arsenic in Hat Yai, Thailand is a result of its particular hydrogeology and geochemistry. Thus, while general principles apply, site-specific conditions must always be considered.

3.2.4 Managing Urban Groundwater

Urban groundwater has been, and for the most part remains, unmanaged and unregulated (Shanahan and Jacobs, 2007). The United Kingdom is typical: extraction of groundwater was considered a right under common law (Walters, 1936) and was essentially unregulated until passage of the National Water Resources Act of 1963 and initiation of a system to license water abstraction (EA, 2002). With no central control prior to 1963, groundwater was overexploited and the groundwater resource diminished as water levels fell. In the 1960s, controls on groundwater abstraction along with reduced industrial activity halted the decline in water levels and, eventually, led to rising water levels.

A lack of rules for preventing overexploitation or contamination of groundwater, or a failure to enforce the rules, is the norm, particularly during periods of rapid economic development. Groundwater is perhaps the ultimate "out of sight, out of mind" resource. It is difficult and expensive to monitor and manage, and there is usually no oversight until some crisis intervenes. In London, it was the extreme fall in water levels that led to action in the 1960s to prevent overexploitation. Today, the issue is rising water levels and only informal controls are in place to manage it. In 1992, London Underground, the National Rivers Authority and Thames Water formed the GARDIT to address the rising groundwater that threatened subsurface infrastructure (Stedman, 1999). The Environment Agency, in turn, still operating under the 1963 Water Management Act, instituted in 2002 a comprehensive system for managing water abstraction known as the Catchment Abstraction Management Strategy (CAMS) (EA, 2002, 2006). GARDIT now cooperates with the Environment Agency within the CAMS framework to pump groundwater from selected wells in London so as to manage groundwater levels (Cherry, 2007). The result is a comprehensively but informally managed aquifer.

London's comprehensive management structure is the exception and most other jurisdictions around the world still leave groundwater unmanaged. For the most part, regulatory systems are absent, but even when present, are often unenforced. In a study of a groundwater basin in West Java, Indonesia, Braadbaart and Braadbaart (1997) blame groundwater overexploitation on a lack of enforcement rather than a lack of regulatory structure. Thus, in addition to the technical challenges discussed above, managing urban groundwater also presents a special political challenge.

How can we improve the management of ground water and increase the resilience of the groundwater component of the urban water system? Clearly, prior system management has not been resilient: groundwater has boomeranged through different problems at different times, even changing from an important resource in pre-industrial cities to a nuisance in post-industrial cities. In large measure, groundwater is a problem because it is "out of sight and out of mind." The public and even decision makers know little about the state of the groundwater resource. Thus, adverse changes in groundwater quality or elevation can fester for years or decades without action or even notice.

Technology offers one potential remedy. While several cities with critical groundwater issues, including London and Boston, monitor groundwater elevations on a regular basis, we have identified no instances in which cities provide web-based depiction of spatial groundwater conditions. The U.S. Geological Survey (USGS, 2008) offers real-time data from individual monitoring wells—something potentially very useful for evaluating trends—but does not provide maps on the scale of an individual city. Such information could remedy the out-of-sight, out-of-mind problem and allow for more resilient management of the resource. For example, homeowners in landfilled sections of Boston worry about construction dewatering as a potential cause of compromise to the integrity of the wood piles that support their homes. Easy access and visualization of groundwater data could allow them (and construction managers) to evaluate impacts and make adjustments in real time. In water-short areas such tools could permit adaptive management of the conjunctive use of ground and surface water.

The out-of-sight, out-of-mind issue is also raised by Drangert and Cronin (2004) as partly responsible for past pollution of groundwater resources. They advocate a new management paradigm that focuses on households as a source of groundwater pollutants and involves residents in preventing toxic chemicals, detergents, oil, and pharmaceuticals from being discharged with wastewater. Here as well, more accessible information could play an important part in educating the public and thereby securing a more resilient groundwater supply. Indeed, better groundwater management and control of pollution can create a new resource. As pointed out by Foster and Chilton (2004), urban wastewater is probably the only water resource that is becoming more available. Thus, intelligent management of water reuse and artificial groundwater recharge will inevitably become an increasingly important part of urban water management.

While it is expensive and time-consuming to develop information systems that could enable better management of groundwater resources, pollution prevention, and conjunctive use, such tools offer great potential for adaptive management of the urban water system, including its little noticed groundwater component. The need for such systems can only grow as the dual forces of climate change and growing urban population increase pressure on the groundwater resources of cities.

References

Aldrich, H. P., and J. R. Lambrechts. 1986. Back Bay Boston, Part II: Groundwater levels. Civil Engineering Practice **1**:31–64.

Al-Rashed, M.F., and M.M. Sherif, 2001. Hydrogeological aspects of groundwater drainage of the urban areas in Kuwait City. Hydrological Processes **15**: 777–795.

Barrett, M. H., K. M. Hiscock, S. Pedley, D. N. Lerner, J. H. Tellam, and M. J. French. 1999. Marker species for identifying urban groundwater recharge sources: a review and case study in Nottingham, UK. Water Research **33**:3083–3097.

Barton, N. 1992. The Lost Rivers of London, Second Edition. Historical Publications Ltd., London.

Brassington, F. C. 1990. Rising groundwater levels in the United Kingdom. Proceedings of the Institution of Civil Engineers, Part 1 **88**:1037–1057.

Cherry, H. 2007. Groundwater Levels in the Chalk-Basal Sands Aquifer of the London Basin. Environment Agency, Bristol. http://www.environment-agency.gov.uk/ common-data/acrobat/gw_report_june2007_1831580.pdf. Accessed September 18, 2007.

Clow, D. G. 1989. Rising groundwater in urban areas—implications for British Telecom. British Telecommunications Engineering **8**:176–181.

DEFRA, 2007. e-Digest Statistics about: Inland Water Quality and Use, Water levels. Department for Environment, Food and Rural Affairs, London. http://www.defra.gov.uk/environment/statistics/inlwater/iwlevels.htm. Accessed September 28, 2007.

Domenico, P. A., and F. W. Schwartz, 1998. Physical and Chemical Hydrogeology, Second Edition. John Wiley & Sons, New York.

Downing, R. A. 1994. Keynote paper: Falling groundwater levels—a cost-benefit analysis, pp. 213–236. In: W. B. Wilkinson, editor. Groundwater Problems in Urban Areas. Proceedings of the International Conference organised by the Institution of Civil Engineers and held in London, 2–3 June 1993. Thomas Telford, London.

Drangert, J. O., and A. A. Cronin, 2004. Use and abuse of the urban groundwater resource: Implications for a new management strategy. Hydrogeology Journal **12**: 94–102.

Driscoll, F. G., 1986. Groundwater and Wells, Second Edition. Johnson Division, St. Paul, Minnesota.

EA, 2002. Managing Water Abstraction: The Catchment Abstraction Management Strategy process. Environment Agency, Bristol, United Kingdom. July 2002. (http://www.environment agency.net/commondata/acrobat/mwa˙english.pdf)

EA, 2006. The London Catchment Abstraction Management Strategy Final Strategy Document. Environment Agency, Bristol, United Kingdom. April 2006. (http://publications.environment-agency.gov.uk/pdf/GETH1205BJYD-e-e.pdf)

Fetter, C. W., 1999. Contaminant Hydrogeology, Second Edition. Prentice Hall, Upper Saddle River, New Jersey.

Fetter, C.W. 2001. Applied Hydrogeology, Fourth Edition. Prentice Hall, Upper Saddle River, New Jersey.

Foster, S. S. D., and P. J. Chilton, 2004. Downstream of downtown: urban wastewater as groundwater recharge. Hydrogeology Journal **12**:115–120.

Freeze, R. A., and J. A. Cherry, 1979. Groundwater. Prentice Hall, Englewood Cliffs, New Jersey.

Halliday, S., 1999. The Great Stink of London: Sir Joseph Bazalgette and the Cleansing of the Victorian Capital. Sutton Publishing, Gloucestershire, UK.

Heath, R. C., 1982. Basic Ground-Water Hydrology. Water-Supply Paper 2220. U.S. Geological Survey, Washington, D.C. (http://water.usgs.gov/pubs/wsp/wsp2220/)

Hiscock, K. 2005. Hydrogeology: Principles and Practice. Blackwell Publishing, Malden, Massachusetts.

Hutson, S. S., N. L. Barber, J. F. Kenny, K. S. Linsey, D. S. Lumia, and M. A. Maupin. 2004. Estimated Use of Water in the United States in 2000. Circular 1268. U.S. Geological Survey, Reston, Virginia. (http://pubs.usgs.gov/circ/2004/circ1268/)

IPCC, 2007. Climate Change 2007. The Physical Science Basis. Contribution of Working Group I to the Fourth Assessment Report of the Intergovernmental Panel on Climate Change. Cambridge University Press, Cambridge, United Kingdom. (http://ipcc-wg1.ucar.edu/wg1/wg1-report.html)

Johnson, S., 2006. The Ghost Map: The Story of London's Most Terrifying Epidemic—and How it Changed Science, Cities, and the Modern World. Riverhead Books, New York.

Johnson, S. T. 1994. Keynote paper: Rising groundwater levels: engineering and environmental implications, pp. 285–298. In: W. B. Wilkinson, editor. Groundwater Problems in Urban Areas. Proceedings of the International Conference organised by the Institution of Civil Engineers and held in London, 2–3 June 1993. Thomas Telford, London.

Lambrechts, J. R. 2000. Investigating the Cause of Rotted Wood Piles, pp. 590–599. In: K. L. Rens, O. Rendon-Herrero, and P. A. Bosela, editors. Forensic Engineering, Proceedings of the Second Conference, San Juan, Puerto Rico. American Society of Civil Engineers, Reston, Virginia.

Lawrence, A. R., D. C. Gooddy, P. Kanatharana, W. Meesilp, and V. Ramnarong. 2000. Groundwater evolution beneath Hat Yai, a rapidly developing city in Thailand. Hydrogeology Journal **8**:564–575.

Lerner, D.N., D. Halliday, and M. Hoffman, 1994. The impact of sewers on groundwater quality, pp. 64–75. In: W. B. Wilkinson, editor. Groundwater Problems in Urban Areas. Proceedings of the International Conference organised by the Institution of Civil Engineers and held in London, 2–3 June 1993. Thomas Telford, London.

Lundy, D., and T. Gogel. 1988. Capabilities and limitations of wells for detecting and monitoring liquid phase hydrocarbons, pp. 343–362. In: Proceedings of the Second National Outdoor Action Conference on Aquifer Restoration, Ground Water Monitoring and Geophysical Methods. National Water Well Association, Dublin, Ohio, Las Vegas, Nevada.

Mercer, J. W., S. D. Thomas, and B. Ross, 1982. Parameters and Variables Appearing in Repository Siting Models. NUREG/CR-3066. U.S. Nuclear Regulatory Commission, Washington, D.C. December 1982.

Prescott, S. C., and M. P. Horwood. 1935. Sedgewick's Principles of Sanitary Science and Public Health. The Macmillan Company, New York.

Richards, H. C., and W. H. C. Payne. 1899. London Water Supply being a Compendium of the History, Law, & Transactions relating to the Metropolitan Water Companies from Earliest Times to the Present Day, Second Edition. P.S. King & Son, London.

Shanahan, P., and B. L. Jacobs. 2007. Ground water and cities. In: V. Novotny and P. Brown, editors. Cities of the Future: Towards Integrated Sustainable Water and Landscape Management. IWA Publishing, London.

Sharma, H. D., and K. R. Reddy. 2004. Geoenvironmental Engineering. John Wiley & Sons, Hoboken, New Jersey.

Simpson, R. W. 1994. Keynote paper: Quantification of processes, pp. 105–120. In: W. B. Wilkinson, editor. Groundwater Problems in Urban Areas. Proceedings of the International Conference organised by the Institution of Civil Engineers and held in London, 2–3 June 1993. Thomas Telford, London.

Snow, B. F. 1936. Tracing loss of groundwater. Engineering News Record **117**:1–6.

Stedman, L., 1999. Businesses act in attempt to keep London dry. Water & Waste Treatment **40**(5): 16 (http://www.edie.net/library/view˙article.asp?id=1845).

Todd, D. K., and L. W. Mays, 2005. Groundwater Hydrology, Third Edition. John Wiley & Sons, Hoboken, New Jersey.

USGS, 2008. USGS Ground-Water Data for the Nation (http://waterdata.usgs.gov/nwis/gw. Accessed June 9, 2008).

Ward, R. 2003. London's New River. Historical Publications Ltd., London.

Walters, R. C. S., 1936. The Nation's Water Supply. Ivor Nicholson and Watson Limited, London.

Winter, T. C., J. W. Harvey, O. L. Franke, and W. M. Alley, 1998. Ground Water and Surface Water: A Single Resource. Circular 1139. U.S. Geological Survey, Denver, Colorado (http://water.usgs.gov/pubs/circ/circ1139/).

WWAP, 2003. Water for People, Water for Life, The United Nations World Water Development Report. World Water Assessment Programme, UNESCO.

Chapter 4
Urban Infrastructure and Use of Mass Balance Models for Water and Salt

Paul Westerhoff and John Crittenden

4.1 Introduction

Urban infrastructure supplies water to urban areas and drains away sewage and stormwater. These services are critical to the health and prosperity of modern cities. The built infrastructure includes reservoirs, concrete channels, canals, pipes, pumps, and treatment facilities. This infrastructure is usually owned by public entities (cities or water and sewer agencies). Some smaller sized systems are operated by private companies who can adequately train personnel and achieve economies of scale through operating facilities for several localized treatment and distribution systems. Privatization for operations of large-sized water and wastewater systems (i.e., serving >100,000 people) is slowly expanding in the USA, although the cities still own the infrastructure. Public entities bill private residences, commercial and industrial users, etc. to repay the enormous capital investment of this infrastructure, reoccurring replacement and repair and continuous operating expenses. In fact, the electrical demand for pumping water is usually among the highest single recurring expenses for cities, just behind the high costs related to human resources (salaries, benefits, etc). Some of these facilities are above ground, but hundreds of miles of pipes are buried 3 to 20 ft underground. Most components of this built urban water environment are constructed with materials designed to last 25 to 50 years and represent enormous capital investments (i.e., billions of dollars). Creation and operation of urban water systems fundamentally changes the natural hydrologic flow across landscapes (Chapter 2). Consequently, studying water and material fluxes in urban settings requires a basic understanding of the key components comprising the urban water infrastructure.

This chapter describes the typical water consumption/production patterns within urban societies and the infrastructure required to support the efficient use and movement of water. The first part of the chapter provides an overview on critical components of the urban water system. The second part of the chapter is a case

P. Westerhoff (✉)
Arizona State University, Tempe, AZ, USA
e-mail: p.westerhoff@asu.edu

study into modeling fluxes of water and pollutants through an arid-region water infrastructure system. Water in arid regions exemplifies the degree to which urban water systems can become almost completely managed, with very little remnants of natural hydrologic flow or aqueous material fluxes.

4.2 Urban Water Infrastructure Components

Urban water systems include four major categories of engineered infrastructure components: (1) water supply, (2) drinking water, (3) wastewater/sewage, and (4) stormwater. Water supply, drinking water and wastewater systems are interdependent upon each other. Information is presented below for each of the major urban water infrastructure components.

4.2.1 Water Supply

Urban areas require water for agricultural, industry, commercial, and residential uses (Fig. 4.1). Agriculture and several types of large industries (e.g., power production facilities) use large quantities of water which is usually separate from water that is centrally treated at a water treatment plant and pumped into pipe networks for commercial or residential water users. A common unit for measuring volumes of water is *acre-feet*, or the volume of water that when spread out covers 1 acre of land to a depth of 1 foot. (1 acre-ft equals 43,560 ft^3 ~ 325,800 gallons ~ 1230 m^3). For the last 20 years in the USA, water usage has been approximately 400 billion gallons per day (USGS, 2004). Agricultural crops use approximately 34% of this

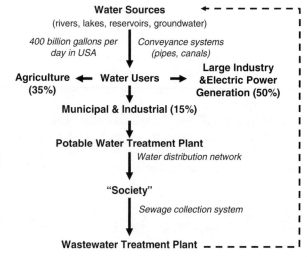

Fig. 4.1 Connectivity of water supply and water usage systems. Details provided for the urban water system. *Lines* indicate flow direction of water. *Dashed line* indicates discharge of treated wastewater effluent back to water sources

water, through application of 6 to >10 ft of water per acre per year. Power generation is the largest water user with the national industrial sector, approximately 50% of the domestic water usage. The most common form of electricity production, thermo-electric power generation from coal, oil, nuclear, consumes approximately 21 gallons per kilo-watt-hour (kWhr) of electricity produced; a typical residential house uses on the order of 20 kWhr of electricity per day, which equates to using over 400 gallons per day per household of water via electricity production. Approximately half of this water comes from fresh water supplies and half from brackish water source unsuitable for human consumption. A portion of the water withdrawn from surface waters for power production evaporates, and is termed *consumptive use*. However, most water passes through the power plant, returns as warm water to the river or lake, and is termed *non-consumptive use*. The returning warmer water can pose risks to downstream aquatic organisms and fisheries.

In addition to industrial water use, a typical "person in society" is responsible for consumption of approximately 90 to 320 gallons per capita per day (*gpcd*) of treated potable water. Potable drinking water is usually centrally treated (water treatment plant, WTP) and pumped out into the urban area under pressure in a vast underground pipe network. Potable water uses include direct drinking and cooking water, toilets, bathing, clothes washing, outdoor irrigation, restaurants, car washes, commercial buildings, malls, and industries. Figure 4.2 illustrates the distribution of annual water use by households across the country. Indoor water use, which is non-consumptive because it returns as sewage to centralized wastewater treatment plants (WWTPs), averages 63,000 gallons per household per year (roughly 69 gpcd) and has very little geographic dependence. However, outdoor water use (consumptive use) varies geographically from <8000 gallons per household per year to >200,000 gallons per household per year.

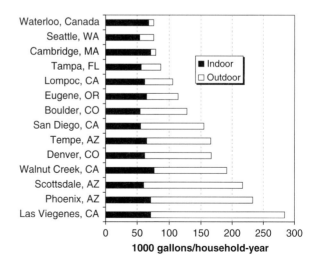

Fig. 4.2 Geographic distribution of indoor and outdoor water use patterns for households served with public drinking water (adapted from Mayer et al. 1999)

Water demands exhibit seasonal patterns. Agriculture needs more water during the summer. Thermo-electric power plants require more cooling water during peak energy use periods (i.e., summer time). Potable water demand is roughly two to three times greater during summer than winter months, and is almost regardless of geographic location.

Rivers, lakes, reservoirs, and groundwater all can store and serve as water supply sources for urban uses. There are roughly 55,000 public water supply systems in the USA. Approximately 64% of public water supplies are surface waters and 36% are groundwaters. Smaller communities (<50,000 residents) are more often served by groundwater. Many urban areas import water from water sources located tens to hundreds of miles away to meet the seasonal and total water demands of the urban population. This is true in both temperate areas (e.g., New York City imports water from the Catskill Mountains over 50 miles away), mountainous areas (Denver, located on the east side of the continental divide, imports water from the wetter western side of the divide) and arid areas (Southern California imports water hundreds of miles from both the Sacramento River in Northern California or Colorado River in Arizona). Surface water impoundments (i.e., lakes and reservoirs) are critical for storing water during wetter periods of the year(s) for drying periods. The average time water spends in a reservoir (i.e., *hydraulic residence times*, HRT) can vary from a few months in Eastern US reservoirs to years (western US reservoirs) or decades (Great Lakes). HRT is calculated by dividing the reservoir volume (ft^3) by the outflow (ft^3 per day, month, or year). Reservoirs are usually located on rivers, where a dam is constructed to allow the reservoir to fill (Fig. 4.3). Outlet structures have "gates" at different depths to permit discharge/release of water from the reservoir downstream into the river.

The water quality of surface waters can be adversely affected by natural or anthropogenic means. During warmer months of the year reservoirs will thermally stratify as warmer, less dense water "floats" above colder, denser water in the depths

Fig. 4.3 Photographs of a typical outlet structure located in a reservoir (Lake Pleasant, Arizona)

of a lake. In the upper sunlight strata of a lake, algae grow and can produce organic chemicals that are neural toxins or cause off-flavor or odors (Baker et al. 2000). Increases in nutrient levels from land surfaces or industrial discharges can lead to increased algae activity, which detrimentally impacts the quality of the water as a drinking water supply and would necessitate more costly water treatment processes to be implemented later. In the lower strata of a lake decomposition of organic matter that settles to the bottom causes bacteria to grow and consume oxygen. When all the dissolved oxygen is consumed, biogeochemical processes can transform metals present in sediments (e.g., arsenic, manganese, iron) into forms that are soluble in the water and would require treatment to remove them prior to use as drinking water. Land development within watersheds lead to numerous adverse water quality effects on rivers and lakes that impact not only recreational and fishing activities (i.e., nutrients or bacteria in stormwater runoff), but also leads to more extensive treatment in drinking water treatment plants before it can be safe to drink.

Groundwater is also an important water supply source for several reasons: (1) groundwater does not require construction of large above ground reservoirs, (2) groundwater can be pumped during periods of drought when reservoirs are low, and (3) groundwater usually is of fairly high quality. Groundwater is "stored" and flows in the pores of the sediment. Wells are drilled into the ground to depths of tens to hundreds of feet to access the water. Most larger municipal groundwater systems require pumping to bring the water to the surface for use, as the aquifers are not "artesian" wells. Artesian wells are in-confined aquifers and water will flow upwards out of the well without need for pumping.

Historically groundwater supplies required little, if any, treatment before potable use. Groundwaters may require treatment to remove naturally occurring pollutants (e.g., arsenic), unpleasant odors (e.g., hydrogen sulfide), high mineral content (hardness or other salts), chemicals that cause staining (e.g., iron or manganese) or anthropogenic pollutants (e.g., gasoline additives such as MTBE or solvents such as TCE). Nitrate is among the most pervasive pollutant found in groundwaters across the USA, primarily due to use of nitrogen fertilizers for agriculture. Nitrate percolates through the soil and into the groundwater. In drinking water nitrate causes methemoglobinemia, or "blue baby" disease. Many groundwater supplies have been abandoned as drinking water supplies because of nitrate and the costly treatment necessary to remove nitrate. New federal regulations enacted in 2006 (USEPA Groundwater Disinfection Rule) also aims to improve the safety of groundwaters used as drinking water supplies by mandating disinfection to kill microorganisms in the water. Because of pollution and new regulations there is an increasing demand to treat groundwater prior to its use as a drinking water.

Pumping groundwater consumes electricity. The amount of electricity required to lift water out of the ground is a function of the desired flowrate (gallons per minute, gpm), height to which the water has to be lifted out of the ground (feet) and efficiency of the pump. Pumping a series of wells from a depth of 100 ft at a rate of 1 million gallons of water per day (1 MGD), enough to supply roughly 6700 people, would consume roughly 145,000 kWhr of electricity per year. At a cost of $0.10 per kWhr, pumping 1 MGD costs roughly $14,500 per year just for pumping the water.

Unrestricted pumping of groundwater can lead to significant declines in groundwater levels (i.e., mining of groundwater) because the rates of withdrawal are greater than the rates of natural recharge. *Safe yield* is a term that refers to the maximum amount of groundwater that can be withdrawn from an aquifer without long-term decline in its water level because the withdrawal rate is equal to or lower than the rate of natural groundwater recharge. As groundwater levels drop (i.e., water level in wells is deeper) a number of problems emerge: (1) less resilience during droughts as wells run dry, (2) increased cost of pumping water from deeper depths, (3) increased risk of land subsidence, or (4) more variable water quality.

Many municipalities in arid regions intentionally recharge surface water or treated wastewater into the ground, using recharge basins (Fig. 4.4), in an effort to increase future groundwater pumping capabilities without exceeding an aquifers safe yield. Many cities (Palm Springs, metro-Phoenix, Tucson, Colorado Springs, Las Vegas, and southern Florida) use surface recharge, or subsurface injection of water (i.e., aquifer storage and recovery) and subsurface aquifers for long-term storage and even treatment of chemicals in the water (i.e., soil aquifer treatment). Another growing trend is referred to as riverbank filtration. Groundwater wells are drilled adjacent to or laterally beneath rivers and water is "sucked" through sediments

Water is conveyed from surface and ground water supplies to locations of water demand (e.g., farms, industry, potable WTPs). River channels serve as a common conveyance system, but many urban areas also employ large-diameter (8 ft to over 12 ft in diameter) buried pipes or aqueducts (e.g., New York City, Boston, Las Vegas) or concrete-lined canals (Phoenix, San Diego, Los Angeles) to move water. Figure 4.5 illustrates an example of a concrete-lined canal in central Arizona. Concrete linings are used to prevent loss of water from the canal (i.e., seepage into the groundwater).

Fig. 4.4 Recharge basin located near a river channel that has high soil permeabilities. A wastewater treatment plant is located in the background, and treated wastewater flows into this recharge basin to supplement regional groundwater supplies

Fig. 4.5 Typical concrete canal in the western USA delivering water from distant reservoirs and rivers to urban regions in the desert

Potable WTPs purify water to make it suitable for human consumption. Typical *unit treatment processes* involve addition of metal salts that aid in removal of inorganic clay-like particles and biological pathogens (bacteria, virus, protozoa) in sedimentation and filtration systems (i.e., pathogen removal via physical processes), followed by application of chemical disinfectants (chlorine or chloramines) for chemical disinfection. Additional processes (e.g., granular activated carbon in deep filters) can be added to the typical processes to improve removal of organic pollutants or compounds that impart off-flavors or odors to water. Figure 4.6 is an aerial photograph of a drinking water treatment plant with several of the unit treatment processes labeled. It takes roughly 4 to 6 hours for water to pass through a WTP.

There are hundreds of miles of buried pipes in urban areas to deliver water from water sources (WTPs, groundwater wells) to houses, commercial buildings, schools, etc. High pressure pumps pressurize the water to roughly 80 pounds per square inch (psi) in buried pipe networks. The pressure of 80 psi is roughly enough pressure to pump water to an elevation of 180 ft above ground level ($\sim$ a 15 story building). Pressure is needed to move water across the topography of a city, provide pressure to flow water out of showers and other domestic uses, and most importantly provide sufficient pressure to fight fires. In fact, the size of pipes buried in the ground

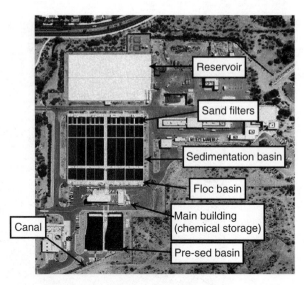

Fig. 4.6 Aerial photograph of a water treatment plant in Arizona. Water flows from the canal, into the presedimentation basin, through the flocculation system where coagulants are added to help remove particles and organics, into the sedimentation basins where particles settle out into sand filters (3 to 8 ft deep) and particle removal also occurs, then chlorine is added before water flows into the reservoir. Drinking water is pumped from the reservoir into the pipe network into "society". Note the facility is surrounded by desert and encroaching urban development from the top of the picture frame

and minimum required pressures in water distribution system networks of pipes is designed around the ability to meet required fire flow and fire fighting pressure requirements. Pipes range from 8 ft in diameter leaving WTPs, with many common pipes being 24 to 48 inches in diameter, and pipes in residential neighborhoods on the order of 2 to 6 inches in diameter. Most pipes are constructed from some type of steel, except for the larger pipes which are lined concrete pipes. There are hundreds to thousands of miles of buried pipes buried in the ground and used to move around treated water, and thousands of valves to control the flow (i.e., water distribution system). Water moves through these buried pipes at approximately 1 foot per minute, resulting in frictional losses which decrease the pressure of water in the pipes. Because of these pressure drops and changes in elevation, most cities have multiple "pressure zones" where water pumped again to increase pressure to serve the public. Storage tanks scattered around cities help regulate pressures and available water, such as the large metal tanks typically located atop mountains or "lollipop" metal tanks elevated into the air. Average travel times of water from a centralized WTP to your house may range from 1 to 3 days, and in some areas take over 7 days.

Water distributions systems plan an important role in understanding the overall sustainability of urban water systems. First, water is heavy and pumps required to move the water consume significant amounts of electricity. Approximately 30% to

50% of the entire electrical use for all city services comes from pumping of water. Second, there are nearly 240,000 water main breaks per year in the USA. These breaks and slow leaks lead to a loss of nearly 1.7 trillion gallons per year (USEPA 2007).

4.2.2 Sewage Collection and Treatment

Depending upon the region of the country, between 50% and 75% of the water distributed from the WTP (~ 100 gpcd) is discharged to sanitary sewers. The remainder is used for outdoor irrigation or evaporated in commercial cooling systems (i.e., consumptive uses). Domestic water used for showers, toilets, and washing clothes mixes with sewage from restaurants and industry in a network of buried pipes (sewage collection systems). In contrast to the potable water system which contains pressurized water in pipes (i.e., pipes are full of water), sewage collection systems flow by gravity in partially full-flowing pipes. As such the pipe diameters are larger. Common materials are clay or lined concrete. Manholes are located every 300 to 500 ft for *every* pipe within sewage collection systems in order to provide (1) locations to re-align (i.e., turn) pipes (2) access for maintenance, and (3) ventilation of volatile gases (odors, methane, etc).

Sewage flows from households and commercial sources through the sewage collection system to centralized WWTPs. WWTPs include sedimentation systems to remove sand, grit and other heavy debris, but rely primarily on biological treatment processes to purify the water (Fig. 4.7). Bacteria consume nutrients (carbon and nitrogen) in the sewage (i.e., proteins, carbohydrates, soaps, etc) and convert them into gases (methane, carbon dioxide, nitrous oxides) and biomass (bacterial cells) (see Chapter 5). The gases are discharged to the atmosphere, although methane can be captured and burned to heat water and keep the bacteria warm. All the gases pose

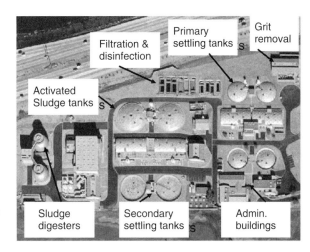

Fig. 4.7 Aerial photo of a wastewater treatment plant. Sewage flows into the grit removal to remove large objects, then into primary settling tanks, activated sludge tanks, secondary settling tanks, filtration and disinfection systems. Settled materials (biosolids) are digested in sludge digesters where they produce methane gas that is recovered to help operate the facility

somewhat of a global warming concern. Biomass is dewatered (i.e., squeezed through filter presses) and then (1) sold as fertilizer, (2) incinerated to produce electricity, or (3) landfilled. To reduce the volume of biomass solids that have to be trucked off and disposed, larger WWTPs operate anaerobic digesters to produce methane, which is collected and burned to provide electricity and/or heating at the facility.

Treated wastewater is briefly disinfected using chlorine or ultraviolet light before being discharged into surface waters (rivers, lakes, streams) or groundwater (i.e., Fig. 4.3). The quality of wastewater effluent is often characterized by (1) the organic carbon strength (i.e., biochemical oxygen demand – see Chapter 5), because bacteria growing in receiving waters consume this carbon and simultaneously consume dissolved oxygen (DO) in water, leading to fish kills, (2) the amount of nitrogen and/or phosphorous because algae growth from these nutrients lead to biomass in rivers which consumes oxygen as it decays, (3) levels of bacteria that may inhibit recreational use of the rivers, and (4) known toxic, regulated pollutants (heavy metals, pesticides, etc).

In addition to known toxins, there has been growing concern for the presence of trace-level (parts per trillion) of pharmaceutical compounds (e.g., synthetic estrogen from birth control) in sewage effluents which can affect the gender of fish in rivers. Therefore, where these or other issues are a concern advanced wastewater treatment processes are added to the WWTP. There is also growing concern about salt pollution for the southwestern USA, because household detergents and chemicals, food, water softeners and industries add salt to sewage. Salt pickup is usually 200 to 400 mg/L, which can make it difficult to reuse treated wastewater for beneficial uses (public outdoor irrigation) if the potable water also has a high level of salts. Salt concentrations, measured as total dissolved solids (TDS), exceeding 1000 to 1200 mg/L will not be suitable for irrigation of most grasses (i.e., golf courses, parks, etc). WWTPs that need to remove salts and trace organics often use *reverse osmosis* (RO) technologies. Because a lot of energy is needed to pressure water for RO systems (e.g., operating pressures are 200 to 300 psi), they are very energy intensive. Inclusion of RO treatment at WWTPs more than doubles the initial infrastructure and operating costs. Figure 4.8 illustrates a mass balance on water and salt during RO treatment. Treating 10 MGD of wastewater containing 1130 mg/L of total dissolved solids (i.e., salts) by RO will produce approximately 8.5 MGD of low-salt (27 mg/L), clean, permeate water and 1.5 MGD of high-salt (7380 mg/L) reject water. Disposal of the reject water can be challenging, as it poses risks to aquatic life because of the high salt levels.

4.2.3 Stormwater Collection

Removing rainwater from streets and land surfaces is critical to prevent localized flooding and associated damages. Therefore, urban water systems include a third network system comprised of stormwater drains, localized treatment, and buried pipe collection systems. Localized treatment usually involve grass lined swales or

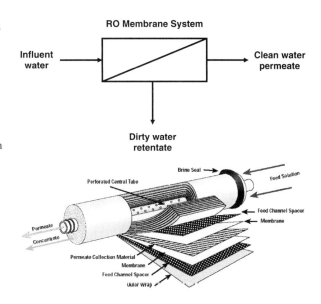

Fig. 4.8 Representation of a mass balance on water and salt in a reverse osmosis (RO) system and image of RO membranes at a wastewater treatment plant. Membrane systems include high pressure pumps, chemical feed facilities and membrane elements. Bottom photograph is spiral-wound RO membrane element (source: www.pyrocrystal.it/sito%20secondario2/osmosys.htm)

small retention basins located in neighborhoods, parking lots, and commercial sites. Roughly 50% of many pollutants in stormwater (e.g., metals, nutrients, organic toxins, bacteria) are associated with particulate matter. Providing time for particles to settle our or filter out dramatically improves the quality of stormwater. However, 50% of the pollutants are not associated with particles (i.e., soluble) and remain in the water. Some pollutants, such as ions from road salts are 100% soluble and are not removed by filtration processes. Common stormwater control measures include grass buffer strips, grass-lined swales, retention ponds, or infiltration basins (See Chapter 5).

Rainfall patterns vary widely across the USA, and as such stormwater systems are designed around local hydrologic considerations. Rainfall frequency and intensity historic data is used to design the number and size of stormwater drains and pipe systems. Roads are beveled/sloped to prevent flooding, but must be drained. Drains are usually located every 500 ft along roads, and more frequent around intersections. Stormwater pipes are large diameter (12 to over 96 inches in diameter) and located beneath nearly every street. Stormwater often flows without any type of treatment to improve water quality into rivers or lakes. In older communities such as Pittsburg, New York and Chicago, stormwater and sewage systems remain connected; these sewers are termed *combined sewers*. This is undesirable, because during larger flow events not all this water can be treated, and consequently is discharged to rivers and streams without treatment (i.e., *combined sewer overflow*). The surface waters become polluted with untreated sewage, nutrients, and microorganisms. There are up to 75,000 sanitary sewer overflow events per year in the USA, resulting in discharge of 3 to 10 billion gallons of untreated wastewater (USEPA 2007). This leads to up to 3700 illnesses annually due to exposure in recreational waters contaminated by sanitary sewer overflows (USEPA 2007).

4.3 Case Study of Modern Integrated Urban water Modeling

The text above describes the components of modern urban water systems. However, integration of these systems across the urban landscape leads to profound differences in urban hydraulics from natural hydrologic processes. A case study is described here to provide an example of how several urban water components become interconnected. The example is from a city in the Sonoran Desert of Arizona because less water falls within city boundaries than is required by the >250,000 residences. Therefore, it serves as a representative model for many other cities around the country that face similar water issues. In addition to "wet" water alone, the quality of the water is modeled because this can affect the beneficial uses of the water.

4.3.1 Modeling Approach

A modeling approach that integrates water sources/uses, water treatment and distribution, wastewater collection and treatment and stormwater will be discussed in terms of both the quantity and quality of water for one city in the southwestern USA (Scottsdale, Arizona). Per capita water use is over 150 gpcd and there are more than 20 golf courses that exclusively use reclaimed wastewater for irrigation. Salt levels are naturally high in the southwestern USA, and as an inland basin (no connection to the ocean) salt accumulation in soils and economic impact on in-home and commercial plumbing appliances are serious concerns. Salts preclude many beneficial uses of water. When water evaporates during outdoor irrigation, the salt in the water stays behind and accumulates in the soil and/or groundwater. This poses important risks to long-term sustainability of soils for growing plants since salts in the root zone inhibit or prevent plant growth, and salts in groundwater leads to undrinkable water unless RO technologies are employed to desalt the water. Because salts move through the urban infrastructure without evaporating, they are good tracers of other pollutants in the water. For these reasons, salts are used in this case study model as an indicator of water quality.

A single city is modeled for one year, but the concepts and models developed herein are transferrable to other cities, over any timescale, and can be run under any desirable scenario (e.g., long-term droughts, floods, urban development growth patterns, etc). These activities are currently underway and are providing deep insight as long-term management planning techniques.

The City of Scottsdale, Arizona was modeled. It has a population of nearly 250,000, spans 180 square miles, and annually receives 9 inches of precipitation. Potable water supplies for Scottsdale include Salt River and Verde River water delivered through Salt River Project (SRP), Colorado River water delivered through

Central Arizona Project (CAP) canal, and ground water wells. There are between 20 and 25 golf courses in the city. Wastewater is treated and reclaimed for golf course irrigation as well as groundwater recharge. Most stormwater runoff is conveyed in the Indian Bend Wash (IBW) south to the Salt River. A few other stormwater drainages exist but constitute minimal flow and most water in these washes and other areas of the cities percolate into the ground. Data on drinking water and wastewater was supplied by the city and other agencies.

A conceptual illustration for the modeling framework is presented in Fig. 4.9. The model illustrates inflow and outflow of water, which carries dissolved salts, through the city boundaries. Within the city water is used. The conceptual model

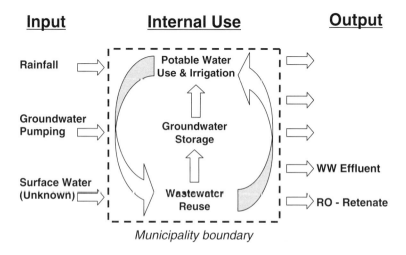

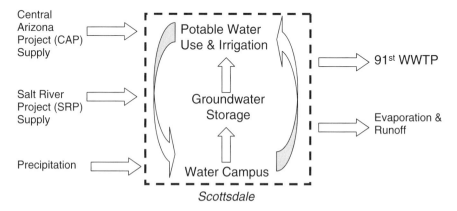

Fig. 4.9 Conceptual framework for integrated urban water modeling of water and salt mass balances. *Upper diagram* is general for an urban area. *Lower diagram* is specific for Scottsdale, Arizona

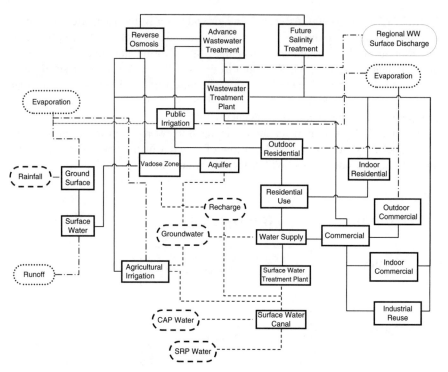

Fig. 4.10 Schematic representation of water and salt flux model. *Ovals* and *dashed lines* represent sources of water into the city or processes that result in water leaving the city. *Rectangles* and *solid lines* represent parts of the urban water system that use water

was mathematically represented using a simulation software program (Powersim Software (www.powersim.com)). PowerSim simulates multi-source fluxes (i.e., movement of materials across time). The PowerSim representation of water quantity fluxes for the City of Scottsdale is illustrated in Fig. 4.10. Water sources (ovals with dashed lines) or sinks (ovals with dotted lines) are linked (arrows) to storage components (rectangles). The storage components include WTPs and WWTPs, and other key aspects for the urban system. The vadose zone (unsaturated soils between the ground surface and groundwater table) and groundwater are represented as separate components because of long retention times in each and because accumulation of water (or salt) in these components have significantly different management implications. "Society" is represented as a single component (labeled "Potable Use") because it fundamentally changes where water is used (i.e., indoor/outdoor use, changes water quality). Overall the mass balance on water moving into or out of a city, and change in groundwater storage is represented by the following relationship: Input Volume = Outflow Volume + Evaporation Volume + Change in Groundwater Storage Volume (See also Chapter 3). The units of water volume are expressed here as acre-ft. One acre-foot is approximately the amount of water that a typical household uses in one year.

The time dependency of water and salt movement (i.e., flux) for different seasons is captured in the model. In most cases these time-dependent relationships are available from the city (e.g., daily water produced each day of the year), streamflow gauging stations (e.g., ft^3/sec), or estimated using standard empirical relationships (e.g., fraction of rainfall converted to runoff). Details of these relationships are provided elsewhere, and all models are run for the calendar year 2005 (Zhang 2007). As one example, based upon a 30-year continuous rainfall data and runoff capture curve for Phoenix metropolitan area it was found that if a rainfall event was less than 2.5 mm, no runoff is predicted; above 2.5 mm of rainfall (P) the volume of runoff (R) can be estimated (R = 0.9P – 2.5) (Guo and Urbonas 2002). Resulting runoff is captured in washes and/or retention basins where the runoff infiltrates into the ground. Only large or longer storm events produce significant runoff (i.e., water not recharged into groundwater).

4.3.2 Water Quantity Mass Balance Modeling

Fluxes of water and salt into/out of one Arizona City over the course of one year is shown in Fig. 4.11. The largest single source of water into the city is actually precipitation, despite being an arid region that receives only 9 inches of rainfall annually. Most of this rainfall occurred over only three or four rainfall events, resulting in "steps" or "pulses" of water into the urban system. Forty nine percent of the rainfall results in percolation into the vadose zone, 40% evaporate, and only 11% leaves as runoff. Potable water use (indoor and outdoor) is second largest water use in the city. The model accounts for seasonal variation in water demand (higher demand in summer than winter months). During the summer nearly all the wastewater is used for outdoor golf course and public parks irrigation, whereas during the winter less irrigation water is needed and it is treated by RO and recharged into the groundwater.

Figure 4.11a shows Scottsdale's annual water balance. The largest single loss of water is evaporation which occurs in response to rainfall and outdoor irrigation practices. Despite common perception, the Sonoran Desert is green because of the ecological function that the rain stimulates. Plants rapidly take up the water and then slowly evaporate the water through photosynthetic processes as the plant grows (i.e., evapotranspiration). In personal yards and public greenways, water is also evapotranspired to keep the community green with living plants. Very little stormwater runoff actually leaves the city boundaries, and only does so at periods related to the intermittent rainfall events. Only 20% of the water imported into Scottsdale from the CAP or SRP canals or fallen as precipitation results in flows out of the city as unused wastewater or stormwater runoff; all other water is internally recycled and 80% of the imported water volume eventually evaporates.

Figure 4.11 also illustrates the importance of subsurface waters for this community. The city pumped 29,000 acre-feet of water per year but only directly recharged 5,800 acre-feet directly back into the aquifer. Another 53,000 acre-feet of water did

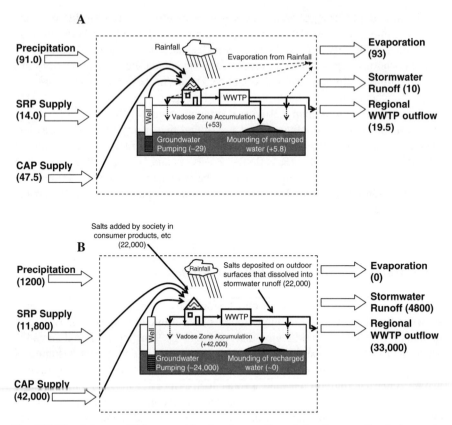

Fig. 4.11 Mass balances of (**a**) water with values in parantheses being the water flux in thousands of acre-feet of water per year (*upper*) and (**b**) salt with values in parentheses being the salt flux in tons of salt per year (*lower*)

not immediately evaporate upon use for irrigation, and is "trapped" in the vadose zone. Downward movement of water through the vadose zone into the groundwater occurs over a time scale of decades. Simultaneously with this downward water migration, gas exchange from the atmosphere and vadose zone will result in slow loss of water vapor from the vadose zone. The current state of knowledge is lacking to predict how much water or how long it will take for water in the vadose to reach the groundwater. This is a major obstacle in predicting the long term sustainability of this community as a significant volume of water is trapped within the vadose zone.

4.3.3 Water Quality Modeling: Salt Mass Balance

Salts occur naturally in many rivers. The geological formations in the southwestern USA consist of sandstones and salt deposits laid down by ancient oceans. As

4 Urban Infrastructure and Use of Mass Balance Models for Water and Salt

Table 4.1 Salinity of water within Scottsdale

Component	Selected TDS (mg/L)	TDS Range (mg/L)
CAP Water Supply	650	500–800
SRP Water Supply	620	450–1200
Groundwater Supply	620	200–1500
Wastewater (including reclaimed wastewater for irrigation)	1130	900–1500
Reclaimed water for recharge	27	20–50
Brackish Water from Reverse osmosis treatment of wastewater from AWT	7380	6000–1000
Runoff to the Salt River	350	60–700

surface and ground waters flow through these deposits, salts dissolve into the water. Salt pose potential economic, aesthetic and potential health issues when present at concentrations above 500 mg/L in drinking waters. To address salt flux issue TDS, which is a measurement of the salt concentration of water, was employed in the model as an indicator of water salinity. TDS levels for different water sources were based from local sources (Table 4.1). Salt fluxes (mass per time) are computed from the TDS concentration (mass per volume) times the water flowrate (volume per time) within the PowerSim modeling program. The following equation represents how the balance between salts moving into and out of a city, and the potential to accumulate salts within the city:

$$\text{Inflow Mass of Salt in Water} + \text{Salt Added by Society}$$
$$= \text{Outflow Mass of Salt in Water} + \text{Salt Accumulated in Soils}$$

Figure 4.11b summarizes the modeling of salt flux through the city for one year. Over the course of a year the single largest source of salt is from the CAP canal. Groundwater pumping moves 24,000 tons per year of salts out of the ground and into the urban water system. Because of household chemical, water softeners and other uses by the public, a significant amount of salts (22,000 tons per year) are added to sewage water. While rain has very low TDS (<10 mg/L) once it contacts the land surface it "dissolves" (i.e., solubilizes) salts from road and land surfaces that were deposited from the atmosphere, fertilizers, etc. This accounts for approximately another 22,000 tons of salt per year entering the urban water system. Consequently, runoff events result in significant mobilization of salts internally within cities boundaries, and in the runoff leaving the city. Because most of the stormwater flow does not leave city boundaries (Fig. 4.11a), stormwater leaving the city only transport 5% of the salt load out of the city.

While evaporation was the major way water exits the urban system, evaporated water is near "distilled" water quality (i.e., carries no salts away). While a small volume of water (15%) leaves the city boundaries as wastewater to a regional treatment facility (91st avenue), this flow accounts for 33,000 tons/year of salt (37% of the salt entering the city) because much of the flow contains RO brine concentrates from the city advanced wastewater treatment facility. The remaining salt, nearly 42,000 tons per year, enters the vadose zone as irrigation water percolates into the soil. As irrigation water is evapotranspired, the salts stay behind. Of the salts entering the vadose zone, 45% is from residential landscape water, 30% from percolation of stormwater runoff, 23% from golf-course irrigation water, 2% from direct groundwater recharge of CAP water and <1% from injection of RO permeate from reclaimed wastewater. Thus the vadose zone is accumulating significant amounts of salts.

Salts in the vadose zone pose other sustainability issues. Salts accumulated in soils make it difficult to grow grass, trees or shrubs because salts in the root zone are toxic to most plants. Therefore, more irrigation water is needed to move salts deeper than the root zones, which increases water use rates. Long-term accumulation of salts in the soil will also decrease permeability of the soils, preventing the vertical movement of water. Salt accumulation in groundwater will require more costly treatment of the water (e.g., RO) in order to use the water for drinking or other purposes. The migration rate of salts through the vadose zone may be on the order of 10 to 20 ft/year in agricultural soils that are heavily irrigated, but is probably significantly slower under residential irrigation practices and is currently unknown. Future research needs to determine rates of vertical water and salt movement in residential settings.

4.3.4 Using Integrated Water Models for Urban Cities

The case study demonstrates for a single city with unique water quantity and quality (salts) issues. Central to the model is the linkage between the many different components of the urban water system (stormwater, imported surface water, groundwater, wastewater), which are present in every city. In other work this modeling platform has been used to investigate several interconnected cities or larger urban regions. The model allows planners and engineers to consider a future with new or different infrastructure, which can easily be represented as "ovals" and "flux lines" in Fig. 4.10. It is quite easy to change the water quality parameter of concern, from salts to nitrate for example.

One year was simulated in the case study, but the model has also been run for 30 consecutive years into the future. The ability to simulate future scenarios is critical to identify non-sustainable trends in the urban water system, and potentially how to prevent them from occurring by addressing the issues now. For example in the model there is a large amount of salt accumulating in the vadose zone. The long term effects of these are unknown. On a timescale of 5 to 15 years this may impact plant growth; over several decades the salts may migrate to groundwater;

over longer time periods impereable clay barriers may form that affect soil properties. Being able to simulate the amount of salt entering the vadose zone can now be used to predict these potential issues. Likewise policies to reduce the amount of salt added by society through regulation (e.g., banning water softeners) or through management practices (e.g., street sweeping to remove soluble salts before rainfall events) can be simulated to determine their cost effectiveness in actually reducing salt loads to the vadose zone.

The model shows that most of the water entering Scottsdale is evaporated. Initially this may seem unsustainable. However, most of that evaporated water is serving an ecological function. Irrigation of parks, golf courses, and personal residences supports plants and trees, which in turn support birds and other animals and improve the human quality of life. Due to evaporative cooling, trans-evaporation by plants and direct evaporation of water also counteracts localized urban heat island effects. Developing water policies that limit such irrigation would change this human relationship with water and the environment it supports.

Modeling allows investigation of not only sustainability issues, but also the resilience of cities to various stressors. What would happen if a surface water supply was lost due to terrorism, natural disaster, contamination or other events? The model could easily provide a tool for evaluating various alternatives, which may be complex and involve multiple alternative sources for varying amounts of time. For example, is there enough groundwater pumping capacity to meet water demands during the summer if the SRP water supply is lost?

Modeling allows urban water managers to look towards the future and identify critical knowledge gaps. In the case study, a large volume of water and mass of salt resides within the vadose zone. There is very little knowledge on the rate of movement of water and salt through vadose zones in urban settings. For agricultural areas where 6 to 10 ft of water is applied per year for crops, the rate of downward migration of water and salt is on the order of several feet per year. Residential yards apply far less water and the downward rate of migration is unknown. Maybe none of the water from residential irrigation ever reaches the groundwater. Another example would be to understand the implications of water demand restrictions on generation of wastewater. Initially, it may seem prudent to encourage water conservation to reduce demand of surface waters (e.g., SRP or CAP supplies). If this is achieved at a household level by reducing outdoor irrigation (Fig. 4.2) then the urban ecosystem and its associated benefits will be reduced. If the water demand is achieved through reduced indoor water use, then less wastewater will be generated. In water stressed communities wastewater is a resource that is sold for public or agricultural irrigation uses (i.e., lower water quality standards than required for drinking water). Thus a water conservation policy would reduce the amount of water available as treated wastewater for irrigation, reducing the ecosystem benefits that could be realized or encouraging the use of higher quality water sources (e.g., groundwater) to irrigate those areas. Either way, implementing water conservation policies would alter the flow of water throughout the urban water system, potentially reduce ecosystem services, and requires an integrated model to understand the secondary and tertiary effects of such management planning.

4.4 Summary

This chapter introduced the major water infrastructure components and demonstrated through a case study that they are intimately connected and affect the movement of water and dissolved chemicals in water. In the past, engineers have modeled each system separately (water supply reservoirs or groundwater, WTP, WWTP, stormwater). The future of urban water systems modeling involves linking these models together across time (days, weeks, months, years, decades) and space (towns, cities, regional centers). The case study illustrated here used historic data. In order to make the models predictive in nature, mathematical algorithms should be incorporated into the models to represent water demand and use, etc. By doing so, long term effects of population growth, water use pattern changes due to pricing of water or economics of water pricing, droughts or floods due to climate change, etc can be simulated.

Acknowledgements This material is based upon work supported by the National Science Foundation under Grant No.DEB-0423704, Central Arizona – Phoenix Long-Term Ecological Research (CAP LTER). Any opinions, findings and conclusions or recommendation expressed in this material are those of the author(s) and do not necessarily reflect the views of the National Science Foundation (NSF). Information from the City of Scottsdale, AZ and discussions with their staff is greatly appreciated. Two graduate students (Chi Choi and Peng Zhang) developed the models presented here. Assistance in PowerSim modeling from Ke Li and Tim Lant at Arizona State University are also appreciated.

References

Baker, L., P. Westerhoff, M. Sommerfeld, D. Bruce, D. Dempster, Q. Hu, and D. Lowry. 2000. Multiple barrier approach for controlling taste and odor in Phoenix's water supply system, pp. 1–11. In: AWWA WQTC Conference, Salt Lake City, UT.
Guo, J. C. Y., and B. Urbonas. 2002. Runoff capture and delivery curves for storm-water quality control designs. Journal of Water Resources Planning and Management **128**:208–215.
Mayer, P. W., W. B. DeOreo, E. M. Opitz, J. C. Klefer, W. Y. Davis, and J. O. Nelson. 1999. Residential End Uses of Water. Awwa Research Foundation, Denver CO.
USEPA. 2007. Aging Water Infrastructure: Addressing the Challenge Through Innovation.
Zhang, P. 2007. Urban Water Supply, Salt Flux and Water Use. Masters of Science Thesis. Arizona State University, Tempe, AZ.

Chapter 5
New Concepts for Managing Urban Pollution

Lawrence A. Baker

5.1 Introduction

5.1.1 Chapter Goals

Most current pollution management in cities is based on treating pollution at the end-of-the pipe, after pollution is generated. This paradigm worked well for treating municipal sewage and industrial effluents – *point sources* of pollutants. Pollution from these sources has been greatly reduced since passage of the Clean Water Act in 1972. However, the remaining pollution problem in post-industrial cities is mostly caused by *nonpoint* sources – runoff from lawns, erosion from construction sites, gradual decomposition of automobiles (e.g., erosion of tire particles containing zinc and brake pad linings with copper), and added road salt from de-icing operations. The next section of this chapter shows why the end-of-pipe paradigm cannot be the primary approach for dealing with these types of pollution and why new approaches are needed.

If the traditional end-of-pipe paradigm doesn't work, what will? I will argue that a new paradigm must be based on analysis of the movement of pollutants through urban ecosystems, an approach sometimes called *materials flow analysis* (MFA). MFA can be used to identify where a potential pollutant enters an urban ecosystem, how it is transferred from one ecosystem compartment to another, and ultimately, it enters surface or groundwater to become an actual pollutant. As we will see, MFA can be used to guide the development of novel approaches for pollution management: reducing inputs of potentially polluting materials to urban ecosystems, improving the efficiency by which they are utilized for their intended purposes, and recycling them. MFA can also be used to develop strategies to mitigate legacy pol-

---

L.A. Baker (✉)
Minnesota Water Resources Center, University of Minnesota, and WaterThink, LLC, St. Paul, Minnesota, USA
e-mail: baker127@umn.edu

lution. The second goal of this chapter is to develop the basic concepts of MFA and demonstrate its application in case studies.

The new paradigm for urban pollution management must also recognize the role of individuals and households in generating pollution – and in reducing it. In modern, post-industrial cities, much of the pollution is generated by the activities of individuals and households, and reducing pollution requires that people change their behaviors. Therefore, the third goal of this chapter is to examine several conceptual models that social scientists use to understand environmental behaviors and several case studies to illustrate how "soft" policy approaches can succeed in reducing pollution or achieving other environmental goals.

Finally, the new paradigm for urban pollution management will make greater use of *adaptive management*. Adaptive management is the concept of using feedback from the environment to managers (individuals; government agencies, etc.) to modify ongoing management practices. Expanded use of adaptive management, made possible by advances in sensor technology and by vast increases in our capacity to acquire, transmit, and store data, could lead to much greater use of adaptive management for managing urban pollution. The final goal of this chapter is to examine the application for adaptive management for managing urban pollution.

Several case studies – managing lawn runoff, reducing the legacy of urban lead pollution, and managing road salt – are developed to illustrate how these concepts can be applied to problems of urban water management and pollution.

5.2 Limitations to End-of-Pipe Pollution Control

5.2.1 Success at Treating Point Sources of Pollution

The traditional focus of urban water pollution management in cities has been treatment of point sources of wastewater – mainly municipal sewage and industrial wastes. Modern wastewater treatment plants are essentially high-speed biogeochemical reactors where pollutants are degraded or transformed into manageable end products, such as sludge or harmless gases (Table 5.1). Physical and biological processes remove pollutants within a few hours in sewage treatment plants, whereas a similar level of pollutant conversion would take days to weeks in a river, during which time pollutants would impact the river. Some pollutants are actually degraded by biological processes. For example, organic matter (measured as *biological oxygen demand*, or BOD) is converted to CO_2 (Table 5.1). Many other pollutants do not actually disappear, but are "removed" only in the sense that they are transferred from the polluted water to sludge (*biosolids*) by sedimentation. The resulting sludge then contains the pollutants.

Fifty years ago, many sewage treatment systems would have been some version of sewage ponds, sometimes aerated. These were moderately effective at reducing the BOD in sewage, but did little to remove nutrients (Table 5.1). Sewage lagoons are still common in many small towns in rural areas of the United States and in

Table 5.1 Treatment reactions, end products, and typical treatment efficiencies for several types of municipal sewage treatment

			Typical treatment efficiencies, as % of inflow concentrations		
Pollutant	Reaction	End product	Sewage ponds[2]	Typical secondary treatment[3]	Advanced treatment[3]
Biological oxygen demand (BOD)	Respiration	CO_2	50–95	95	95
Nitrogen	Sedimentation, mineralization, nitrification-denitrification[1]	Sludge N_2 gas	43–80	50	87
Phosphorus	Concentration by microbes or chemical reaction with alum	Sludge	50	51	85
Suspended solids	Sedimentation	Sludge	85	95	95
Metals	Chemical precipitation and sedimentation.	Sludge	Variable	Variable	Variable

[1] Mineralization converts organic nitrogen to ammonium (NH_4^+); nitrification converts NH_4^+ to nitrate (NO_3^-); and denitrification converts nitrate to nitrogen gas (N_2). All three processes are mediated by bacteria.

[2] Values for lagoon systems are from Metcalf and Eddy (1991) and Reed (1995). Variability reflects wide variations in types of sewage pond (lagoon) designs.

[3] Values for BOD, N and P removal are averages for secondary and advanced treatment systems in the St. Paul-Minneapolis metropolitan region. Suspended solids removal is a "typical" value.

larger cities in lesser developed countries, but the majority of wastewater systems in U.S. cities use at least "secondary" treatment, high tech systems that can remove up to 95% of organic matter. What was once called "advanced treatment" – additional processes used to achieve higher removal efficiencies for nitrogen and/or phosphorus – is becoming standard for large wastewater treatment plants that discharge to waters susceptible to eutrophication.

Industrial waste treatment employs a broader range of treatment technologies, depending on the type of pollutants in the waste stream. Some common treatment technologies include ion exchange, reverse osmosis, filtration, floatation, and sedimentation. Industrial wastewater containing toxic or non-biologically degraded contaminants is generally treated on-site, before it is discharged to public sanitary sewers or waterways. This pre-treatment is necessary to protect sewage treatment plants from toxic chemicals, and to prevent transfer of toxic chemicals to effluent or sludge.

Treating municipal and industrial point sources was a necessary but not sufficient pollution management strategy. The Clean Water Act of 1972 encouraged, through regulation and economic subsidies, massive improvements in sewage treatment throughout the United States (see Chapter 9). From 1968 to 1996, the amount of BOD entering rivers from sewage treatment plants was reduced by 45%, even as the sewered population increased by 50 million (USEPA 2000a). Industrial wastewater treatment, also mandated by the Clean Water Act, has reduced the inputs of many toxic pollutants to surface waters.

5.2.2 The Nonpoint Source Problem

Sewage treatment has probably had relatively little impact on nutrient inputs to rivers, partly because municipal sewage is a fairly small source of nutrients to rivers, accounting for roughly 6% of the total N and 4% of the total P inputs to U.S. surface waters (Puckett 1995). Even without sewage treatment, point sources probably never accounted for more than 10–15% of N and P inputs to U.S. rivers. The rest comes from *nonpoint sources*, such as urban stormwater and agricultural pollution. Nonpoint sources are also the major sources of salts, sediments, lead, pesticides and many other pollutants. According to the EPA (USEPA 2000b), 45% of all assessed lakes, 39% of rivers and 51% of coastal waters are classified as *impaired*, meaning that these waters do not meet one or more *designated uses* under the Clean Water Act, with most of this impairment caused by nonpoint sources of pollution. We now need to contend with a broad set of problems for which new solutions are needed.

Urban stormwater: Nearly all large, older U.S. cities originally developed *combined sewers* that accepted both urban drainage and sanitary sewage (Chapter 1). The problem with combined sewers is that the volume of water that enters them during rain events is so large that it can't be treated in sewage treatment plants and was often discharged directly into rivers during peak flow periods (*combined sewer overflows*), severely polluting rivers. Over the past 40 years, most cities with combined sewers have re-built their sewage infrastructure into separate sanitary sewers and storm sewers.[1] Even so, until recently, urban storm water, though highly polluted, was not treated. Starting in the late 1990s, large U.S. cities were required to seek permits for discharging stormwater into rivers under the National Pollution Discharge Elimination System (NPDES). This has spurred the construction of thousands of end-of-pipe *best management practices* (BMPs) such as detention ponds, infiltration basins, constructed wetlands and rain gardens to treat stormwater. Pollution removal efficiencies are typically 30–80%, depending on the pollutant and

---

[1] Despite this effort, there are still more than 700 sewage systems serving 40 million people in the U.S. which still have combined sewer systems. Most of these are in older cities in the east and mid-west (see http://cfpub.epa.gov/npdes/cso/demo.cfm)

type of BMP (Weiss et al. 2007). They are particularly ineffective (<50% removal) at removing salts, soluble phosphorus, and bacteria. None are sustainable for prolonged periods without maintenance that generally involves transporting polluted sediments to a disposal site. Finally, the overall costs of structural BMPs (capital costs + operation and maintenance costs) are quite high. Weiss et al. (2007) estimated that the total operation and maintenance costs of wet ponds located in the Minneapolis-St. Paul area is $10,000 to 15,000 per hectare of watershed (values are expressed as "current worth"). Analysis of pollutant inputs to streets indicates that source reduction has considerable potential for reducing urban stormwater pollution (Baker 2007).

The upstream problem: In the lexicon of modern industrial ecology, pollution generation occurs in the *pre-consumption system*, the *consumption system*, and the *post-consumption system*. Some of the pollution caused by urban living is generated in the pre-consumption agricultural systems that provide food and other materials to be consumed in cities. For example, in an analysis of N flows through the Twin Cities, runoff and leaching of N in the upstream agricultural system needed to provide protein to the city was four times higher than the amount of N discharged as treated sewage (Baker and Brezonik 2007).

Agriculture is the main source of N and P to surface waters in the United States today. Over the past several decades, reduction of erosion and more efficient use of phosphorus has probably reduced P loading to rivers, but the recent surge in corn production, driven by ethanol demand (and subsidies!) may reverse this decline. Nitrate concentrations in U.S. rivers have generally increased in recent decades (Smith et al. 1994, Goolsby and Battaglin 2001).

Salt pollution: Salt pollution has emerged as a major urban pollution problem that is not amenable to end-of-pipe treatment. In cold weather regions, road salt has become a major urban pollutant. Road salt use was not used much prior to the 1960s; its use has more than doubled since the 1980s, resulting in widespread salt contamination in the Northeastern United States (Kaushal et al. 2005). In the arid Southwestern United States, some urban landscapes are becoming contaminated with salt from irrigation water (Miyamoto et al. 2001, Baker et al. 2004).

Resource conservation: Finally, resource conservation will become an increasingly important driver for pollution reduction. Of particular concern may be phosphorus, which is obtained almost entirely from mining of scarce phosphate deposits. Several analyses (Hering and Fantel 1993, Smil 2000) have suggested that exhaustion of phosphate deposits over the next 100 years may be problematic for human food production. Although the main goal of reducing P pollution has been to ameliorate ecological effects of P enrichment of surface waters, conservation of P via recycling may become an even more important objective in the foreseeable future. Even now, recycling of many metals is driven as much by high prices caused by resource depletion as by pollution reduction goals.

Legacy pollutants: Finally, we need to deal with legacy pollutants in cities – pollutants that were widely used and released to the environment in earlier times and remain today, either in soils (such as lead, creosote, and other persistent organic chemicals) or in groundwater (such as organic solvents).

5.3 Materials Flow Analysis for Cities

5.3.1 The Basics

Single compartment model: The concept of a materials balance is fairly simple in principle. First, consider a simple *system*, which may be the whole ecosystem or part of it (a compartment), defined by a distinct physical or conceptual boundary. The movement of material across the system boundary is called a *flux*, measured in units of mass/time (Fig. 5.1). Inputs and outputs are not necessarily equal. Losses of material can be caused by *reactions*. The difference between inputs, outputs and reactions is *accumulation*:

$$\text{Accumulation} = \text{input} - \text{reaction} - \text{output} \quad (5.1)$$

Accumulation in (5.1) can be positive (the system gains material over time) or negative (the system loses material over time). This definition is different from the colloquial definition, where accumulation is always a gain of material. The term "retention" is used synonymously with positive accumulation. As an example, when phosphorus fertilizer is applied at rates higher than can be removed by crops, some of the excess P will accumulate in soils.

Multi-compartment models: Most MFA studies of urban ecosystems involve more than one ecosystem compartment. Outputs from one compartment may be inputs to another compartment. A given compartment might receive inputs from several other compartments. In turn, outputs from a given compartment may enter any one of several other compartments. Figure 5.2 shows the flow of nitrogen (N) associated with wastewater treatment plants in Phoenix. Inputs to the wastewater treatment plant are lost through treatment (denitrification, producing N_2, and sedimentation, producing biosolids). Wastewater then transports the remaining N to irrigated crops, to a nuclear power plant, where the wastewater is used for cooling, to the Gila River, and to groundwater (via infiltration basins).

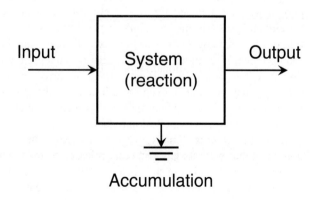

Fig. 5.1 Materials balance for a one-compartment box model

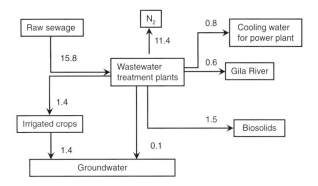

Fig. 5.2 Fluxes of sewage nitrogen in Phoenix, Arizona. Values have units of gG/yr. Source: Lauver and Baker (2000)

Estimating fluxes: Fluxes to and from ecosystem compartments can be estimated in one of several ways (1) by direct measurement of all fluxes to and from a given compartment; (2) by a combination of direct measurements and estimated mass transfer coefficients; (3) by process-based models; and (4) by direct measurement of some fluxes and inferred values for others. Few ecosystem compartments have detailed measurements of all inputs and outputs. A sewage treatment plant might be an exception: Most have continuous records of flow and routine measurements of many pollutants for influent (raw sewage), effluent, and biosolids. Usually, one or more terms cannot be measured and must be inferred by other means. If only one flux value is missing, it can be estimated "by difference". For example, denitrification (conversion of nitrate to N_2 gas) is rarely measured directly in most ecosystem mass balances. In Fig. 5.2, the denitrification flux was estimated as the difference between inputs (wastewater) and measured outputs (biosolids and effluent), on the assumption that there is no net accumulation.

Another common way to estimate fluxes is estimation based on a measured value and a mass transfer coefficient. A simple example is nitrogen inputs from human food consumption in a city. The number of people in a region, by age and sex, can be estimated from Census data. However, measuring food consumption directly (through 24-hour dietary recall surveys) is expensive and requires specialized food composition databases for interpretation. Urban ecologists can take advantage of the fact that national food consumption studies are routinely conducted, providing us with readily accessible data on nutrient consumption rates for each age and sex subpopulation. In this case, the mass transfer coefficient is daily protein consumption, estimated from national studies (Borrud et al. 1996). A simple conversion is then needed to convert protein consumption to N consumption (N = protein/6.25). For each subpopulation, the annual N input from food is:

$$\text{N input} = \text{subpopulation} \times \text{average protein consumption for subpopulation, kg/yr} \div 6.25 \quad (5.2)$$

The total food N input is calculated by summing inputs for all subpopulations.

A third approach is to use outputs from one compartment as inputs to another. For the example of N in human food, we know that about 90% of consumed N is excreted, so the amount of sewage N produced per person is 0.9*N consumption. This output from humans becomes an input to the sewage system.

Detailed process-based models can also be used to develop urban materials balances. Process-based models explicitly recognize biological and chemical processes, and are "dynamic", which means they can represent changing conditions. We are still some years away from a detailed process-based model of an entire urban ecosystem, but process-based models have been used to estimate some processes in urban ecosystems, for example, sequestration of C and N in urban lawns (Milesi et al. 2005, Qian et al. 2003).

Mobility of chemicals in the urban environment: The mobility of chemicals in urban ecosystems is highly variable. Some chemicals are readily adsorbed to soil particles (Table 5.2). For organic chemicals, the tendency to be adsorbed tends to be inversely related to solubility. Readily adsorbed chemicals (like PCBs) are immobilized (trapped) in the upper layer of soils, so they are rarely found in groundwater. However, these chemicals can be transported downstream by erosion and then trapped in sedimentation basins, wetlands, or lakes when the suspended particles settle out. Most metals become adsorbed to some extent. In addition, metals are often immobilized by chemical precipitation, in which metal ions react to form insoluble carbonates, hydroxides and other compounds.

Transformations and decay: Some nutrients – particularly nitrogen and phosphate – are removed from water by assimilation (nutrient uptake) by algae or terrestrial plants (Table 5.2). These nutrients are also released from plants during decomposition, recycling them back to soils and water. Most natural organic com-

Table 5.2 Major transformations of some important pollutants in urban environments. ••• = very important; •• = moderately important; • = somewhat important; ○ = unimportant

	Decay	Adsorption or precipitation	Assimilation by plants	Gaseous endproduct
BOD	•••	○	○	CO_2
Surfactants	•••	○	○	CO_2
NO_3^- (nitrate)	○	•	•••	N_2
NH_4^+ (ammonium)	○	••	•••	NH_3
Phosphate	○	•••	•••	○
Sodium	○	•	○	○
Chloride	○	○	○	○
Zinc	○	•••	○	○
Copper	○	•••	•	○
Arsenic	○	•	○	○
Cadmium	○	•••	○	○
Lead	○	•••	○	○
Glyphosphate (herbicide)	•••	••	•••	CO_2
2,4 D (herbicide)	•••	•	•••	CO_2
PCBs	○	•••	○	○

pounds in urban environments decompose via microbial processes, forming CO_2, and releasing nutrients. Many common herbicides, such as 2,4 D and glyphosphate (Roundup) decay in soils with days to weeks (Table 5.2). Nitrate is readily converted to nitrogen gas (N_2) under anaerobic conditions, which may occur in soils, wetlands, and lake sediments. Note that many pollutants do not decay – they are simply transformed from one form to another (e.g., dissolved in water → bound to sediments).

5.3.2 Data Sources

Public data sources: New tools and databases now make the development of materials flow analysis for large urban regions fairly practical. In particular, the widespread use of geographic information systems (GIS) by all levels of government mean that many types of data are readily available on a spatially discrete basis, allowing a facile GIS analyst to create numerous data layers within a selected watershed or other discrete region.

5.3.3 System Boundaries

An important consideration for analyzing materials flows through cities or parts of cities is definition of system boundaries. One commonly used boundary is a watershed, defined by topographic features. One advantage of using a watershed boundary is that it is relatively easy to define the boundaries. However, for analyzing materials flows through cities, there are several disadvantages: (1) water moves across topographically-defined watershed boundaries, particularly through storm sewers and sanitary sewers (Chapter 2), and (2) urban regions are not neatly bounded by watersheds. There also may be significant vertical movement of solutes via groundwater, which may necessitate delineation of a separate groundwater system. Political entities (such as counties or cities) can also be used to define boundaries. These have the advantage that many types of data are organized by political boundaries (Table 5.3), but suffer the disadvantage that political boundaries are generally not closely related to anything that might be considered an "ecosystem". It is not necessary that "boundaries" be spatial. For example, in our analysis of household ecosystems (Baker et al. 2007a), we developed a conceptual boundary that included all activities of household members. By this definition, all driving and air travel done by household members occurred within the boundary of the household ecosystem.

5.3.4 Scales of Analysis

MFA studies can be conducted at scales from individual households to entire urban regions. Vast amounts of publicly available data are readily accessible for materials

Table 5.3 Some types of data used in development of urban materials balances

	Original source of data	Typical unit of government	Common spatial resolution for reporting
Impervious surface	LandSat images	State	30′ × 30′
Land cover	" "	State	30′ × 30′
Land use	Air photo	State or regional	
Crop production	Surveys of farmers	State agricultural statistics services (also compiled nationally)	County
Agricultural fertilizer use	Surveys of farmers	National Agricultural Statistical Survey	State
Population characteristics	Nationwide census	U.S. Census	Census blocks
Housing characteristics	Surveys	U.S. Census, American Housing Survey	Metropolitan areas
Watershed boundaries	Digital topographic maps		Variable
Sewersheds	Ground-based mapping	Municipality or regional sewage authority	Delineated by individual hookups
Sewage flow and quality	Direct measurement at sewage treatment plants	Municipality or regional sewage authority	Individual sewage treatment plants
Land parcel information	Ground based mapping and reporting	Local governments	Individual properties
Animal feedlots (type and size)	Ground-based reporting	State government (Minnesota)	Varying – about 30 animal units in Minnesota
Animal production	Ground-based reporting	State government	County
Human nutrient consumption	National surveys based on 24-hour dietary recall	Federal government – Continuing Survey of Foods Study	Federal
Lawn fertilizer use	Lawn surveys (individual studies)	Various (not systematic)	Various

flow analysis, compiled by governments at all levels (see examples in Table 5.3). Some of these data are spatially discrete. For example, land cover data is often based on LandSat imagery, with a resolution of 30 × 30 feet. Plat data, which contains a wealth of information about individual properties, is now nearly always available in digitized form from local units of government. Other types of data, however, are available only in aggregated form, often compiled at the county, state or even federal level (Table 5.3).

Urban regions: Public sector data are particularly useful for analyzing metropolitan regions (>1000 km^2). For example, in an ongoing analysis of N and P balances for the Twin Cities, Minnesota, we estimated the population of all U.S. Census blocks located within the regional watershed, allowing a high degree of precision. Because the watershed comprised a large fraction of five counties, we could also confidently utilize a wealth of data collected at the county level, on the assumption that per capita fluxes collected at the 5-county or regional level would be similar to per capita fluxes within the watershed.

Household scale: We have used a hybrid approach to study fluxes of materials through individual households (Baker et al. 2007a). In this study, we used a combination of an extensive questionnaire, plat data, data acquired from homeowners (energy bills and odometer readings) and landscape measurements. This is probably the only way to collect information needed for materials balances for individual households with a reasonable degree of reliability.

Intermediate scales: Intermediate scales of analysis pose greater problems. For example, small watersheds ($< \sim 100$ km^2) may be a small part of a county or regional governmental unit. In this case, it may not be reasonable to assume that average characteristics of the watershed are similar to county to tabulated county or regional characteristics. For example, the American Housing Survey compiles a wealth of data on household characteristics within metropolitan areas, but one could not assume that average values for these characteristics (e.g., household size or energy cost) are averages for a particular study watershed. These problems may overcome in the future, as more and more data are compiled and reported at finer spatial resolutions.

5.3.5 Indirect Fluxes

The issue of "indirect" fluxes of materials presents a problem that requires careful boundary delineation. Indirect fluxes of materials are those that occur outside the system boundary, but are affected by activities within the system. For example, the carbon flux used to manufacture a car may occur outside an urban ecosystem, but is affected by the purchase of cars within the system. Moreover, additional carbon fluxes occur outside the factory that manufactures the car – for example, as food used to feed the workers at the plant. There is no entirely satisfactory solution to this problem, but when indirect fluxes are being included in a MFA, the boundary of indirect fluxes needs to be carefully defined.

5.3.6 Prior Studies

Methodologies for developing urban materials balances and MFA are rapidly improving as more researchers undertake these projects. Table 5.4 shows that most of these studies have dealt with water, nutrients, and salts. Studies on urban metabolism, which generally focus on energy and water, are reviewed by Kennedy et al. (2007).

Table 5.4 Examples of material flow analysis studies in urban ecosystems

Location	Description	References
Phoenix, Arizona	Whole-city N balance.	Baker et al. (2001)
Phoenix, Arizona	Regional salt balance.	Baker et al. (2004)
Five Southwestern U.S. cities	Salt balances for public water supplies.	Thompson et al. (2006)
Hong Kong	N and P balances for Hong Kong.	Boyd et al. (1981)
Sydney, Australia	Whole-city P balance.	Tangsubkul et al. (2005)
Bangkok, Thailand	Partial (food-wastewater) N and P balances.	Faerge et al. (2001)
Paris, France	Early industrial era N balances.	Barles (2007)
Individual households (Falcon Heights, MN)	C, N and P fluxes for 35 households.	Baker et al. (2007a)
Urban residential water systems	Contaminant flows for 12 pollutants in residential water systems.	Gray and Becker (2002)
Hong Kong	Urban N balance.	Boyd et al. (1981)
Gavle, Sweden	P in food and wastes.	Nilsson (1995)
Swedish households	Direct and indirect household energy requirements	Carlsson-Kanyama et al. (2005)
Several U.S. cities	Copper mass balances for municipal water systems.	Boulay and Edwards (2000)

5.4 The Human Element

Studies of MFA often show that much of the generation of pollutants in post-industrial cities occurs as the result of actions by individuals and households, either directly or indirectly. For example, we have estimated that about 70–80% of the N and P entering the Twin Cities region comes directly through households. Other studies have shown that direct energy consumption by households accounts for 41% national energy use in the United States (Bin and Dowlatabadi 2005), an average of 57% of all municipal water use (USEPA 2004) and 28–84% of the salt input to municipal wastewater reclamation plants (Thompson et al. 2006).

Because many homeowner behaviors can not (and most would say, should not) be regulated extensively, we need to expand the use of "soft" (non-regulatory) policy approaches to manage the flow of pollutants through urban systems by changing homeowners' behaviors voluntarily.

Several conceptual frameworks have been used to analyze how environmental behaviors can be changed. Traditional supply-demand economic models have been widely used to develop policies regarding directly marketed goods, either through subsidies, "demand-side" pricing and other means (see Chapter 13). Some environmental psychologists use the framework provided by the Theory of Planned Behavior (TPB) and its derivatives (Ajzen 1991). The TPB seeks to understand underlying attitudes that affect the intent to behave in a particular way. It has been has been used to understand behavioral controls involved in recycling (Gamba and Oskamp 1994), household energy use (Lutzenhiser 1993), urban homeowners' riparian vegetation (Shandas 2006), and lawn management (Baker et al., 2008). Communications theory, which examines how new ideas diffuse from innovation through broad adoption, has also been used to understand the adoption of agricultural erosion (Rogers 1995).

Experience has shown that soft policy approaches can change environmental behaviors. For example, curbside recycling in the U.S. increased from 7% of municipal waste in the 1970s to 32% today (USEPA 2007). Farmland erosion declined by 40% since the early 1980s (USDA 2003). Agricultural P fertilizer use has declined by about 15% since the late 1970s, while yields have simultaneously increased (e.g., corn yields over this period have more than doubled). Many urban water conservation efforts have been successful (Renwick and Green 2000). Most of these programs have included a mixture of policy tools. Typical elements include education, financial subsidies and/or disincentives, outright regulations (e.g. zoning, product bans restrictions) and social marketing. Some also target sensitive areas (e.g., erodible land) or disproportionate consumers or polluters, such as flagrant household water consumers. These examples suggest that in the future, soft policies might also be used to reduce sources of urban stormwater pollution, alter diets to reduce agricultural pollution, and reduce CO_2 emissions.

5.5 Adaptive Management

As we seek solutions for nonpoint source pollution, we will also need new models for decision-making. In the traditional top-down management approach to point source pollution control, the impact of various levels of pollution reduction needed to achieve an environmental goal can readily be predicted using mathematical models. For example, engineers can now readily predict the increase in oxygen levels that will occur in a river as the result of decreased inputs of BOD from sewage. We also have the ability to design sewage treatment plants to precisely achieve a specific BOD load reduction – no more and no less than required to achieve specific oxygen level (usually a legally mandated standard). Decision-making is therefore relatively straightforward, with predictable outcomes and well-defined costs.

Managing the diffuse movement of pollutants through urban ecosystems is more problematic for several reasons: (1) pollution removal efficiencies for BMPs are highly variable (2) mathematical models to represent diffuse pollution are not highly developed, and (3) even if they were, obtaining accurate input data would be expensive and in some cases, nearly impossible (e.g., we have no good way to measure how much fertilizer individual homeowners apply to lawns). One potential solution is *adaptive management*, which involves making environmental measurements over long periods, providing *feedback* to those involved in management, who change management practices in response to feedback (Gunderson and Holling 2002). An example of adaptive management that we are all familiar with is our weather forecasting system. Our government policy to protect citizens from severe weather is to provide continuous forecasts and measurements. The actual response by citizens is entirely unregulated: Each citizen listening to weather forecasts decides whether to carry an umbrella if the forecast shows a 70% chance of rain. A person may choose not to carry an umbrella when heavy rain is predicted, but over time, most people learn that this is a good idea – they adapt. Generally the response is not based on raw data, but requires some level of interpretation and recommendation. The public is interested in weather forecasts, not real-time barometer readings.

Very little water quality management in cities is currently based on adaptive management, but several technological and cultural developments favor broader use of adaptive management in the future. Some of these include: (1) rapidly improving, less expensive monitoring approaches, including citizen-based programs, to provide feedback; (2) increasing internet connectivity, which allows participants to receive feedback and management recommendations, (3) the advent of "Web 2.0", which allows online dialogues, (4) growing environmental awareness and willingness to change behaviors, and (5) a resurgent interest in citizen participation in governance.

5.6 Applications

This section examines several case studies of pollution management that involve MFA, social tools, and adaptive management to illustrate new directions in management of urban pollution.

5.6.1 Case Study 1: Urban Lawns and P Pollution

Lawn runoff typically contains 0.5–2.0 mg P/L, compared with levels around 0.1 mg P/L that typically result in lake eutrophication. Hence, lawns are probably the major source of P to stormwater in residential areas (Baker et al. 2007b). It would therefore be desirable to develop policies that would reduce P export from residential lawns. Targeted lawn management strategies could be based on MFA of lawn P cycling (Fig. 5.3). Logical system boundaries would be the horizontal borders of the lawn (x–y axis) and the lawn surface down to the bottom of the root zone (z-axis). Some lawn P fertilizer is lost immediately as surface runoff, but this loss is typically <5% of applied P (reviewed by Baker et al. 2007b). Most of the applied P is assimilated by grass or enters the soil. Because most urban lawns have been fertilized with P for many years, many have accumulated a large pool of active P that can be "mined" by growing turf for many years following cessation of fertilizer P inputs. When grass is mowed, some enters streets as P-containing particles and the rest decomposes on the lawn, releasing soluble P, which can become runoff or leach downward, replenishing the active P pool.

One policy option would be to restrict the use of lawn P fertilizers, as the Minnesota and several counties and cities have done. In the first few years since the

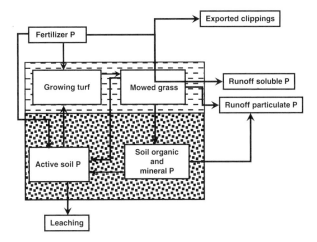

Fig. 5.3 Diagram of P cycling in lawns

restriction has become law, it appears that P export from lawns has declined ~15%. Most lawns will continue to mine stored active P. Most of the P assimilated by grass will be recycled, but a small fraction will be exported as grass particles and soluble P to streets. Eventually, the mass of the active P pool will decline. When it gets very low, the amount being mined by grass will decline, and the amount of P in runoff will decline.

One important policy question is: How long will it take to reduce the P concentration in lawn runoff? This depends on several factors: edaphic features of the lawn (slope and soil texture), the initial concentration in the active soil P pool, how fast the grass is growing, and whether the clippings are bagged and removed or mulched in place. Increasing the growth rate would increase the rate at which P is mined from the soil P pool. Ironically, this could be done by adding N fertilizer to stimulate growth! Removing clippings would prevent them from decomposing on the lawn and releasing P. This would reduce P export in runoff and reduce the rate of recycling to the active soil P pool. The fastest way to deplete the soil P would be to fertilize with N *and* remove the clippings. Another consideration is that depleting the active P pool too much could impair turf growth, thereby increasing erosion. An optimum level of P in the active soil pool is ~25 mg/kg. When soil P falls below this level, grass growth slows down and turf quality deteriorates. Under these conditions, erosion would increase, very likely increasing the rate of P export to streets.

Our analysis indicates that simply restricting P fertilizer inputs would not have a large short-term effect, and may backfire over the long run, if soil P levels fall below the optimum. Furthermore, because lawn management preferences among homeowners vary, strategies for reducing P runoff from lawns should be *tailored* to specific typologies of homeowners. For example, the "casual" homeowner whose soil active P level is below the optimal level for turf growth would be encouraged to add enough fertilizer P to maintain soil P levels high enough to maintain a healthy turf, to reduce the potential for P losses caused by erosion. The "perfectionist" homeowner with very high soil P levels needs to know that there is no benefit in maintaining the active soil P level above the optimal level. In both cases, P fertilization should be adaptive, based on actual measurements of their lawn's soil P levels. This testing is inexpensive, generally provided by university extension services.

Taken to the next level, quantitative modeling of lawn runoff could be used to determine specific circumstances (slope, soil texture) where runoff P is most likely problematic. Watershed managers could then target their efforts on these areas. Homeowners in these areas would learn that they happen to be located on vulnerable areas. The TPB (discussed above) suggests that this type of specific, credible knowledge would likely cause a greater change in homeowners' behaviors than general guidance. Thus, a homeowner who is asked to change his lawn management practices because his lawn is particularly vulnerable – for example, on a steep slope with clayey soils, where runoff can be 25% of precipitation – is more likely to adopt specific advice than a random homeowner given generic advice.

5.6.2 Case Study 2: Using MFA to Devise Improved Lead Reduction Strategy

Lead has probably been as harmful to human health in cities as any pollutant except perhaps fine particulate matter in air. Prior to the 1970s, two main inputs of lead to urban environments were leaded gasoline and lead-based paints, which together accounted for ~10–12 million tons of lead entering the environment. Both uses of lead were banned in the 1970s, leading to dramatic blood lead levels that parallel the decline in use of leaded gasoline (Needleman 2004). Even so, the Center for Disease Control's latest survey (1999–2002) shows that 1.6% of all children and 3.1% of black children in the United States had blood lead levels (BLLs) greater than the "action limit" of 10 μg/dL. Symptoms of moderate lead poisoning include irritability, inattention, and lower intelligence in children and hypertension in adults. The problem today is largely concentrated in poor, non-white urban populations. For example, more than 30% of the residents (all ages) in Baltimore City, Maryland had elevated BLLs in the early 1990s (Silbergeld 1997). Urban lead poisoning therefore remains a critical human health problem (Table 5.5).

Currently, lead hazard interventions to reduce blood lead levels in children have focused on either abatement (removal of lead-based paint or contaminated soil) or

Table 5.5 Trend in BLLs > 10 μg/dL. Source: Centers for Disease Control

	Percentage of children, ages 1–5, with BLLs >10 μg/dL		
	All races	White, non-Hispanic	Black, non-Hispanic
1976–1980	88.2	–	–
1988–1991	8.6	6	18
1991–1994	4.4	2.3	11.2
1999–2002	1.6	1.3	3.1

5 New Concepts for Managing Urban Pollution

interim controls (paint stabilization, dust control). These programs have been only moderately successful, at best reducing blood lead levels by 25% and never reducing them to $\mu 10$ μg/dL (PTF 2000).

A MFA of lead in the urban environment might lead to more effective ways to eliminate lead poisoning. Figure 5.4 shows likely fluxes of lead in cities. The major inputs of lead, paint and leaded gasoline, were phased out in the 1970s. Today, virtually no new lead is deliberately imported into the landscapes of U.S. cities (except car batteries, which must by law be recycled). Declining use of leaded gasoline immediately reduced BLLs, suggesting that much of the lead in gasoline must have been exported from cities quickly. Currently, the main source of "new" lead to urban environments is leaded paint from old buildings, which enters the environment as paint chips. In MFA terminology, this represents a transfer between ecosystem compartments (buildings → soil; Fig. 5.4). Kids probably ingest lead both directly from paint dust in buildings and from external soil and will continue to do so, as long as lead-based paint continues to erode.

There are two likely routes of lead export from the inner city urban environment. One is atmospheric transport. Laidlaw et al. (2007) showed that BLLs of children vary in relation to wind speed, concentrations of fine particles in air (PM-10), and soil moisture. This indicates that fine particles containing lead are dispersed by wind. If this is the case, it is likely that the net direction of lead is outward from inner-urban areas. The reason for this is that soil lead levels are highest in inner-urban areas and lower in the surrounding suburbs and in rural areas (Mielke et al. 1997). The lack of relationship between soil lead levels and land use in Baltimore (Pouyat et al. 2007) also suggests a wind dispersal mechanism. The second export

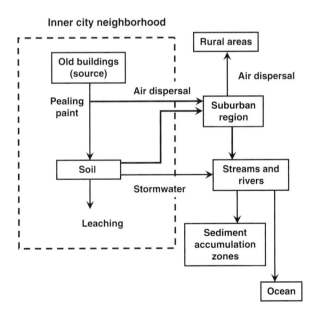

Fig. 5.4 MFA diagram of lead cycling in a city

route is wash-off of wind-blown lead-laden dust that has settled on local impervious surfaces (e.g., streets and sidewalks). Rain events transports lead to storm water drains. Kaushal et al. (2006) found that the mean lead concentration in storm water from some Baltimore streams was 75 μg/L decades after new lead inputs had ceased, indicating that lead is slowly being exported by storm water. Storm sewers would transport lead to low-lying sediment accumulation zones, such as stormwater detention basins and stream riparian areas, where it would be deposited. Particle-bound lead would be transported further downstream during flood events to be re-deposited in regional rivers or transported to the ocean (Fig. 5.4). The one export route that is not likely is downward leaching: Lead is tightly adsorbed to soil and is unlikely to leach.

The obvious means of reducing lead in the urban environment is careful removal or containment of lead-based paint, or removal or burial of entire buildings. Over time, existing environmental lead would be reduced by export via wind and water. More detailed study of export mechanisms could lead to focused management to accelerate removal of environmental lead. For example, frequent high-efficiency street sweeping and removal of highly contaminated soils might accelerate removal of accumulated lead, while also controlling the export route (to hazardous waste landfills, rather than dispersal to outlying residential areas or to downstream aquatic ecosystems).

5.6.3 Case Study 3: Managing Road Salt with Adaptive Management

De-icing roads using salt has accelerated over the past 40 years, increasing ten-fold since the 1960s. Road salting has resulted in widespread contamination of streams and groundwater in cold regions of the United States (Kaushal et al. 2005). Chloride concentrations in some urban areas can reach several thousand mg/L during peak snowmelt periods, and can remain elevated above 250 mg/L even during base flow periods, exceeding chronic water standards for protection of aquatic life. Chloride also is highly mobile and can readily move through soils, contaminating aquifers. In the Shingle Creek watershed in Brooklyn Park, Minnesota, groundwater chloride concentrations have increased 10-fold, presumably due to road salt. Over the long term road salt could potentially contaminate some urban aquifers to the point that the quality of groundwater would no longer be suitable for municipal water supply. The potential problem with reducing road salt use is increased potential for accidents, so reduction has to be done in a way that assures safe driving conditions. Some common strategies used to reduce the amount of salt applied include the use of premade brines rather than dry salt and the use of "pre-wetting" – application of salt or brine to the road before a major winter storm, rather than afterwards. Chloride use can be reduced through the use of alternative de-icers, such as calcium acetate, but these are generally far more expensive and may cause other problems (Novotny et al. 1999).

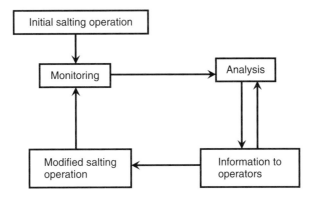

Fig. 5.5 Scheme of adaptive management for road de-icing

Adaptive management might be used to further reduce the amount of road salt used. Management of road salt is well-suited to an adaptive management, for the following reasons: (1) road salt crews are a relatively small, captive audience, which enables frequent communication, (2) because road salt is often overused, there is potential for reduced use, hence savings, (3) several technologies can reduce the amount of salt needed, including pre-wetting, application of brine rather than dry salt and use of non-chloride alternative, and (4) chloride can be readily measured indirectly, as conductivity – a method that is simple, inexpensive, reliable and readily automated, enabling real-time monitoring on pavements, in storm sewers and in streams (Baker 2007). The essential elements of an adaptive management scheme for road de-icing are shown in Fig. 5.5.

Requisite data to guide road de-icing management could readily be obtained via environmental sensors. Road surface temperature and specific conductance (a surrogate measurement for chloride) could readily be measured at the road surface and in streams. Precipitation amount for each event could be interpolated from measured precipitation at weather networks. Newer salting trucks have computerized equipment to record the mass of salt used per mile. Analysis of data from a series of de-icing events might include simple statistical analysis, hydrologic modeling or other methods. In addition, adaptive management often uses human feedback – in this case, the perceptions and knowledge of the salt truck operators. This analysis would be used to guide subsequent de-icing operations, which in turn would result in new data for analysis. This cycle would continue, with the goal of improving de-icing operations until environmental and safety goals are met.

5.7 Summary

Creating sustainable, resilient urban ecosystems requires that we understand and manage the flow of polluting materials through them. From the time of early industrialization through the 20th century, we rarely analyzed the flows of these materials, often with disastrous, or at least, unpleasant effects. For example, allowing urban lead pollution to occur, and allowing it to persist through the 1970s, will probably be recorded by historians as one of the poorest environmental decisions in the 20th century. Today, MFA methods are sufficiently well developed that they can be used to guide pollution management strategies. As a minimum, we can now develop reasonable a reasonable view of the movement of polluting materials through urban regions. We can also expect that new databases, with more types of data and at finer resolution, will enable even broader applications.

Many of the findings from MFA analysis will likely point toward "soft" and policies to change environmental behaviors of ordinary citizens. To do this, we also need to understand how people make environmental decisions. This will require a new degree of interdisciplinary collaboration to develop transdisciplinary knowledge of social–ecological systems. This type of thinking has largely been missing from most pollution management policy, with the exception of economic cost-benefit analysis. One very promising approach that integrates the biophysical and social realms is adaptive management. New technologies enable a whole new realm of feedback mechanisms to inform policy makers at all levels, including individual citizens. We examined case studies involving lawn runoff, urban lead pollution and road salt to illustrate these concepts.

MFA may have particular importance for cities in arid lands, because contaminants in water tend to be retained in arid cities, rather than flushed from cities, as would be the case for wetter regions. Water conservation measures in general, and recycling of wastewater in particular, tend to exacerbate accumulation of solutes such as salt and nitrate within the urban ecosystem, especially in underlying aquifers (Chapter 4, Baker et al. 2004). Although cities in arid lands have persisted for thousands of years, historical per capita water use rates were probably very low, so accumulation of solutes would have been slow. In the modern, post-industrial city, solute accumulation is accelerated by very high water use, often several hundred gallons per person per day and the long-term consequences have not been studied. Because most of the world's population increase is occurring in cities in arid lands, managing the flow of materials through them may be essential for sustainability.

Acknowledgements Research reported in this chapter was supported by NSF Biocomplexity Projects 0322065 and 0709581 to L. Baker.

References

Ajzen, I. 1991. The theory of planned behavior. Organizational Behavior and Human Decision Processes **50**:179–211.
Baker, L. A. 2007. Stormwater pollution: getting at the source. Storm Water **8** (Nov. 2007), http://www.stormh2o.com/november-december-2007/bmps-ms4-pollution.aspx
Baker, L. A., B. Wilson, D. Fulton, and B. Horgan. Disproportionality as a framework for urban lawn management. Cities and the Environment. Cities and the Environment 1:2 Article 7.
Baker, L. A., and P. L. Brezonik. 2007. Using whole-system mass balances to craft novel approaches for pollution reduction: examples at scales from households to urban regions, pp. 92–104. In: V. Novotny and P. R. Brown, editors. Cities of the Future: Toward Integrated

Sustainable Water and Landscape Management. Proceedings of a conference held at the Wingspan Center, Racine, WI (USA), July 12–14, 2006. IWA Publishing, London.

Baker, L. A., P. Hartzheim, S. Hobbie, J. King, and K. Nelson. 2007a. Effect of consumption choices on flows of C, N and P in households. Urban Ecosystems **10**:97–110.

Baker, L. A., R. Holzalksi, and J. Gulliver. 2007b. Source reduction, Chapter 7. In: J. A. Gulliver and J. Anderson, editors. Assessment of Stormwater Best Management Practices. Minnesota Water Resources Center, St. Paul.

Baker, L. A., A. Brazel, and P. Westerhoff. 2004. Environmental consequences of rapid urbanization in warm, arid Lands: case study of phoenix, Arizona (USA), pp. 155–164. In: Marchettini, N., C. A. Brebbia, E. Tiezzi, and L. C. Wadhwa, editors. The Sustainable City III, WIT Press, Sienna, Italy.

Baker, L. A., Y. Xu, D. Hope, L. Lauver, and J. Edmonds. 2001. Nitrogen mass balance for the Phoenix-CAP ecosystem. Ecosystems **4**:582–602.

Barles, S. 2007. Feeding the City: Food Consumption and Circulation of Nitrogen, Paris, 1801–1914. Science of the Total Environment **375**: 48–58.

Bin, S. and H. Dowlatabadi. 2005. Consumer lifestyle approach to U.S. energy use and related CO_2 emissions. Energy Policy **33**: 197–208.

Borrud, L., E. Wilkinson, and S. Mickle. 1996. What we eat in America: USDA surveys food consumption changes. Foods Review 14–19.

Boulay, N., and M. Edwards. 2000. Copper in the urban water cycle. Crit. Reviews in Environmental Science and Technology **30**:297–326.

Boyd, S., S. Millar, K. Newcombe, and B. O'Neill. 1981. The Ecology of a City and its People: The Case of Hong Kong. Australian National University Press, Canberra, Australia.

Carlsson-Kanyama, A., R. Engstrom, and R. Kok. 2005. Indirect and direct energy requirements of city households in Sweden. Journal of Industrial Ecology **9**:221–235.

Faerge, J., J. Magid, and W. T. Penning de Vries. 2001. Urban nutrient balance for Bangkok. Ecological Modeling **139**:63–74.

Gamba, R. J., and S. Oskamp. 1994. Factors influencing community residents participation in comingled curbside recycling programs. Environment and Behavior **126**:587–612.

Goolsby, D. A., and W. A. Battaglin. 2001. Long-term changes in concentrations and flux of nitrogen in the Mississippi River Basin, USA. Hydrological Processes **15**:1209–1226.

Gray, S. R., and N. S. C. Becker. 2002. Contaminant flows in urban residential water systems. Urban Water **4**:331–346.

Gunderson, L. H., and C. S. Holling. 2002. Panarchy: understanding transformations in human and natural systems. Island Press, Washington.

Hering, J. R., and J. R. Fantel. 1993. Phosphate rock demand into the next century: impact on world food supply. Natural Resources Research **2**:226–246.

Kaushal, S. S., K. T. Belt, W. P. Stack, R. V. Pouyat, and P. M. Groffman. 2006. Variations in heavy metals across urban streams. EOS Transactions, Jt. Assem. Suppl. (abstract) 87.

Kaushal, S. S., P. M. Groffman, G. E. Likens, K. T. Belt, W. P. Stack, V. R. Kelly, L. E. Band, and G. T. Fisher. 2005. Increased salinization of fresh water in the northeastern United States. Proceedings of the National Academy Sciences **102**:13517–13520.

Kennedy, C., J. Cuddihy, and J. Engel-Yan. 2007. The Changing Metabolism of Cities. Journal of Industrial Ecology **11**:43–59.

Laidlaw, M. A. S., H. W. Mielke, G.M. Filippelli, D. L. Johnson, and C. R. Gonzoles. 2007. Seasonality and children's blood lead levels: developing a predictive model using climatic variables and blood lead data from Indianapolis, Indiana, Syracuse, New York, and New Orleans, Louisiana (USA). Environmental Health Perspectives **113**:793–800.

Lauver, L., and L. A. Baker. 2000. Nitrogen mass balance for wastewater in the Phoenix-Central Arizona Project ecosystem: implications for water management. Water Research **34**: 2754–2760.

Lutzenhiser, L. 1993. Social and behavioral aspects of energy use. Annual Review of Energy and the Environment **18**:247–289.

Metcalf, and Eddy. 1991. Wastewater engineering: treatment, disposal, and reuse. McGraw-Hill, New York.

Mielke, H. W., D. Dugs, P. W. Mielke, K. S. Smith, S. L. Smith, and C. R. Gonzoles. 1997. Associations between soil lead and childhood blood lead in urban New Orleans and rural Lafourche Parish of Louisiana. Environmental Health Perspectives **105**:950–954.

Milesi, C., C. D. Elvidge, J. B. Dietz, B. T. Tuttle, R. N. Ramkrishna, and S. W. Running. 2005. Mapping and modeling the biogeochemical cycling of turf grasses in the United States. Journal of Environmental Management **36**:426–438.

Miyamoto, S., J. White, R. Bader, and D. Ornelas. 2001. El Paso Guidelines for Landscape Uses of Reclaimed Water with Elevated Salinity. El Paso Public Utilities Board, Water Services, El Paso.

Needleman, H. 2004. Lead poisoning. Annual Review of Medicine **55**:209–222.

Nilsson, J. 1995. A phosphorus budget for a Swedish municipality. Journal of Environmental Management **45**:243–253.

Novotny, V., D. W. Smith, D. A. Duemmel, J. Mastriano, and A. Bartosova. 1999. Urban and Highway Snowmelt: Minimizing the Impact on Receiving Water. Water Environment Research Foundation, Alexandria, VA.

Pouyat, R. V., I. D. Yesilonis, J. Russell-Anelli, and N. K. Neerchal. 2007. Soil chemical and physical properties that differentiate urban land-use and cover types. Soil Science Society of America Journal **71**:1010–1019.

PTF. 2000. Eliminating Childhood Lead Poisoning: A Federal Strategy Targeting Lead Paint Hazards. President's Task Force on Environmental Health Risks and Safety Risks to Children, Washington, DC.

Puckett, L. J. 1995. Identifying the major sources of nutrient water pollution. Environmental Science Technology **29**:408–414.

Qian, Y. L., W. W. Bandaranayake, W. J. Parton, B. Mecham, M. A. Harivandi, and A. Mosier. 2003. Long-term effects of clipping and nitrogen amendment on management in turfgrass on soil organic carbon and nitrogen dynamics: the CENTURY model simulation. Journal of Environmental Quality **32**:694–1700.

Reed, S. C., R. W. Crites, and E. J. Middlebrooks. 1995. Natural Systems for Waste Management and Treatment. McGraw-Hill, New York.

Renwick, M. E., and R. D. Green. 2000. Do residential water demand side management policies measure up? An analysis of eight California water management agencies. Journal of Environmental Economics and Managment **40**:37–55.

Rogers, E. 1995. Diffusion of Innovations (4th ed.). Free Press, New York.

Shandas, V. 2006. An empirical study of streamside landowners' interest in riparian conservation. Journal of the American Planning Association **73**:173–183.

Silbergeld, E. K. 1997. Preventing lead poisoning in children. Annual Review of Public Health **18**:187–210.

Smil, V. 2000. Phosphorus in the environment: natural flows and human interferences. Annual Review Energy and Environment **25**:53–88.

Smith, R. A., R. B. Alexander, and K. J. Lanfear. 1994. Stream Water Quality in the Conterminous United States – Status and Trends of Selected Indicators During the 1980's. National Water Summary 1990–91 – Stream Water Quality Water-Supply Paper 2400, U.S. Geological Survey, Washington, DC.

Tangsubkul, N., S. Moore, and T. D. Waite. 2005. Incorporating phosphorus management considerations into wastewater management practice. Environmental Science & Policy **8**:1–15.

Thompson, K., W. Christofferson, D. Robinette, J. Curl, L. A. Baker, J. Brereton, and K. Reich. 2006. Characterizing and Managing Salinity Loadings in Reclaimed Water Systems. American Water Works Research Foundation, Denver, CO.

USEPA. 2000a. Progress in Water Quality: An Evaluation of the National Investment in Municipal Wastewater Treatment. U.S. Environmental Protection Agency, Washington, DC.

USEPA. 2000b. National Water Quality Inventory: 2000. U.S. Environmental Protection Agency, Washington, DC.
USDA. 2003. 1997 erosion rates. Natural Resources Conservation Service, U.S. Department of Agriculture, Washington, DC.
USEPA. 2004. How we use water in the United States. U.S. Environmental Protection Agency, Washington, DC.
USEPA. 2007. Recycling-basic facts. http://www.epa.gov/msw/facts.htm
Weiss, P. T., J. S. Gulliver, and A. J. Erickson. 2007. Cost and pollutant removal of storm-water treatment practices. Journal of Water Resources Planning and Management **133**:218–229.

Chapter 6
Streams and Urbanization

Derek B. Booth and Brian P. Bledsoe

"Urbanization" encompasses a diverse array of watershed alterations that influence the physical, chemical, and biological characteristics of streams. In this chapter, we summarize lessons learned from the last half century of research on urban streams and provide a critique of various mitigation strategies, including recent approaches that explicitly address geomorphic processes. We focus first on the abiotic conditions (primarily hydrologic and geomorphic) and their changes in streams that accompany urbanization, recognizing that these changes may vary with geomorphic context and climatic region. We then discuss technical approaches and limitations to (1) mitigating water-quantity and water-quality degradation through site design, riparian protection, and structural stormwater-management strategies; and (2) restoring urban streams in those watersheds where the economic, social, and political contexts can support such activities.

6.1 Introduction and Paradigms—How Do Streams "Work"?

6.1.1 Channel Form

The term *stream channel* means different things to different people. To an engineer, it is a conduit of water (and perhaps, sediment). To a geologist, it is a landscape feature typically constructed by the very flow of water and sediment that it has carried over many years or centuries. To an ecologist, it is an interconnected mosaic of different aquatic and riparian habitats, and the organisms that populate it. To a government regulator, it is a particular landscape feature that may impose adjacent land-use constraints and whose flow should meet certain standards for chemical composition. And to the urban public, it can be an aesthetic amenity, a recreational focus, or an eyesore (and sometimes, all three).

D.B. Booth (✉)
Quaternary Research Center, University of Washington, Seattle WA 98195; Stillwater Sciences Inc., 2855 Telegraph Avenue, Berkeley, CA 94705
e-mail: dbooth@stillwatersci.com

In this chapter, we approach the stream channel primarily from its physical perspective, namely as the product of the primary watershed processes of water runoff and sediment delivery, together with the secondary components of large woody debris and trace (but locally critical) chemical constituents. Of course, the effects of human activity on stream channels cannot be ignored in the context of the urban water environment. Our goal, however, is to provide a basis from which to understand the influences of watershed urbanization, deliberate channel manipulations, and climate change. This is best achieved by approaching the topic through the perspective of the multi-scale processes that normally give rise to these features, and that in turn have supported the suite of biota that have evolved to thrive in these dynamic environments (Frissell et al. 1986, Church 2002).

Before embarking on a discussion of river-channel form and behavior, we must draw a distinction between two fundamentally different types of channels. *Alluvial channels* are those that have been carved by the water flow into deposits of the very sediment carried by that flow in the past, and that presumably could be carried by that flow in the future. These "self-formed" channels are free to adjust their shape in response to changes in flow, because their flows are capable (at least episodically) of moving the material that forms their boundaries (Fig. 6.1). The detailed

Fig. 6.1 View of an alluvial channel, whose boundaries are composed of the sediment previously transported by the flow under its current hydrologic regime

Fig. 6.2 View of a non-alluvial channel, whose boundaries cannot be modified under the current discharge regime (Los Angeles River, California)

hydrodynamics of how these channels establish their preferred dimensions and shape are complex and still not fully understood. However, we can recognize remarkable similarities in the behavior of these channels worldwide, readily expressing the net result of processes only imperfectly understood.

In contrast, *non-alluvial channels* are unable to adjust their boundaries, or at least not over relatively short time periods. A variety of channels express this condition to varying degrees: bedrock ravines, channels choked with landslide sediment or the debris of a catastrophic flood, channel sediment dominated by immovable boulders derived from the surrounding hillside deposit, or channels with thick and deeply rooted bank vegetation. In the urban environment, the most common non-alluvial channel is a piped or concrete-lined conduit (Fig. 6.2). In nearly all such instances, any degree of sediment movement or deposition within a non-alluvial channel will encourage that channel towards a more "alluvial" behavior. Thus these categories are not absolute but instead are gradational in both space (i.e., up and down the channel) and in time. Nevertheless, the distinction is a useful one and its recognition can save the planner or engineer from much fruitless analysis in certain types of channels and stream systems.

6.1.2 Water Discharge

In every setting, the most obvious role of a stream channel is to convey water from the contributing watershed. Flows rise and fall relatively rapidly in response to rainfall during storms or snowmelt, and they maintain a more steady discharge from the slow release of groundwater. With small contributing areas or in arid climates, stream channels may not carry any flow at all during dry weather.

A useful distinction is between the components of runoff that reach the stream channel quickly and those that arrive more slowly, often days (or longer) after the rain has stopped. If hillslope runoff reaches a stream channel during or within a day or so of rainfall, commonly following a flow path over or close to the ground surface, it causes high rates of discharge in the channel and is usually classified as *storm runoff* or *direct runoff*. Water that percolates to the groundwater moves at much lower velocities by longer paths and so reaches the stream slowly, over long periods of time. Water that follows these paths sustains streamflow during rainless periods and is usually called *base flow*. A formal distinction between these types of runoff is needed for certain computational procedures, but for our purposes a qualitative understanding is sufficient.

The relative importance of these flow paths in a region (or more particularly on each hillslope) can be affected by climate, geology, topography, soil characteristics, vegetation, and land use. The dominant flow path may vary between large and small storms. The most important discrimination, however, is based on which is larger: the rate of precipitation (known as the "rainfall intensity") or the rate at which water can be absorbed by the soil (the "infiltration capacity"). Where runoff primarily occurs in regions (or during particular storms) in which the rainfall intensity exceeds the infiltration capacity of the soil, surface runoff occurs because the ground cannot absorb all of the rainfall. This characterizes areas of "Horton overland flow regime." In contrast, a "subsurface flow regime" predominates where the rainfall intensity is typically low and so all precipitation typically infiltrates. Runoff can still occur in areas dominated by subsurface flow, but measured discharges have a much more attenuated response to rainfall because flow paths are primarily via the subsurface.

In most humid regions where the soil's infiltration has not been locally impacted, a subsurface flow regime commonly predominates. In arid and semi-arid regions, infiltration capacity is commonly limiting and rainfall, when it occurs, can be quite intense; Horton overland flow is thus the dominant storm runoff process. One common expression of these different regimes is the persistence of dry-weather (i.e., "perennial") flow in humid regions, because subsurface water is abundant and groundwater discharges continue to occur between storms.

The changing discharge in a stream is commonly displayed as a *hydrograph*, a graph of the rate of discharge at a point in a stream (or runoff from a hillside) plotted against time. Discharge is usually expressed as a volume of water per unit time (as cubic meters per second (cms) or cubic feet per second (cfs)) (Fig. 6.3).

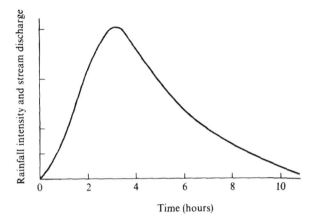

Fig. 6.3 An example of a hydrograph, showing the variation of discharge with time (modified from Dunne and Leopold (1978))

If this volume per unit time is divided by the area of the catchment in appropriate units, the runoff can be expressed as a depth of water per unit time (e.g., centimeters per hour or inches per day), which is very convenient for comparing with similarly expressed rates of rainfall, infiltration, and evaporation.

6.1.3 Sediment Transport

Precipitation falling on the landscape, together with the action of biological agents, breaks down rocks by weathering. Surface runoff and streamflow carry this load and transport the weathered debris. These various actions gradually move the rock debris toward the oceans, ultimately lowering the continents and depositing the materials in the sea. Successive periods of uplift ensure that the leveling process never becomes complete. But the downcutting or denudation of the land masses proceeds inexorably on all continents.

The rate of denudation seems slow but the amount of debris moved is immense. The rate is variously expressed as the spatially averaged speed at which the land surface is being lowered (e.g., in millimeters per 1,000 years), the annual amount of sediment being delivered into stream channels produced per unit area of watershed area (the sediment delivery, e.g., in tonnes per square kilometer per year), or the amount of sediment being carried past a point in a river in a given day under a given discharge (the sediment yield, e.g., in kilograms per day).

The average sediment load of a channel thus comprises the average rate at which hillslope sediment is delivered into stream channels, combined with the amount of sediment that is eroded from the bed and banks of the channel itself. Although not nearly as self-evident to the urban planner or city dweller as the water flow within the channel, the sediment load is a critical contributor to the physical, chemical, and biological conditions of an urban stream.

Fig. 6.4 A view of a stream and its adjacent floodplain, the recently constructed surface adjacent to the channel that is still episodically inundated by high flows

6.1.4 Floodplains

Most alluvial river channels are bordered by a relatively flat area or valley floor. When the water fills the channel completely (and so is at "bankfull stage"), the water level matches the elevation of this ground surface, which is called the *floodplain* (Fig. 6.4). This term is also used by both planners and engineers to identify the area adjacent to a channel that is inundated by floods of a given recurrence interval (e.g., "the 10-year floodplain"), but here we mean a distinct, observable land feature itself.

Geomorphically, a floodplain is defined as the flat area adjoining a river channel, constructed of alluvium by the river under the present climatic and land-use regimes. In natural settings, floodplains commonly are constructed by the lateral migration of channels and the subsequent deposition of sediment over a period of many hundreds and thousands of years without significant change in that channel's width or depth. This definition of a floodplain includes the concept, very difficult for the public and their elected officials to grasp, that the floodplain is an integral part of the river channel itself. It is not occupied by water as often as is the identifiable (low-flow) channel, but as a part of the river's "high-flow channel," its inundation is virtually assured over time, and its modification almost always has significant downstream consequences.

6.1.5 Water Chemistry

Just as the flow and sediment load of a stream integrates the contributions from the upstream watershed, so the chemical composition of the water reflects the contributions of both natural constituents and human-generated compounds throughout the watershed. Urbanization invariably results in a net increase in surface runoff because of soil compaction and new impervious surfaces, and so a great proportion of the water delivered to streams bypasses the cleansing influence of soil and plants. Because human activities in urban areas also increase inputs of nutrients, metals, organic compounds, and other potential pollutants to the land surface, urban storm runoff normally results in larger loads and more variable concentrations of chemical pollutants than runoff from undisturbed watersheds.

6.1.6 Biota

River water supports a world of its own. The microorganisms alone comprise a surprising variety and number of forms, while freshwater fish are often one of the most prized natural resources of a region. The *biotic health* of a stream is indicated by the variety and the composition of the population of organisms, both visible and microscopic. Although this chapter does not fully explore the details of stream ecology in the urban environment, we recognize that biology is commonly the overriding goal that drives much of the present activity in stream enhancement. The environmental planner has a large stake in the biotic health of the watercourse because it affects the perceived value of the amenity, the potential for recreation, the degree of regulatory attention, and the health of the surrounding community.

Using measures of plant and animal populations is also a particularly attractive way to assess aquatic health because organisms tend to integrate the effects, both known and unknown, of stream and watershed conditions (Karr and Chu 1999). However, a sole reliance on measures of biotic health can also limit our ability to act promptly and effectively to solve socially important problems. If freshwater fish are a major resource value, for example, then measuring their abundance will surely tell us the status of that resource, but any decline in that population will come only when degradation has already occurred and may be too late to correct.

6.1.7 Social Amenities of Urban Streams

Stream corridors in urban areas range from repulsive, polluted drainage ditches to verdant oases of biodiversity, recreation, and renewal. There is an emerging perspective that urban stream corridors should be much more than engineered conduits for fast conveyance of runoff and other discharges. Indeed, many communities are now focusing on stream and river corridors as high-value amenities not only for recreation, but as focal points for providing social, aesthetic, and educational benefits. Stream corridors are increasingly viewed not only as a "right-of-way" for floodwaters, but also as places where urban dwellers can access pedestrian and bicycle paths, go boating, experience a renewing environment, learn more about local animals and plants and whole ecosystems, and even swim. Accordingly, the management of urban stream corridors is most effective when multiple uses and functions are recognized, and policies balance human uses with practices necessary for sustaining the ecological health of the stream.

6.2 How Development Affects Stream Processes

6.2.1 Hydrologic Effects

The urbanized landscape: Modifications of the land surface during urbanization change the type and the magnitude of runoff processes. These changes in runoff processes result from vegetation clearing, soil compaction, ditching and draining, and finally covering the land surface with impervious roofs and roads. The infiltration capacity of these covered areas is lowered to zero, and many areas that remain soil-covered are trampled to an almost impervious state. Thus Horton overland flow is introduced into areas that formerly may have generated runoff only under the subsurface flow regime. Resulting increases in storm runoff rates and total volumes lead to difficulties with storm-drainage control, stream-channel maintenance, groundwater recharge, and water quality.

This fundamental change in runoff-generating processes, then, is the major hydrologic consequence of urban development. Even where Horton overland flow occurred in the undeveloped landscape, runoff rates and volumes will increase further as a result of urban development. Although the downstream impacts of those increases are not expected to be as great as where subsurface flow once occurred, they can also be quite significant.

Besides eliminating soil-moisture storage and increasing imperviousness, urbanization affects other elements of the drainage system. Gutters, drains, and storm sewers are laid in the urbanized area to convey runoff rapidly to stream channels. Natural channels are commonly straightened, deepened, or lined with concrete to make them hydraulically smoother. Each of these changes increases the hydraulic efficiency of the channel, so that it transmits the flood wave downstream more quickly and with less storage in the channel. Higher downstream flood peaks typically result.

The increase of storm runoff has many costly consequences in urban areas. Frequent overbank flooding damages houses and gardens, or disrupts traffic. The capacities of culverts and bridges may be overtaxed. Channels become enlarged in response to the larger floods, and building lots suffer erosion and reduction of their value. Biological communities are disrupted by both these physical changes and the altered flow regime itself.

The measurement and prediction of hydrologic response: The human activities accompanying development produce measurable effects in the hydrologic response of a drainage basin. Most dramatic, and most often studied, is the increase in the maximum ("peak") discharge associated with floods. Other hydrologic changes also accompany watershed urbanization, but they require relatively sophisticated methods to recognize their effects and predict their magnitude. Hydrologic models are the most common tools by which runoff changes are studied; they allow us to understand the changes wrought by urbanization and show why many of the efforts to control runoff problems have not been terribly successful.

Decades of direct hydrologic measurements and simulation models quantify several related consequences of watershed urbanization: For any given intensity and duration of rainfall, the peak discharge is greater (by factors of 2 to 5; Hollis 1975), the duration of any given flow magnitude is longer (by factors of 5 to 10; Barker et al. 1991), and the frequency with which sediment-transporting and habitat-disturbing flows move down the channel network is increased dramatically (by factors of 10 or more; Booth 1991).

More recent assessments of hydrologic change have recognized other aspects of an altered flow regime, however, that are not expressed by traditional hydrologic metrics such as these but that may have even more significant geomorphic and ecological consequences. These include various attributes of non-extreme flows, such as the relative distribution of runoff between wet-season base-flow periods and high-flow periods (Konrad and Booth 2002) or the rate of rise or fall of individual storm hydrographs (Poff et al. 1997). As such, they may provide useful criteria for identifying flows, and entire flow regimes, that may have significant geomorphic or ecological effects on streams.

The influence of urban development on base flow will change by location and with the season, because base flow derives from different sources in different places and at different times of the year. During the wet season, base flow includes slow drainage from soils, which is likely to be lower in urban areas. During the dry season, base flow is fed from groundwater discharging from deeper aquifers, whose recharge may or may not be affected by the land-surface modifications associated with urban development. Human use of shallow groundwater or surface-water resources can reduce base flow during the dry season, whereas using water from a deep aquifer or imported from another basin to irrigate landscape during a dry season can actually increase base flows in urban streams (Konrad et al. 2005). Thus this attribute of stream hydrology, critical to both ecological and aesthetic functions, does not have a uniform response to urbanization.

Characterizing imperviousness: Although we commonly invoke "impervious surfaces" as a prime determinant of runoff changes in urban areas, not all imperviousness is created equal. Most important is the distinction between total impervious area (TIA) and effective impervious area (EIA). TIA is the "intuitive" definition of imperviousness: that fraction of the watershed covered by constructed, non-infiltrating surfaces such as concrete, asphalt, and buildings. Hydrologically, however, this definition is incomplete for two reasons. First, it ignores nominally "pervious" surfaces that are sufficiently compacted or otherwise so low in permeability that the rate of runoff from them is similar or indistinguishable from pavement (Burges et al. 1998). The second limitation of using TIA as a metric of hydrologic response is that it includes some paved surfaces that may contribute nothing to the storm runoff into the downstream channel. For example, rooftops that drain onto splashblocks that disperse the runoff onto a garden or lawn may not create any change in flow in the downstream channel at all. This metric, therefore, cannot recognize any contribution to stormwater mitigation that may result from alternative runoff-management strategies using, for example, pervious pavements or rainwater harvesting.

The first of these TIA shortcomings, the production of significant runoff from nominally pervious surfaces (Burges et al. 1989), is typically ignored in the characterization of urban development. The reason for such an approach lies in the difficulty in identifying such areas and estimating their contribution, and because of the credible belief that pervious areas will shed water as overland flow in proportion, albeit imperfectly, with the amount of impervious area. The second of these TIA shortcomings, the inclusion of non-runoff-contributing impervious areas, is formally addressed through the concept of EIA, defined as the impervious surfaces with direct hydraulic connection to the downstream drainage (or stream) system. Thus, any part of the TIA that drains onto pervious (i.e., "green") ground is excluded from the measurement of EIA. This parameter, at least conceptually, captures the hydrologic significance of imperviousness. EIA is the parameter normally used to characterize urban development in hydrologic models, although its direct measurement is difficult and commonly accomplished only by correlation to TIA.

6.2.2 Geomorphic Effects of Urbanization

Historically, human-induced alteration of stream channels was not universally seen as a problem. Dams and other stream-channel "improvements" were a common activity of municipal and federal engineering works of the mid-20th century (Williams and Wolman 1984); "flood control" implied a betterment of conditions, at least for streamside residents (Chang 1988); and fisheries "enhancements,"commonly reflected by massive infrastructure for hatcheries or artificial spawning channels, were once seen as unequivocal benefits for fish populations. Today, however, these alterations are widely recognized as commonly degrading the physical function, the biological integrity, and the aesthetic appeal of urban streams.

Even when not subjected to direct manipulation, however, urban-induced channel changes commonly do occur. As a result of hydrologic changes, channel widths and depths commonly increase throughout urban areas, and heterogeneous channel morphology becomes more simplified and uniform. Channels expand gradually in response to progressive increase in the flow regime (e.g., Hammer 1972, Booth and Jackson 1997, Bledsoe and Watson 2001). Yet this relationship, although common and intuitive, is not universal. A few studies note a reduction in channel width or depth with increases in watershed urbanization and, presumably, the discharge that accompanies it (e.g., Leopold 1973).

Although channel dimensions do commonly increase in response to gradual increases in the flow regime, changes in channel dimensions are usually sporadic and abrupt, often happening during particular storms when a single large flow can annul periods of stability that may have spanned many years (Booth and Henshaw 2001). Channels can also experience rapid and nearly uncontrolled downcutting of the stream bed, usually in response to an increase in the flow rate combined with specific combinations of gradient, substrate, and reduced in-channel roughness (Booth 1990).

The flow increases themselves can also increase the washout of in-stream woody debris or erosion of riparian vegetation, critical components of both channel stability and ecological health in forested (or once-forested) watersheds. Even under the best of circumstances, accelerated wood removal cannot be compensated by natural rates of regrowth and replacement. More commonly, however, urbanization eliminates the riparian corridor altogether, which means that in-channel wood is not replaced at all. This can result in further acceleration in rates of urban-induced channel expansion.

Change in the rate of sediment delivery into the channel network is another common consequence of urban development with potentially significant consequences for channel form. The broad relationship between stages of watershed development and resulting sediment loads have long been recognized and presented in studies such as Wolman (1967). In general, an initial phase of increased sediment delivery is associated with land clearing and soil disturbance during watershed development. As impervious surfaces such as road networks, parking lots, buildings, and compacted areas increase their footprint, sediment yield from upland areas is diminished as runoff is simultaneously increased. In terms of stream processes, the capacity to transport sediment is significantly increased even as the supply of sediment for transport may be concurrently decreased. In subsequent stages of the process, channel erosion from increased flows can provide a new source of sediment that can account for more than half of the total sediment load of an urban stream (Trimble 1997).

The observed sequence of channel responses, however, can be complex. Increased sediment loads, generated at particular stages in the forest–agriculture–urban sequence of North American land development, exert a tendency for channel aggradation that opposes the tendency for erosion that accompanies increasing discharge. The time-varying interplay of these contradictory factors probably explains much of the channel narrowing or shallowing that is sometimes measured.

Efforts to integrate the generally similar, but locally disparate, observations of channel change (see Schumm 1977, Park 1997, Thorne et al. 1998) into a unified model generally articulate a sequence of anticipated changes over time. Simon (1989), for example, evaluated the consequences of channelization and described a widely used evolutionary sequence of undercutting, bank failure, channel widening, and restabilization that closely resembles that of urbanization. Arnold et al. (1982) also recognized the interplay of spatial factors, notably upstream stream erosion and downstream deposition, that can result in multiple "responses" along the same channel, a theme of complex spatial and temporal response that is echoed by many careful studies of urban channels.

Such changes to channel morphology are among the most common and readily visible effects of urban development on natural stream systems (Walsh et al. 2005). The actions of deforestation, paving of the uplands, and channelization can produce tremendous changes in the delivery of water and sediment into the channel network. In channel reaches that are alluvial, subsequent responses can be rapid, dramatic, and readily documented: channels widen, deepen, and in extreme cases may downcut many meters below the original level of their beds. Alternatively, they may fill with sediment derived from farther upstream and braid into multiple rivulets threading between gravel bars. In either case, they are transformed far beyond the range of conditions displayed at any time during their pre-urban period. They can become hazardous to any surrounding human infrastructure, and they no longer can support their once-natural populations of benthic invertebrates and fish.

6.2.3 Chemical Effects

The chemical constituents of natural streams vary widely with climatic region, stream size, soil types, and geological setting. However, small natural streams typically have relatively low levels of both dissolved and particulate constituents. As urbanization alters the pathways by which water passes over and through the ground surface, and as we introduce new chemical constituents into the near-surface environment, the chemical composition of surface and ground waters change. The worst of these problems have historically emanated from discrete sources such as a municipal sewage outfall or the cooling-water discharge of a thermal power plant. In the United States, large expenditures on existing sources and new regulations on future sources have yielded dramatic reductions in this type of "point-source" pollution during the 1970s and 1980s. Yet these gains are slowly being lost to more diffuse nonpoint sources of contaminants, which continue to change the quality of surface and ground waters almost unabated.

These changes in water quality are nearly inescapable byproducts of modern land-use development and human activities in both agricultural and urban settings. The spatial pattern of such increases, however, is quite irregular, and simple correlations between any measure of urbanization (e.g., percent watershed imperviousness) and concentrations of chemical pollutants are generally poor. Furthermore, the linkages between chemical constituents and beneficial uses are very poorly known, particularly at low but chronic levels, and the natural variability of many of these constituents often makes the identification of human effects ambiguous or very time-consuming. In areas of low or even moderate urban development, water-chemistry parameters often do not exceed water-quality standards (Horner et al. 1997). Other constituents, particularly manmade compounds with unknown but potentially significant biological activity at very low concentrations, have no health or water-quality standards at all.

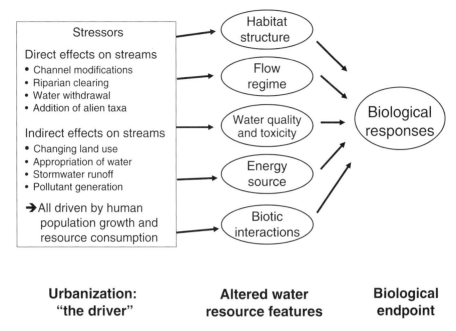

Fig. 6.5 Five features that are affected by urban development and that affect biological conditions in urban streams (modified from Karr (1991); Karr and Yoder (2004))

6.2.4 Ecological Implications

Stream biota evolves over millennia as a result of the complex interactions of chemical, physical, and biological processes. These processes and interactions can be grouped into five major classes of environmental "features" to form a simple conceptual framework (Fig. 6.5; Karr 1991, Karr and Yoder 2004). When one or more of these features is affected by human activities, the result is ecosystem degradation (Allan 2004, Paul and Myer 2001). No one feature, however, is always the limiting factor for biological condition; conversely, improving any one feature does not guarantee corresponding improvement in biology. An important corollary for our subsequent consideration of stream enhancement is that correcting or "restoring" one altered feature does not necessarily eliminate the need to correct another.

In the urban environment, changes are imposed on these features by a wide variety of human activities, via a number of pathways that operate at multiple spatial scales. So, for example, watershed-scale changes in land cover alter hydrology through stormwater inflows to streams and reduced groundwater recharge. Adjacent to stream channels, local changes to land cover can affect the input of energy via organic material and sunlight; and, at a single site, direct modification of the channel itself can disrupt the habitat structure.

Although any of the five features of Fig. 6.5 can be responsible for the loss of biological health in an urban stream, changes in flow patterns are commonly recognized as a particularly important and ubiquitous pathway by which urbanization influences biological conditions. This primacy reflects the magnitude of hydrologic change commonly imposed by urbanization (e.g., Booth and Jackson 1997, Konrad and Booth 2002) and the close correlations reported between biological health and various metrics of hydrologic alteration (e.g., Poff and Ward 1989, Poff and Allan 1995, Roy et al. 2003). Such metrics reflect interactions between flow regime and the physical characteristics of the channels upon which they are imposed. Because the frequency and erosive potential of flows that shape in-stream habitats are amplified by imperviousness, the overall intensity of habitat disturbance experienced by stream biota is often more severe after watershed development. The resulting disequilibrium between flow regime and channel form alters habitat "dynamics" and degrades biological health by reducing the quantity, quality, and diversity of available habitats.

Even where urban-modified flows have been managed and downstream channels have adjusted (or been directly modified), a "stabilized" channel should not be mistaken for a return of the channel to its natural state (Henshaw and Booth 2000), and a "stream-stabilization project" should never be mistaken for ecological restoration. A re-stabilized channel will typically be larger and less geomorphically complex than the pre-urbanization channel form. It will also have altered habitat and flow patterns, water velocities, sediment flux, and organic inputs (e.g., Jacobson et al. 2001, Roesner and Bledsoe 2002), and it may carry an ecological legacy of extirpations that precludes the return of pre-disturbance biota (Harding et al. 1998). Additional assessment and rehabilitation actions are almost always required to restore the biological integrity of the stream even after geomorphic stability is achieved, and the success of such additional efforts is by no means assured.

The inherent complexity of watershed processes makes it difficult to isolate the effects of urbanization on ecological health. Interactions between stream water quality and quantity, and year-to-year climate variability, can confound predictions regarding the ecological implications of urbanization. At present, it is usually not possible to accurately predict the specific ecological changes that will occur under alternative watershed-management scenarios. Nevertheless, the last few decades of research and management experience provide a very useful knowledge base and suite of science-based strategies for managing urban watersheds.

6.3 Management Principles

Channels are problematic for people, because they are attractive but resist our efforts to manage them—they flood, they migrate, they deposit sediment, they downcut—in short, they are dynamic systems, but they spend long periods of time in quiescence that lull the unwary into approaching too closely and developing too permanently. People are problematic for channels too—we alter them directly for our own purposes, and our manipulation of the watershed's land surface affects every aspect of what combines to form a natural stream or river. As a result, channels can lose both their physical and biological functions without any intentional (but no less influential) actions on our part.

We recognize that pervasive watershed changes, notably during urbanization, fundamentally alter the rates and processes by which water and sediment are delivered to the stream channels. The channel form, in turn, changes in response to the altered delivery regime. Yet rather than address the problem at its source, namely the watershed area, most remedial efforts are expended at the final point of expression, namely the stream channel. Clearly, this is not rational.

Complete restoration of an already urbanized watershed, however, is rarely judged feasible (because of astronomical expense, daunting logistics, and limited effectiveness of available tools). However, the success of in-channel mitigation, however feasible and conscientiously applied, also is limited. This is the conundrum that faces even the most well-intentioned efforts at stream protection or enhancement in the urban water environment.

Even if achievable goals are of necessity limited, effective actions do exist and typically follow certain key underlying principles:

- hydrologic alteration is profound; hydrologic mitigation is critical;
- hydrologic mitigation must reflect both geomorphic and ecological principles;
- protecting riparian zones provides synergistic benefits; and
- goals, objectives, and evaluation are all needed for successful urban-stream enhancement.

These principles are enumerated in the following sections.

6.3.1 Hydrologic Alteration Is Profound; Hydrologic Mitigation Is Critical

As a consequence of urban-induced runoff changes, which in turn cause flooding, erosion, and habitat damage, jurisdictions have long required some degree of stormwater mitigation. The most common historic approach has been to convey stormwater runoff as rapidly and efficiently as possible away from developed areas to minimize the consequences of standing water. As this conveyance becomes more effective, however, the receiving downstream channels become subject to increasing peak discharges and consequent flooding of their own.

Thus, the first recognized hydrologic consequences of urbanization were those associated with peak-flow increases (i.e., "more flooding"). Careful analysis, culminating in a synthesis of many separate studies (Leopold 1968, Hollis 1975), showed how two factors, watershed percent imperviousness and watershed percentage with storm sewers, increased the peak discharges of floods. Large, infrequent floods were increased less than smaller, more common events; in general, Hollis found peak-flow increases of two- to three-fold are common for the moderate-sized floods in moderately urbanized watersheds. These general results have been replicated in both empirical and modeling studies, on many dozens of watersheds throughout the United States. Although there is a consistent pattern of peak-flow increase associated with increased watershed imperviousness, differences in "styles" of development (e.g., connectivity of imperviousness surfaces and drainage infrastructure) as well as climatic and geologic contexts contribute to high variability among regions and watersheds (e.g., Bledsoe and Watson 2001, Poff et al. 2006).

The first (and still most common) approach in reducing the magnitude of peak discharge has been through the use of detention ponds (Fig. 6.6a), which are placed downstream of the developed area (from which runoff is drained rapidly and efficiently) and upstream of areas prone to urban-increased flooding or erosion from high flows. These facilities can be designed to various levels of performance, depending on the desired balance between achieving downstream protection and the cost of providing that protection. A "peak" standard, the classic (and least costly) goal of detention facilities, seeks to maintain post-development peak discharges at their pre-development levels (Fig. 6.6b). This approach addresses the concern of flooding, for which the "peak" discharge is the only important parameter. Even if this goal is achieved successfully, however, the aggregate duration that such flows occupy the channel must increase because the overall volume of runoff is greater,

Fig. 6.6a A detention pond, designed to capture and temporarily store runoff from the adjacent residential development before releasing the water to the downstream channel (King County, Washington)

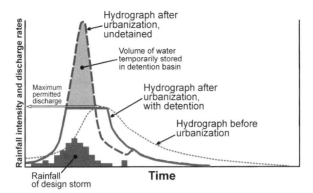

Fig. 6.6b Idealized hydrograph of a detention pond, showing the presumed rainfall from a chosen "design storm" (low gray bars) and comparing the three alternative hydrographs (pre-development, post-development without detention, and post-development with detention) that result. The maximum permitted discharge from the detention pond is normally set by the peak discharge from the pre-development watershed (modified from Dunne and Leopold (1978))

resulting in substantial stream-channel erosion (McCuen 1979, Booth and Jackson 1997, Roesner et al. 2001). If the channel is erosive, or if it supports biota with a particular suite of flow-related needs, significant damage may still result.

Thus, mitigating the erosive potential of increased runoff requires control of the *duration* (not just the magnitude) of flows across a wide spectrum of sediment-transporting discharges. A "duration" standard for detention-pond performance thus was developed in several jurisdictions to maintain the post-development duration of all discharges at pre-development levels (e.g., King County 1990, MacRae 1997). Duration standards are motivated by a desire to avoid potential disruption to the downstream channels by not allowing any flow changes that might increase sediment transport beyond pre-development levels. Without infiltration of runoff, however, the total volume of runoff must still increase in the post-development condition, and so durations cannot be matched (or reduced) for all discharges—below some discharge rate, the "excess" water must be released. This is accomplished by determining (or otherwise assuming) a threshold discharge below which sediment transport, or any other disruptive conditions, in the receiving channel is presumed not to occur.

The flow-duration control approach is a significant improvement over the "peak-shaving" standard, but it is not a panacea. Reductions in sediment delivery to stream channels may result in accelerated channel erosion and, therefore, habitat degradation, even if the pre-development flow characteristics are largely maintained (Bledsoe 2002). This occurs because the flow becomes more "hungry" for channel-forming sediment and the stream consequently compensates for the reduction in the watershed sediment supply through local boundary erosion. Moreover, additional analyses have shown that other measures of flow variability with likely biological importance, such as the seasonality of peak discharges or the time between sediment-transporting events, are not well maintained by such flow-mitigation approaches in the face of watershed urbanization (e.g., Konrad and Booth 2005). In other words, maintaining sediment-transport capacity is not an adequate surrogate for protecting the full universe of flow-related attributes of a stream.

6.3.2 Hydrologic Mitigation Must Reflect both Geomorphic and Ecological Principles

Native biological communities are adapted to and tolerate a range of aquatic habitat conditions that may become less available or completely disappear as a consequence of land-use changes. Watershed urbanization alters the interactions between flow, sediment and channel form that fundamentally control the quality, quantity, and spatial distribution of stream habitats (e.g., Jacobson et al. 2001, Roesner and Bledsoe 2002, Walsh et al. 2005). As such, there is broad consensus among river scientists that sustaining biological communities, and especially sensitive biota, requires maintaining flow and habitat dynamics within some range of natural variability (e.g., Bunn and Arthington 2002). Thus, hydrologic-mitigation practices derived from an understanding of both geomorphic and ecological processes are a prerequisite for maintaining stream ecological integrity. Because flow regime is the "master variable" controlling erosion, habitat availability, and ecological processes, the stormwater-management practices that are the most protective of stream health are those that minimize changes in the magnitude, frequency, duration, and variability of streamflows.

6.3.3 Protecting Riparian Zones Provides Synergistic Benefits

Urban development not only increases rates of water and sediment delivery but also encroaches on the riparian corridor. With the clearing of streamside vegetation, less wood enters the channel, depriving the stream of stabilizing elements that help dissipate flow energy and usually (although not always) help protect the bed and banks from erosion (Booth et al. 1997). Deep-rooted bank vegetation is replaced, if at all, by shallow-rooted grasses or ornamental plants that provide little resistance to channel widening. Furthermore, the overhead canopy of a stream is lost, eliminating the shade that controls temperature and supplies leaf litter that enters the aquatic food chain (Roberts et al. 2008).

Abundant research has demonstrated the ecological importance of preserving riparian zones, even where other measures have not been taken to mitigate the effects of urbanization. For example, Morley and Karr (2002) documented an increase in biological health from "very poor" to "fair" over <2 km along a single suburban Puget Lowland stream channel, finding that the variability was strongly explained by riparian land cover but not by overall catchment land cover. Good correlations between physical condition of channels and frequency of stream-road crossings are shown by a variety of studies (e.g., Avolio 2003, McBride and Booth 2005). Over two decades ago, Steedman (1988) showed the importance of both watershed disturbance and riparian corridor integrity in supporting a healthy fish population in streams of the American Midwest.

6.3.4 Goals, Objectives, and Evaluation Are Needed for Successful Urban-Stream Enhancement

Most urban streams are managed in a piecemeal, reactionary fashion. Managers often find themselves in a perpetual cycle of treating the symptoms of urban degradation with small-scale "band-aids" that are largely divorced from any sort of strategic planning for streams within their watershed context. Many managers perceive that social attitudes and values around urban-stream amenities are rapidly evolving in their jurisdictions, but most programs and activities are not rooted in stakeholder preferences or clearly defined goals.

Goals for enhancing and sustaining urban stream amenities are generally most useful and achievable when they grow out of an envisioning process that proactively garners input from the full spectrum of watershed stakeholders. The envisioning process necessarily involves planners, engineers, ecologists, and social scientists to connect alternative management strategies to probable future states defined in terms of valued amenities. That is, the process involves developing predictive scientific assessments (in the sense of Reckhow 1999) that integrate modeling, expert judgment, extensive communication, and developing the institutional commitments requisite for achieving a long-term vision for the stream systems within a particular jurisdiction. They also require a list of tangible activities that can make concrete progress towards these overarching goals.

This progress must be measured and assessed, lest the entire effort become meaningless. Even where basin-planning programs have identified and implemented practices aimed at achieving a particular long-term vision, stream-enhancement activities are rarely monitored and assessed (Wohl et al. 2005, Palmer et al. 2005). As such, the practice of managing urban streams suffers from a paucity of information and, therefore, knowledge, regarding which policies and tools are effective for placing a given type of urban stream on a desired trajectory.

Despite substantial knowledge gaps, a critical examination of the last few decades of research and monitoring suggests that it is plausible that integrated programs of hydrologic mitigation, riparian zone conservation, and pollution controls can potentially sustain aquatic biodiversity and valued social amenities in urban streams. Systematic monitoring and assessment of pre- and post-urban processes and conditions are essential for understanding the extent to which integrated management can maintain ecosystems that closely resemble pre-impact structure and function, as opposed to yielding new types of regional stream ecosystems (Westman 1985). Without such information, the goal of identifying sustainable management strategies becomes unattainable, and the rapidly growing population of urban streams will never reflect the aspirations of the people who inhabit their watersheds.

6.4 Technical Approaches to Urban Stream Enhancement

6.4.1 Hydrology and Geomorphology

Although the physical "channel design" is a common element of stream enhancement or stream restoration, a true alluvial channel is ultimately the product of its water and sediment regime. Although a set of design drawings or engineering plans can establish the initial template for channel form, the long-term morphology will reflect the hydrologic regime and the sediment load that passes down the channel. Conversely, if the channel is not designed to adjust then the interplay of channel form, flow regime, and sediment load will determine whether or not the outcome is "stable" or "successful."

In each of these circumstances, the role of hydrology is paramount, and there is no substitute for accurate hydrologic predictions. Current computer models use hourly (or more frequent) precipitation data as input to simulate many years of hydrologic response, keeping a running account of the amount of water within various hydrologic storage zones, both surface and subsurface. Individual storm "events" are not discriminated; the actual rainfall record, over time, determines how the hydrologic system responds. This approach is necessary to achieve the overarching goal of recognizing relationships between flow and biota, because much of the biotic response depends not on the characteristics of an individual storm but on the timing and the relationships of flows arising from multiple storms, and the sequence and distribution of those flows throughout the year. These critical factors cannot be explored in any other way.

One-size-fits-all practices based on "single-factor" ecology or extrapolation across all stream types is not likely to protect stream amenities. Streams differ in their resilience and response to the effects of urbanization (Montgomery and MacDonald 2002). A channel that naturally contains extensive bedrock control or very resistant boundary materials, for example, will be less physically susceptible to the hydrologic changes typical of urbanization than a fully alluvial stream in relatively erodible material. This suggests that stream-management activities aimed at mitigating the effects of hydrologic modifications will be most effective when tailored to different stream types.

Identifying simple thresholds that can be used to broadly prescribe stormwater policy will continue to be an attractive goal (e.g., $\leq 10\%$ total watershed imperviousness, Schueler 1994)), but the outcomes of such an approach will be constrained and difficult to predict. Instead, a linked modeling framework that combines continuous hydrologic simulation, sediment delivery, and channel erosion models is probably necessary to protect fully the physical habitat characteristics of streams that are susceptible to geomorphic impacts (Richards and Lane 1997). Such a framework can provide a process-based, albeit uncertain, foundation for envisioning alternative future states of streams. Identifying appropriate predictive and assessment tools, and designing management practices that are demonstrably effective in conserving ecological integrity is an ongoing challenge that begs for improved interdisciplinary collaboration between engineers and ecologists.

6.4.2 Riparian-Zone Conservation and Restoration

Protecting and restoring riparian zones is a cornerstone of stream conservation. Although riparian corridors often constitute less than 5% of the total watershed area, they have a profound and disproportionate influence on the ecological integrity of streams (Gregory et al. 1991, Naiman et al. 2005). Protected streamside zones, sometimes called "buffers" in a regulatory setting, support stream health by moderating temperatures, filtering pollutants, providing food and cover, and preventing excessive channel erosion. There are many excellent resources that describe management strategies for riparian zones, including information on multi-purpose designs, model ordinances, and overcoming implementation issues (Lowrance et al. 1995, Schueler 1995, Center for Watershed Protection (CWP) 2008).

One of the key challenges in urban watersheds is restoring riparian zones along streams that have been engineered for drainage conveyance, or where channels have incised and have become disconnected from their floodplain, lowering the water table of the surrounding landscape (Fig. 6.7). Restoration of riparian corridors in these contexts requires careful prioritization of activities and multidisciplinary design teams of geomorphologists, engineers, and ecologists. Increases in channel and floodplain roughness associated with reestablishment of vegetation, debris inputs, and adjustments in stream morphology are generally at odds with the traditional approach to drainage infrastructure that emphasized "fast conveyance" of floodwaters. In many contexts, enhancing stream riparian corridors will require engineers and environmental planners to transcend the tension between encouraging fast conveyance versus establishing functional (and hydraulically rough) riparian corridors, in part by strategically identifying locations where riparian enhancement is feasible within the constraints of existing infrastructure and floodplain encroachment.

6.4.3 Low Impact Development and Land-Use Planning

Low Impact Development (LID) is a strategy for stormwater management that uses on-site natural features integrated with engineered, small-scale hydrologic controls to manage runoff by maintaining, or closely mimicking, pre-development watershed hydrologic functions (U. S. Environmental Protection Agency (USEPA) (1999), Puget Sound Action Team (PSAT) (2005). It is achieved most effectively at multiple scales—land-use planning at the scale of an entire watershed to identify and preserve key elements of the hydrologic system, together with engineering and site-design elements that are implemented at the scale of individual parcels, lots, or

Fig. 6.7 Modestly incised urban channel, with likely lowering of the water table beneath the adjacent and now-disconnected floodplain (Juanita Creek, Kirkland, Washington)

structures. In combination, these actions seek to store, infiltrate, evaporate, or otherwise slowly release stormwater runoff in a close approximation of the rates and processes of the pre-development hydrologic regime.

Most applications of LID have several common components:

- Preserving elements of the natural hydrologic system that are already achieving effective stormwater management, recognized by assessment of a site's watercourses and soils; channels and wetlands, particularly with areas of overbank inundation; highly infiltrative soils with undisturbed vegetative cover; and intact mature forest canopy.
- Minimizing the generation of overland flow by limiting areas of vegetation clearing and soil compaction (Arendt 1997); incorporating elements of urban design such as narrowed streets, structures with small footprints (and greater height, as needed), use of permeable pavements as a substitution for asphalt/concrete surfaces for vehicles or pedestrians; and using soil amendments in disturbed areas to increase infiltration capacity.
- Storing runoff with slow or delayed release, such as in cisterns or distributed bioretention cells, across intentionally roughened landscaped areas, or on vegetated roofing systems ("green roofs"). Runoff storage in LID differs from traditional stormwater management, notably the latter's use of detention ponds, primarily

by its scale—namely small and distributed in LID, large and centralized in traditional approaches.

These objectives are achieved through five basic elements that constitute a "complete" LID design (Coffman 2002):

1. Conservation measures—maintaining as much of the natural landscape as possible.
2. Minimization techniques—reducing the impacts of development on the hydrologic regime by reducing the amount of disturbance when preparing a site for development.
3. Flow attenuation—holding runoff on-site as long as possible, without causing flooding or other potential problems, to reduce peak discharges in the downstream channel.
4. Distributed integrated management practices—incorporating a range of integrated best management practices throughout a site, commonly in sequence.
5. Pollution prevention measures—applying a variety of source-control, rather than treatment, approaches.

Although these five elements can be applied to virtually any development, the specific manner in which they are used must be determined by the local climate and soils. Native soils, in particular, play a critical role in storage and conveyance of runoff. In humid regions, one to several meters of soil, generally high in organic material and relatively permeable, commonly overlie less permeable substrates of largely unweathered geologic materials. While water is held in this soil layer, solar radiation and air movement provide energy to evaporate surface-soil moisture and contribute to the overall evapotranspiration component of the water balance. Water not evaporated, transpired or held interstitially moves slowly downslope or down gradient as shallow subsurface flow over many hours, days, or weeks before discharging to streams or other surface-water bodies. In arid regions with relatively lower organic-content soils and vegetation cover, precipitation events can produce rapid overland flow response naturally; however, the principles of LID remain: retain native soils, vegetation, topography, and the various elements of the hydrologic system to preserve aquatic ecosystem structure and function.

6.5 Next Steps

6.5.1 Rivers and Streams Are Focal Points for Urban Renewal: These Are Systems Worth Restoring

Over one hundred years ago, urban designs were deliberately linked to water systems. As discussed later in this book, we have only recently rediscovered the fundamental idea that cities can express the multiple purposes of the urban water environment. Urban streams are neighborhood amenities that inspire passion and ownership from their nearby residents, and they can support self-sustaining biotic communities, even though those communities depart significantly from pre-disturbance conditions. This combination is particularly timely as we address the dual challenges of climate change and sustainability of our modern cities.

6.5.2 Define Realistic Goals for Urban-Stream Restoration

Functioning stream systems and watershed urbanization are not mutually exclusive, but seeking a direct analog to undisturbed aquatic systems ignores the profound alteration to water and sediment fluxes that are the hallmark of urban watersheds and the streams that result. Based on nearly a half-century of studies of urban streams, the challenges of establishing a self-sustaining trajectory towards aquatic function and health are seemingly insurmountable. Even if a natural flow regime could be reestablished through effective, watershed-wide application of site-scale runoff management, natural geomorphic processes of sediment delivery and channel change are incompatible with most adjacent urban land uses. These processes, however, are the very agents of habitat creation and rejuvenation, and they ensure the persistence of the channel form through dynamic, short-term adjustments to floods and droughts. These adjustments are rarely tolerable in urban landscapes. Particularly in climatic regimes such as the American Southwest, where large discharges are many times larger than "typical" flows, the immediate consequences on the surrounding terrain can be quite dramatic.

As a result, we expect that the paradigm for a "restored" urban stream must combine the recovery of certain natural processes with a respect for the unyielding constraints and multiple objectives of the urban setting. Channels will not meander across the landscape, and so entire categories of key habitat features may not exist. Sediment will not pass down the channel as freely or as efficiently as in pre-development time, because the morphology of the channel will be constrained, and urban infrastructure (e.g., road crossings) will impose immutable constraints (Chin and Gregory 2001). A riparian corridor may (and should) be present, but its species composition will probably not mimic pre-human conditions, and the exclusion of people and domestic animals cannot be assured—indeed, their active use of this space will probably be encouraged to achieve other goals set for these watercourses.

Short-term, local-scale actions can improve the condition of urban streams and are generally feasible under many different management settings. They are unlikely to produce permanent effects, however, because they do not incorporate the reestablishment of self-sustaining watershed processes. Such actions include riparian fencing and planting, water-chemistry source control, fish-passage projects, and certain in-stream structures. Short-term actions address acute problems typical to stream channels in urban and urbanizing catchments; they are commonly necessary, but not sufficient, to restore biotic integrity.

In contrast, if restorative actions are intended to achieve sustainable ecological goals, they would need to effectively address all five elements of disturbed stream ecosystems (Fig. 6.5). These actions might include various types of land-use planning (e.g., preserves and zoning), avoiding road and utility crossings of the channel network or minimizing their footprint, upland hydrologic rehabilitation (e.g., stormwater infiltration or on-site retention) and erosion control, re-establishing the age structure of riparian vegetation communities, and reconnection of floodplains with their associated channels (Booth 2005). Because streamflow is a key element of ecological conditions and driver of habitat-forming processes, reestablishing streamflow patterns is almost certainly necessary for restoration of an aquatic ecosystem. Given the constraints common to cities and the challenges of watershed-scale hydrologic rehabilitation in a built-out catchment, this goal may not be realistic for many urban watersheds. However, nearly half of the urban development projected for the United States for the year 2030 has not yet been built (Nelson 2004), and so opportunities to achieve better stormwater management through both new development and redevelopment still abound.

Short-term actions alone, and even some well intentioned and well-reasoned long-term actions, will not achieve broad ecosystem protection in the urban environment. At best, biological communities in urban streams may be diverse and complex, but they will depart significantly from pre-development conditions. These streams can be neighborhood amenities and provide their nearby residents with a connection to a place, and they can support a self-sustaining and self-regulating biological community. If we articulate these goals and work towards them, such outcomes for urban streams should be achievable even without fully reestablishing natural hydrologic processes or hydrologic conditions.

6.5.3 Climate Change and the Uncertain Coupling Between Human and Environmental Systems

Climate change is an impending threat to aquatic ecosystems, urban and non-urban alike, but the particular constrains of the urban water environment are likely to amplify some of the most serious consequences. Increases in water temperatures as a result of general warming will alter the geographic distribution of aquatic plant and animal species. Although some species can migrate as the climate changes, the barriers to migration and fragmentation of habit at that commonly accompanies urban development will likely result in local and regional extirpation, absent extensive and innovative restoration approaches.

Changes in precipitation will alter streamflows, with the most commonly anticipated change being an increase in extreme events and a corresponding increase in channel-scouring flows and flooding. The urban infrastructure is generally not tolerant of increased magnitudes or frequency of flooding, and the most common responses to increased flood risk are costly and further damaging the aquatic ecosystems. Future actions will need to do better! Those actions that will improve the resiliency of urban streams to such changes include maintaining riparian forests, reconnecting floodplains and other overbank areas, reducing pollution, restoring already-damaged systems, and minimizing groundwater withdrawal (Poff et al. 2002).

6.5.4 Lessons from Prior Efforts, Guidelines for the Future

Failure of the last century's management of hydrologic alteration should not condemn us to the same future. Instead, it underscores the need for new approaches to stormwater management that integrate multiple scales of watershed planning, site layout, and infrastructure design. Full, or at least partial, long-term restoration of some hydrologic and geomorphic processes, with subsequent biological recovery, may be possible even in highly disturbed urban environments. The absence of abrupt thresholds in biological responses to urbanization (e.g., Thomson et al. 1996, Morley and Karr 2002) suggests that even incremental improvements can have direct, albeit modest, ecological benefits. Urban streams can be self-sustaining to biotic communities, even though those communities depart significantly from pre-disturbance conditions. Last, urban streams should also retain the possibility, however remote, of one day benefiting from the long-term actions that can produce greater, sustainable improvements. Current costs, uncertainties, or sociopolitical constraints are no excuse to continue building urban developments or traditional rehabilitation projects that permanently preclude future long-term stream improvements.

The scientific literature and numerous case studies demonstrate the value of following ten principles to achieve sustainable stream health and resiliency in urbanizing watersheds. Conversely, our many failures can commonly be traced back to ignorance of one or more of these elements (Williams et al. 1997, Frissell 1997). We offer them as a summary of this chapter's lessons and a checklist for the management and enhancement of streams in the urban water environment:

1. Address problem causes, not just symptoms: focus on ecosystem processes rather than a specific, tangible form.
2. Recognize many scales, in both time and space. A long-term, large-scale, multidisciplinary perspective that includes both ecological history and future changes is critical.
3. Work with, rather than against, natural watershed processes, and reconnect severed linkages—the only channels that persist on the landscape without continuous human intervention are those with an intact set of watershed processes that sustain their form and features.
4. Clearly define goals and make both sustainability and enhancing ecological integrity explicit goals.
5. Utilize the best available science in predictive assessments that are risk-based and decision-oriented, acknowledging the desired outcomes of interest to all

stakeholders: Human health and safety, clean water, productive fisheries, other valued biota, reliable water supply, recreation, and aesthetics.
6. Honestly identify and openly debate the key knowledge gaps and uncertainties, but adopt an action-oriented principle that ensures that the decision-making exercise will lead to results.
7. Make decisions in a transparent, organized framework that:

    - structures the problem clearly;
    - provides a ranking of the options even though the uncertainties may not be resolved in the foreseeable future;
    - involves affected stakeholders;
    - documents and justifies the decision process to all stakeholders; and
    - provides research priorities by showing whether resolving particular uncertainties would affect the preferred option(s).

8. Watershed-restoration projects are as much a social undertaking as an ecological one; understand social systems and values that support and constrain restoration while establishing long-term personal, institutional, and financial commitments.
9. Some strategies will work, some will not, and some will take many years to assess. Learn through careful long-term monitoring of key ecological processes and biotic elements. Reevaluate and update management strategies based on monitoring, recognizing that every "restoration" effort is actually an experimental treatment that requires evaluation and future modification to achieve its stated goals.
10. The best strategy is to avoid degradation in the first place. The highest emphasis should be placed on preventing further degradation rather than on controlling or repairing damage after it has already occurred.

References

Allan, J. D. 2004. Landscapes and riverscapes: the influence of land use on stream ecosystems. Annual Review of Ecology, Evolution and Systematics 35:257–284.

Arendt, R. 1997. Conservation Design for Subdivisions. Island Press, Washington, District of Columbia, USA.

Arnold, C. L., P. J. Boison, and P. C. Patton. 1982. Sawmill Brook: an example of rapid geomorphic change related to urbanization. Journal of Geology 90:155–166.

Avolio, C. M. 2003. The Local Impacts of Road Crossings on Puget Lowland Creeks. MSCE Thesis. Department of Civil and Environmental Engineering, University of Washington, Seattle, Washington, USA.

Barker B. L., R. D. Nelson, and M. S. Wigmosta. 1991. Performance of detention ponds designed according to current standards, pp. 64–70. In: Puget Sound Research 91 Conference Proceedings, Puget Sound Water Quality Authority, Seattle, Washington, USA.

Bledsoe, B. P., and C. C. Watson. 2001. Effects of urbanization on channel instability. Journal of the American Water Resources Association 37(2):255–270.

Bledsoe, B. P. 2002. Stream erosion potential associated with stormwater management strategies. Journal of Water Resources Planning and Management 128:451–455.

Booth, D. B. 1990. Stream-channel incision following drainage-basin urbanization. Water Resources Bulletin 26(3):407–417.

Booth, D. B. 1991. Urbanization and the natural drainage system—impacts, solutions, and prognoses. Northwest Environmental Journal 7:93–118.

Booth, D. B., and C. R. Jackson. 1997. Urbanization of aquatic systems: degradation thresholds, stormwater detection, and the limits of mitigation. Journal of the American Water Resources Association 33(5):1077–1090.

Booth, D. B., D. R. Montgomery, and J. P. Bethel. 1997. Large woody debris in urban streams of the Pacific Northwest. In: L. A. Roesner (Ed.), Effects of Watershed Development and Management on Aquatic Ecosystems, Proceedings of an ASCE Engineering Foundation Conference, Snowbird, Utah, August 4–9, 1996; published by ASCE, Reston, Virginia, USA.

Booth, D. B., and P. C. Henshaw. 2001. Rates of channel erosion in small urban streams, pp. 17–38. In: M. S. Wigmosta and S. J. Burges (Eds.), Influence of Urban and Forest Land Use on the Hydrologic-Geomorphic Responses of Watersheds, Monograph Series, Water Science and Applications, Volume 2, American Geophysical Union, Washington, District of Columbia, USA.

Booth, D. B. 2005. Challenges and prospects for restoring urban streams, a perspective from the Pacific Northwest of North America. Journal of the North American Benthological Society 24:724–737.

Bunn, S. E., and A. H. Arthington. 2002. Basic principles and ecological consequences of altered flow regimes for aquatic biodiversity. Environmental Management 30:492–507.

Burges, S. J., B. A. Stoker, M. S. Wigmosta, and R. A. Moeller. 1989. Hydrological Information and Analyses Required for Mitigating Hydrologic Effects of Urbanization. University of Washington, Department of Civil Engineering, Water Resources Series Technical Report No. 117, p. 131.

Burges, S. J., M. S. Wigmosta, and J. M. Meena. 1998. Hydrological effects of land-use change in a zero-order catchment. Journal of Hydrological Engineering 3:86–97.

Chang, H. H. 1988. Fluvial Processes in River Engineering. John Wiley and Sons, Inc., New York, USA, p. 336.

Chin, A., and K. J. Gregory. 2001. Urbanization and adjustment of ephemeral stream channels. Annals of the Association of American Geographers 91(4):595–608.

Church, M. 2002. Geomorphic thresholds in riverine landscapes. Freshwater Biology 47(4):541–557.

Coffman, L. S. 2002. Low-impact development: an alternative stormwater management technology. In: R. L. France (Ed.), Handbook of Water Sensitive Planning and Design, Lewis Publishers, Boca Raton, Florida, USA.

CWP. 2008. Model ordinances. Center for Watershed Protection, Ellicott City, Maryland, USA, available from http://www.cwp.org/.

Dunne, T., and L. Leopold. 1978. Water in Environmental Planning. W. H. Freeman and Co., San Francisco, California, USA, 809 pp.

Frissell, C. A., W. J. Liss, C. E. Warren, and M. D. Hurley. 1986. A hierarchical framework for stream habitat classification: viewing streams in a watershed context. Environmental Management 10(2):199–214.

Frissell, C. A. 1997. Ecological principles, pp. 96–115. In: J. E. Williams, C. A. Wood, and M. P. Dombeck (Eds.), Watershed Restoration: Principles and Practices, American Fisheries Society, Bethesda, Maryland, USA.

Gregory, S. T., F. J. Swanson, W. A. McKee, and K. W. Cummins. 1991. An ecosystem perspective of riparian zones. BioScience 41:540–551.

Hammer, T. R. 1972. Stream channel enlargement due to urbanization. Water Resources Research 8:1530–1546.

Harding, J. S., E. F. Benfield, P. V. Bolstad, G. S. Helfman, and E. B. D. Jones III. 1998. Stream biodiversity: the ghost of land use past. Proceedings of the National Academy of Sciences 95:14843–14847.

Henshaw, P. C., and D. B. Booth. 2000. Natural restabilization of stream channels in urban watersheds. Journal of the American Water Resources Association 36(6):1219–1236.

Hollis, G. E. 1975. The effect of urbanization on floods of different recurrence intervals. Water Resources Research 11(3):431–435.

Horner, R. R., D. B. Booth, A. A. Azous, and C. W. May. 1997. Watershed determinants of ecosystem functioning. In: L. A. Roesner (Ed.), Effects of Watershed Development and Management on Aquatic Ecosystems, Proceedings of an ASCE Engineering Foundation Conference, Snowbird, Utah, August 4–9, 1996; published by ASCE, Reston, Virginia, USA.

Jacobson, R. B., S. R. Femmer, and R. A. McKenney. 2001. Land-Use Changes and the Physical Habitat of Streams – A Review with Emphasis on Studies Within the U.S. Geological Survey Federal-State Cooperative Program. U.S. Department of the Interior, U.S. Geological Survey Circular 1175, Reston, Virginia, USA, 63 pp.

Karr J. R. 1991. Biological integrity: a long-neglected aspect of water resource management. Ecological Applications 1:66–84.

Karr, J. R., and E. W. Chu. 1999. Restoring Life in Running Water. Island Press, Washington, District of Columbia, USA.

Karr, J. R., and C. O. Yoder. 2004. Biological assessment and criteria improve TMDL planning and decision making. Journal of Environmental Engineering 130:594–604.

King County. 1990. Surface-Water Design Manual. King County Public Works Department, Surface Water Management Division, Seattle, Washington, USA, 5 chapters.

Konrad, C. P., and D. B. Booth. 2002. Hydrologic Trends Associated with Urban Development for Selected Streams in the Puget Sound Basin, Western Washington. U.S. Department of the Interior, U.S. Geological Survey, Water Resources-Investigations Report 02-4040, Tacoma, Washington, USA, p. 40.

Konrad, C. P., and D. B. Booth. 2005. Hydrologic changes in urban streams and their ecological significance. American Fisheries Society Symposium 47:157–177.

Konrad, C. P., D. B. Booth, and S. J. Burges. 2005. Effects of urban development in the Puget Lowland, Washington, on interannual streamflow patterns: consequences for channel form and streambed disturbance. Water Resources Research 41(WO7009), dx.doi.org/10.1029/2005WR004097.

Leopold, L. B. 1968. Hydrology for Urban Land Planning – A Guidebook on the Hydrologic Effects of Urban Land Use. U.S. Department of the Interior, U.S. Geological Survey Circular 554, Washington, District of Columbia, USA, p. 18.

Leopold, L. B. 1973. River channel change with time: an example. Geological Society of America Bulletin 84:1845–1860.

Lowrance, R., L. S. Altier, J. D. Newbold, R. R. Schnabel, P. M. Groffman, J. M. Denver, D. L. Correll, J. W. Gilliam, J. L. Robinson, R. B. Brinsfield, K. W. Staver, W. Lucas, and A. H. Todd. 1995. Water quality functions of riparian forest buffer systems in the Chesapeake Bay watershed. Report EPA 903-R-95-004/CBP/TRS 134/95, U.S. Environmental Protection Agency, Chesapeake Bay Program Office, Annapolis, Maryland, USA, p. 67.

MacRae, C. R. 1997. Experience from morphological research on Canadian streams: is control of the two-year frequency runoff event the best basis for stream channel protection? pp. 144–162. In: L. A. Roesner (Ed.), Effects of Watershed Development and Management on Aquatic Ecosystems, Proceedings of an ASCE Engineering Foundation Conference, Snowbird, Utah, August 4–9, 1996; published by ASCE, Reston, Virginia, USA.

McBride, M., and D. B. Booth. 2005. Urban impacts on physical stream conditions: effects of spatial scale, connectivity, and longitudinal trends. Journal of the American Water Resources Association 41(3):565–580.

McCuen, R. H. 1979. Downstream effects of stormwater management basins. ASCE Journal of the Hydraulics Division 105(HY11):1343–1346.

Montgomery, D. R., and L. H. MacDonald. 2002. Diagnostic approach to stream channel assessment and monitoring. Journal of the American Water Resources Association 38:1–16.

Morley S. A., and J. R. Karr. 2002. Assessing and restoring the health of urban streams in the Puget Sound basin. Conservation Biology 16:1498–1509.

Naiman R. J., H. Décamps, and M. E. McClain. 2005. Riparia: Ecology, Conservation, and Management of Streamside Communities. Elsevier/Academic Press, San Diego, California, USA, p. 430.

Nelson, A. C. 2004. Toward a New Metropolis: the Opportunity to Rebuild America. The Brookings Institution Metropolitan Policy Program, Washington, District of Columbia, USA, December, 44 pp., available from http://www.brookings.edu/metro.

Palmer, M. A., E. S. Bernhardt, J. D. Allan, P. S. Lake, G. Alexander, S. Brooks, J. Carr, S. Clayton, C. N. Dahm, J. Follstad Shah, D. L. Galat, S. G. Loss, P. Goodwin, D. D. Hart, B. Hassett, R. Jenkinson, G. M. Kondolf, R. Lave, J. L. Meyer, T. K. O'Donnell, L. Pagano, and E. Sudduth. 2005. Standards for ecologically successful river restoration. Journal of Applied Ecology 42:208–217.

Park, C. C. 1997. Channel cross-sectional change, pp. 117–145. In: A. Gurnell and G. Petts (Eds.), Changing River Channels, John Wiley and Sons, Inc., Chichester, UK.

Paul M. J., and J. L. Meyer. 2001. Streams in the urban landscape. Annual Review of Ecology and Systematics 32:333–365.

Poff, N. L., and J. V. Ward. 1989. Implications of streamflow variability and predictability for lotic community structure: a regional analysis of streamflow patterns. Canadian Journal of Fisheries and Aquatic Sciences 46:1805–1818.

Poff, N. L., and J. D. Allan. 1995. Functional organization of stream fish assemblages in relation to hydrological variability. Ecology 76(2):606–627.

Poff, N. L., J. D. Allan, M. B. Bain, J. R. Karr, K. L. Prestegaard, B. D. Richter, R. E. Sparks, and J. C. Stromberg. 1997. The natural flow regime: a paradigm for river conservation and restoration. BioScience 47(11):769–784.

Poff, N. L., M. M. Brinson, and J. W. Day. 2002. Aquatic Ecosystems and Global Climate Change—Potential Impacts on Inland Freshwater and Coastal Wetland Ecosystems in the United States. Pew Center on Global Climate Change, January, 44 pp., available from http://www.pewclimate.org/docUploads/aquatic.pdf.

Poff, N. L., B. P. Bledsoe, and C. O. Cuhaciyan. 2006. Hydrologic alterations due to differential land use across the contiguous United States: geomorphic and ecological consequences for stream ecosystems. Geomorphology 79:264–285.

PSAT. 2005. Low Impact Development – Technical Guidance Manual for Puget Sound. Publication No. PSAT 05-03, Puget Sound Action Team, Olympia, Washington, USA, January, available from http://www.psp.wa.gov/downloads/LID/LID˙manual2005.pdf.

Reckhow, K. H. 1999. Lessons from risk assessment. Human and Ecological Risk Assessment 5:245–253.

Richards, K. S., and S. N. Lane. 1997. Prediction of morphological changes in unstable channels, pp. 269–292. In: C. R. Thorne, R. D. Hey, and M. D. Newsom (Eds.), Applied Fluvial Geomorphology for River Engineering and Management, Chapter 10, John Wiley and Sons, Inc., New York, USA.

Roberts, M. L., R. E. Bilby, and D. B. Booth. 2008. Hydraulic dispersion and reach-averaged velocity as indicators of enhanced organic matter transport in small Puget Lowland streams across an urban gradient. Fundamental and Applied Limnology 171(2):1451–1459.

Roesner, L. A., B. P. Bledsoe, and R. W. Brashear. 2001. Are best-management-practice criteria really environmentally friendly? Journal of Water Resources Planning and Management 127(3):150–154.

Roesner, L. A., and B. P. Bledsoe. 2002. Physical Effects of Wet Weather Flows on Aquatic Habitats – Present Knowledge and Research Needs. Final Report to Water Environment Research Foundation, WERF Project Number 00-WSM-4, p. 250.

Roy A. H., A. D. Rosemond, M. J. Paul, D. S. Leigh, and J. B. Wallace. 2003. Stream macroinvertebrate response to catchment urbanisation (Georgia, USA). Freshwater Biology 48: 329–346.

Schueler, T. 1994. The importance of imperviousness. Watershed Protection Techniques 1(3): 100–111.

Schueler, T. 1995. The architecture of urban stream buffers. Watershed Protection Techniques 1(4):155–163.

Schumm, S. A. 1977. The Fluvial System. John Wiley and Sons, Inc., New York, USA.

Simon, A. 1989. A model of channel response in disturbed alluvial channels. Earth Surface Processes Landforms 14(1):11–26.

Steedman, R. J. 1988. Modification and assessment of an index of biotic integrity to quantify stream quality in Southern Ontario. Canadian Journal of Fisheries and Aquatic Sciences 45:492–501.

Thomson, J. D., G. Weiblen, B. A. Thomson, S. Alfaro, and P. Legendre. 1996. Untangling multiple factors in spatial distributions: lilies, gophers, and rocks. Ecology 77:1698–1715.

Thorne, C. R., C. Alonso, R. Bettess, D. Borah, S. Darby, P. Diplas, P. Julien, D. Knight, L. Li, J. Pizzuto, M. Quick, A. Simon, M. A. Stevens, S. Wang, and C. C. Watson. 1998. River width adjustment, I: processes and mechanisms. Journal of Hydraulic Engineering 124(9):881–902.

Trimble, S. W. 1997. Contribution of stream channel erosion to sediment yield from an urbanizing watershed. Science Magazine 278:1442–1444.

USEPA. 1999. Low-impact development design strategies: an integrated design approach. Report EPA 841-B-00-003, June, available from http://www.lowimpactdevelopment.org/pubs/LID_National_Manual.pdf.

Walsh, C. J., A. H. Roy, J. W. Feminella, P. D. Cottingham, P. M. Groffman, and R. P. Morgan. 2005. The urban stream syndrome: current knowledge and the search for a cure. Journal of the North American Benthological Society 24:706–723.

Westman, W. E. 1985. Ecology, Impact Assessment, and Environmental Planning. John Wiley and Sons, Inc., Chichester, UK.

Williams, G. P., and M. G. Wolman. 1984. Downstream Effects of Dams on Alluvial Rivers. U.S. Department of the Interior, U.S. Geological Survey Professional Paper 1286, U.S. Government Printing Office, Washington, District of Columbia, USA, p. 83.

Williams, J. E., C. A. Wood, and M. P. Dombeck. 1997. Understanding watershed-scale restoration, pp. 1–16. In: J. E. Williams, C. A. Wood, and M. P. Dombeck (Eds.), Watershed Restoration: Principles and Practices, American Fisheries Society, Bethesda, Maryland, USA.

Wohl, E. E., P. L. Angermeier, B. P. Bledsoe, G. M. Kondolf, L. MacDonnell, D. M. Merritt, M. A. Palmer, N. L. Poff, and D. Tarboton. 2005. River restoration. Water Resources Research 41(W10301), doi:10.1029/2005WR003985.

Wolman, M. G. 1967. A cycle of sedimentation and erosion in urban river channels. Geografiska Annaler 49A(2–4):385–395.

Chapter 7
Urban Water Recreation: Experiences, Place Meanings, and Future Issues

Ingrid E. Schneider

7.1 Introduction

City sights ... city smells ... city sounds. In stark contrast to most urban images, water resources afford aesthetic relief among a variety of other individual and community benefits, including recreation. Water-based recreation experiences in urban environments contribute to citizen's quality of life through opportunities for enhanced health and wellness, environmental protection, and stimulated economic development. Beyond the benefits afforded through recreation experiences, however, are the meanings urban dwellers and visitors attach to these recreation resources. Both the recreation experiences and the meanings attached to these recreation environments have short- and long-term implications for water resource managers, policy makers, and urban planners. These urban water-based recreation experiences and resources, their place meanings, and future issues associated with them are the subject of this chapter.

Prior to addressing urban water-based recreation experiences and place meanings, it is appropriate to understand the magnitude of water-based recreation. Beyond serving municipalities as water supplies and modes of transportation, water resources provide significant recreation services. Nationally, the majority of US citizens report visiting a waterside area (60%; NSRE 2000). In addition to visiting a waterside area, a variety of recreation activities depend upon water resources. For example, in 2006, 29.9 million US residents, 16 years and older, participated in recreational fishing; an additional 71 million enjoyed wildlife viewing (USFWS 2007). At the state level, 735 million visitors enjoyed state parks, many of which include water-based activities such as swimming, boating, and fishing (National Association of State Park Directors 2004). In terms of the urban water-based recreation experience, the total area covered by urban parkland in the US exceeds one million acres and recreational use of this area is significant. For example, Lincoln Park in Chicago hosts more than 12 million users each year while New York's

I.E. Schneider (✉)
University of Minnesota, Minneapolis and Saint Paul, MN, USA
e-mail: ingridss@umn.edu

Central Park hosts about 25 million visits annually—more than five times as many to the Grand Canyon (Center for City Park Excellence 2007). Parks such as these contain significant water features, which afford aesthetic enjoyment, wildlife habitat, spaces for contemplation, and other recreation opportunities.

Beyond providing recreational experiences to local residents, urban environments also host a variety of business and leisure travelers, both attracted to water-based features. Destinations with significant water features have the ability to differentiate themselves from their competitors and therefore obtain a greater share of tourism's economic contributions. Nationally, tourism is a $1.3 trillion industry and one out of every eight US non-farm jobs is directly and indirectly created by travel and tourism (TIA 2007). City and urban sight seeing is among the top four activities for domestic tourists in the US. Further, among the ten most visited cities in the US, seven have significant water features (Travelers Digest 2007). Beyond those, consider San Antonio's Riverwalk, Seattle's pier, and Baltimore's Inner Harbor. Each community embraces and capitalizes on their water resources to enhance recreation and tourism opportunities. As the evidence suggests, urban water areas are significant resources for recreation and tourism experiences that also contribute to local economies and protect natural environments.

7.2 Recreation Benefits

Upon initial contemplation of urban water based recreation experiences, a list of activities probably comes to mind: boating, swimming, fishing, and enjoying the beach or wildlife viewing. However, beyond a simple list of activities, recreation experiences provide benefits to individuals, groups and even society as a whole. A significant line of recreation research posits that recreation experiences provide opportunities to realize benefits and suggests that individuals seek out these experiences to attain benefits. Recreation benefits represent (1) a change in a condition or state viewed as more desirable than a previous one; (2) maintenance of a desired condition and thereby prevention of an unwanted condition; and (3) realization of a satisfying recreation experience (Moore and Driver 2005). Purportedly, personal benefits of leisure exist in a chain of causality that goes from antecedent condition through motivation and participation to the realization of end-state benefits (Fig. 7.1). Antecedent conditions include management inputs (e.g., recreation space, labor, and capital) and consumer inputs (e.g., time, discretionary income, equipment, and skills). Consumer outputs are motivations (e.g., be close to nature, to relax physically, to do something with my family, to experience solitude, to test my skills and abilities, and to learn about the natural history of the area) and end-state benefits include individual activities (e.g., cognitive and physiological processes) and individual outputs (e.g., physical and mental health, friendships, knowledge, and self-esteem). For example, individuals who visit recreation areas typically report health improvements (Driver et al. 1996), stress reduction and renewal (Ulrich et al. 1990), and learning outcomes (Roggenbuck

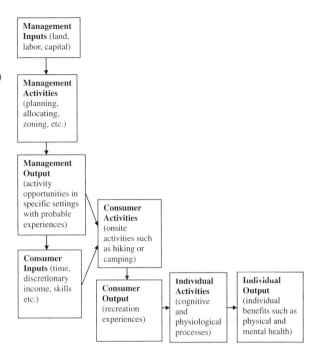

Fig. 7.1 Abbreviated model illustrating the process and subprocesses for producing outdoor recreation benefits (modified from Brown 1984)

et al. 1991), among other benefits. Benefits are not only afforded at the individual level. For instance, at the community level, the use and availability of recreational resources add positively to quality of life perceptions (Stein et al. 1999). Further, recreation environments contribute to economic development opportunities as well as enhance property values (Lindsey et al. 2004). Perhaps of particular interest in the 21st century are recreation's health and wellness benefits, environmental protection, and economic development opportunities.

Health encompasses aspects of physical, mental and social well-being (WHO 2007). Recreation and leisure professionals are aggressively addressing the obesity epidemic in the US through initiatives such as "healthy parks, healthy lives" and "step up to health" (e.g., National Recreation and Parks Association). At the start of the 21st century, nearly one in three US citizens were completely inactive, and only one of four attained enough physical activity to meet recommended physical activity guidelines (defined as at least 30 minutes of moderate physical activity for five or more days of the week) (CDC 2001). On a related note, the percent of overweight and obese US residents exceeded 65% (Hedley et al. 2004). Even though 80% of US citizens participate in outdoor recreation (Roper ASW 2004), almost 25% report no physical activity during leisure time (Ewing et al. 2003). Urban water bodies provide significant opportunities for physical activity through on- and off-water experiences. Given urban water bodies typically incorporate trails, and they serve as magnets for physical activity (Shafer et al. 2000). Invariably, such trails and urban park systems are important for active living in that they accommodate a diverse set of activities (Sallis et al. 1998) and retain a relatively safe and pleasing appearance that attracts recreational use (Humpel et al. 2002). Similarly, swimming and water sports provide opportunities for physical activity for all ages and abilities.

Beyond physical health, the mental health benefits afforded by water-based recreation experiences are similarly important. Arguably, the stress-relief function of merely viewing water, moreover the benefits of recreation experiences in and around water, can lessen the severe impact of mental illness. Data developed by the massive Global Burden of Disease study (http://www.who.int/topics/global burden of disease/en/) reveals that mental illness accounts for more than 15% of disease in established market economies, such as the US. The use of leisure to cope with illness and stress is both documented and important. In the early 1990s, Coleman and Iso-Ahola's (1993) seminal publication conceptualized the functions of leisure as a way to cope with stress and maintain good health. Evidence for this ranges from Caltabiano's (1994) findings that three major groups of activities (i.e., outdoor-active sport, social, and cultural-hobbies leisure) reduce stress to Iwasaki's extensive work (cf. Iwasaki and Mannell 2000, Iwasaki and Smale 1998) on leisure as coping. Given the stress encountered in urban environments, the stress-relief benefits of water-based recreation experiences are of paramount importance.

Importantly, perceived benefits are differentiated by proximity to the recreation area. Anderson et al. (2008) explored distance from a water-based recreation resource and its ability to differentiate visitor perceived benefits. Three benefit factors—enjoy nature, mental and physical health, and social interaction—were important to both proximate (within 15 miles, per management suggestion) and distant visitors (greater than 15 miles), but proximate visitors identified all benefits as more important. In particular, learning and solitude benefit experiences were more important to proximate visitors than distant visitors. Such findings are important to understand the benefits of water-based recreation sites for urban residents, particularly the benefit of solitude, given the continued "noise pollution plague" in urban areas (Goines and Hagler 2007).

When considering urban water policy and planning, the assessment and monitoring of recreation derived benefits to individuals and community's is essential. Since the 1993 Government Performance and Results Act, federal land management agencies have increasingly focused on the documentation of outcome-based performance that seeks to better justify budgets, formulate policies, and make decisions. While quantification of such benefits is complex and daunting, significant progress has been made and continued efforts are advantageous at all planning levels.

Certainly the benefits described above are not mutually exclusive. For example, individuals walking for exercise along waterways can enjoy the wildlife along the way, listen to or read interpretive material about the area as they walk, as well as get immediate stress reduction by taking in the area's viewscapes. This same trail may house several restaurants or cultural exhibits that encourage economic expenditures and benefit the community from a tax revenue perspective. Finally, this water-adjacent trail and its surrounding environment provides habitat for birds, small mammals and fish that urban citizens appreciate and seek out. In essence, the benefits related to urban water-based recreation experiences and resources are important for both individuals and the communities to which they belong: socially, environmentally and economically.

7.3 Place Meanings

Beyond benefits, however, urban water-based resources in which we work and play evolve into places infused with meaning. As planners and geographers consider it, *place* moves beyond space and refers to a combination of "setting, landscape, ritual, other people, personal experiences, care and concern for home ... in the context of other places" (Relph 1976). Williams (2008) summarizes the extensive work on place since the 1970s and identifies four approaches to place: (1) place as an attitude towards a geographic locale or resource, such as sense of place or specialness of a place, (2) place as meaning and relationship that exists at both individual and cultural level where individuals have relationships with places and culture's assign meaning to places by protecting them, (3) place as environmental philosophy where moral reasons demand place protection, and (4) place as sociopolitical processes that, by nature, produce conflicts over appropriate uses and values.

The growing emphasis on collaborative ecosystem management has increased interest in place-based concepts within the natural resources field and the recreation field is no different. Place attachment caught the attention of outdoor recreation academicians in the 1980s. Place attachment has most often been described and measured by two primary dimensions in a recreational context: place dependence and place identity. *Place dependence* is defined as the potential for a place to satisfy the needs of an individual and how that place compares in the satisfaction of needs compared to another place (Stokols and Shumaker 1981). *Place identity* is defined by Proshansky (1978) as "a subculture of the self-identity of the person consisting of, broadly conceived, cognitions about the physical world in which the individual lives". Place meanings can evolve both within individuals and throughout a site's environmental history. Within individuals, as experience with sites builds, so does the meaning of a place. For example, as one has more and varied experiences with a water body, the meaning of and relationship with the water body changes. In terms of a site's history, consider the Cuyahoga River which actually caught on fire repeatedly from 1936 onward due to pollution (see Chapter 1). In the 21st century, however, the River Valley is now protected as a National Park, but portions remain areas of critical concern.

Place identity and place dependence have been reliably assessed in as few as four items each (Williams and Vaske 2003) and therefore, can easily be incorporated into existing visitor or community questionnaire efforts (Table 7.1). Moving beyond identity and dependence, however, requires qualitative assessment of place meaning. Additionally, identification of special places through stakeholder mapping or auto-photography can add rich contextual data layers useful in urban planning and management. Like perceived benefits, differences in place attachment

Table 7.1 Questions regarding place identity and place dependence (from Williams and Vaske 2003). Items are measured on a 5 point scale where 1 = strongly disagree and 5 = strongly agree

Place Identity
I feel "X" is a part of me.
"X" is very special to me.
I identify strongly with "X".
I am very attached to "X".
Visiting "X" says a lot about who I am.
"X" means a lot to me.

Place Dependence
"X" is the best place for what I like to do.
No other place can compare to "X".
I get more satisfaction out of visiting "X" than any other.
Doing what I do at "X" is more important to me than doing it in any other place.
I wouldn't substitute any other area for doing the types of things I do at "X".
The things I do at "X", I would enjoy doing just as much at a similar site.

exist even within the same area. Warzecha and Lime (2001), for example, found differences in place attachment between recreational users of two rivers in the same management area. Specifically, respondents demonstrated significant differences in tolerances for encountering other watercraft based on their level of agreement with attachment statements (place attachment increased, tolerance for encounters decreased). As such, they suggest providing more than one type of recreation opportunity could help prevent visitor displacement and provide opportunities for visitor solitude. Subsequently, identifying the meaning of and relationships with urban water-based recreation areas is essential for effective management and planning.

7.4 Future Considerations

As population and development intensity increase in water-based recreation areas, so do the associated resource management and policy challenges (Budruk et al. 2008, Cordell 2000, Manning 1999, Driver et al. 1996). Of particular interest for the future of urban water-based recreation environments are the diversifying population, an increased interest in cultural and heritage tourism, and conflict management.

Diversifying Population: A diversifying populace will influence urban water-based recreation experience preferences and place meaning. The projected population shift is significant. By 2050, non-Hispanic Whites are expected to comprise about half of the total population, down over a quarter percentage from 2000 (US Census Bureau 2006). The impact of the population shift is and will continue to be marked in urban environments. Hispanics, Asians, and Blacks remain more likely to reside in large metropolitan areas than the population as a whole (Frey 2006). Hispanic populations are the fastest growing group as identified by the Census

Bureau with a tripling projected by 2050. The foreign-born population of the US underwent a 7.9% increase between 1990 and 2000 and, in 2000, half of the 28.4 million foreign-born population migrated from Latin America.

To meet the needs of a diversifying public and create a more resilient community, the values and preferences both within and between groups should be assessed and monitored. For example, Sasidharan et al. (2005) explored inter-ethnic differences in type of urban park use, as well as intra-ethnic variations among Hispanics and Asians in two similar metropolitan areas. Their survey found that ethnic respondents visited in larger groups, Hispanics/Latinos preferred water activities and that the acculturation to the US did not have a significant impact on outdoor recreation choices. Few studies examine the impact of immigration on leisure experiences. Given the urban pattern of immigration, identifying the role of water-based recreation opportunities to adjustment in the US and as an expression of ethnic identity are important areas of inquiry. Similarly, understanding how urban water-based recreation opportunities contribute to sense of place and community among immigrants is important.

Planning and management implications of the diversifying population include assessing urban water-based recreation sites for their (1) ability to accommodate large groups, (2) levels of development in relation to preferred development levels, (3) language use, (4) perceived physical accessibility, as well as (5) perceived sense of welcomeness by newer residents. Similarly, planning processes need to incorporate the knowledge that stakeholders are diversifying and, as such, reconsider typical media outlets and primary languages used to invite public participation to enhance diversity and coverage. Likewise, typical planning processes and forums may not be culturally appropriate for newer residents and participatory planning may be completely new to immigrating urban residents (Chapter 8).

Sidebar #1 Diversifying Population: Cuyahoga Valley National Park (Floyd and Nicholas, 2008)

Illustrating the challenge of the demographic change is visitation to one of the most recent urban National parks: Cuyahoga Valley National Park in urban Ohio (established in 2000). The 32,864-acre park adjoins the urban areas of Cleveland and Akron. Primary activities at the park include hiking, bicycle trails, a historical depot and Brandywine Falls. The 2000 US Census indicates Cleveland's diverse racial makeup as 50% Black, 42% white, and 7% Hispanic or Latino of any race. However, a 2006 National Park Service study indicated a visitor base that is 97% non-Hispanic white. Efforts toward a visitorship that represents the local population might include assessing staff composition in comparison to the population, assessing perceived discrimination among visitors and non-visitors, and auditing marketing and interpretative materials to ensure representativeness.

Cultural/heritage Tourism: Due to their history and diverse stories, urban areas serve as primary culture and/or heritage based tourist destinations. This is particularly true of urban water-based areas which developed as transportation and industrial centers, transitioned through industrial periods, and are experiencing rebirth as sought-after housing and recreation sites. Cultural or heritage based tourism is the fastest growing US leisure travel segment (Nicholls et al. 2004). TIA defines a cultural/historic tourist as someone who engages in cultural, arts, historic or heritage based activities or events. According to national data, a total of 118.1 million US adults participated in cultural or heritage tourism in 2002 (TIA 2003). Tourists interested in cultural and heritage experiences are both an attractive and emerging travel market due to their higher expenditures and longer lengths of stay than non-cultural based tourists (TIA 2003). However, as urban areas seek to enhance or revitalize their culture and heritage around water, assessing the impacts of increased visitation will be mandatory to ensure social sustainability. An increase in the number of tourists can negatively impact sense of place, community cohesion and viability. Further, to maintain the attractions, consideration of the environmental sustainability associated with increased tourist numbers is essential.

Related to cultural/heritage tourism is the concept of sustainable tourism, which encompasses all three pillars of sustainability: ecological, social and economic. According to the World Tourism Organization (WTO), sustainable tourism development "meets the needs of present tourists and host regions while protecting and enhancing opportunities for the future ... management of all resources in such a way that economic, social and aesthetic needs can be fulfilled while maintaining cultural integrity, essential ecological processes, biological diversity, and life support systems". Certainly sustainable cultural/heritage tourism development related to urban water resources will be essential to a destination's long-term success. The WTO has a list of indicators to consider for sustainable tourism planning, including those specific for urban sites (WTO 1996, Table 7.2).

Sidebar #2 Telling River Stories: A Program Highlighting Distinctive Urban Water-Oriented Recreational Heritage

Patrick Nunnally, Executive Director

Telling River Stories, a program led by the University of Minnesota and featuring half a dozen community partners from Minneapolis and St. Paul, capitalizes precisely on the shift from urban industrial to urban recreational landscape. Between 1880 and 1930, the flour mills at St. Anthony Falls led the world in the production of flour. Today, the St. Anthony Falls Heritage Zone, often utilizing the same material fabric and buildings from the glory days of milling, is home to over $2 billion in reinvestment, houses approximately 4,000 new residents, and hosts more than a million visitors a year to the

Table 7.2 Sustainable tourism indicators: core, urban and cultural/heritage (World Tourism Organization 1996)

Core Indicators of Sustainable Tourism	
Indicator	Specific Measures
1. Site Protection	Category of site protection according to IUCN* index
2. Stress	Tourist numbers visiting site (annum/peak month)
3. Use Intensity	Intensity of use – peak period (persons/hectare)
4. Social Impact	Ratio of tourists to locals (peak period and over time)
5. Developing Control	Existence of environmental review procedure of formal controls over development of site and use densities
6. Waste Management	Percentage of sewage from site receiving treatment (additional indicators may include structural limits of other infrastructural capacity on site such as water supply)
7. Planning Process	Existence of organized regional plan for tourist destination region (including tourism component)
8. Critical Ecosystems	Number of rare/endangered species
9. Consumer Satisfaction	Level of satisfaction by visitors (questionnaire based)
10. Local Satisfaction	Level of satisfaction by locals (questionnaire based)
11. Tourism Contribution to Local Economy	Proportion of total economic activity generated by tourism only
Composite Indices	
A. Carrying Capacity	Composite early warning measures of key factors affecting the ability of the site to support different levels of tourism
B. Site Stress	Composite measure of levels of impact on the site (its natural and cultural attributes due to tourism and other sector cumulative stresses)
C. Attractiveness	Qualitative measure of those site attributes that make it attractive to tourism and can change over time

Cultural Sites – Built Heritage		
Issue	Indicators	Suggested Measures
Site Degradation	Restoration costs	- estimated costs to maintain/restore site per annum
	Levels of pollutants affecting site	- acidity of precipitation
	Measures of behavior disruptive to site	- traffic vibration (ambient level) - number of incidents of vandalism reported
Determining Tourism Capacity	Use intensity**	-
Lack of Safety	Crime rate and type	- number and type of crimes against tourists reported**

Table 7.2 (continued)

Urban Environment		
Issue	Indicators	Suggested Measures
Lack of Safety	Crime levels	- number of crimes reported (e.g., theft and assault)
	Types of crimes committed	
	Traffic safety	- traffic injuries as a % of population
Uncleanliness	Site attraction**	- counts of levels of waste on site
Crowding at Key Urban Attributes	Use intensity**	- traffic congestion - length of wait
Degradation of Key Urban Attributes	See Cultural Sites Built Heritage below	
Health Threats	Air pollution measurements	- air pollution indices (e.g., sulfur dioxide, nitrogen oxide, particulates) - number of days exceeding specified pollutant standards
	Drinking water quality	- availability of clean water (e.g., can tap water be consumed on site)
	Type and extent of communicable diseases	- statistics on disease prevalence
	Noise levels	- records on decibel count at key locations

*International union for Conservation of Nature and Natural Resources.
**May be a function of change in level of crime or changes in level of reporting.

region's Central Mississippi Riverfront Park. In St. Paul, which lagged behind its neighbor by some two decades in making the "turn toward the river", the figures are not as startling, but nevertheless the trend is clear: The Twin Cities have turned to face their origins on the Mississippi River, and visitors from all over the world are coming to the cities.

Often, what they are coming to see is the history and culture of the area. Again, water is the draw, and the world-renowned Guthrie Theater moved to the Minneapolis riverfront, opening a new theater in 2006. A host of shops, restaurants, and night spots have grown up around the Guthrie. What is missing, though, is a deeper, richer sense of how this landscape has evolved over time, and what the complex tapestry of human stories that have played themselves out in this landscape have to tell us about being urban, and about the river. This is where Telling River Stories comes in. Using a web-based platform (www.riverstories.umn.edu) that will ultimately feature an array of downloadable materials, the project seeks to repopulate the urban river corridor with the people and the stories of how the landscape has changed over time. Web-based materials will be supplemented by on-ground installations that provide "accidental learning" opportunities for people who have come to the riverfront for other reasons. Finally, materials developed in support of other components of the program will be made accessible to the public through tours, storytelling, and other programs. Again, there's something magical about the urban river that makes this program sing.

Telling **river stories** capitalizes on the person-to-person contact that comes from the act of telling, even mitigated through digital and other media. Ultimately, the project captures something of the elusive quality of voice, rekindling connections between people and enriching the experience for visitors, whether they are new to the region or, long-time residents, or new to the riverfront.

Telling *River* stories highlights stories and the experiences of water. Rivers are dynamic; indeed all waterfronts, whether facing onto rivers, lakes, bays or the ocean, reflect something of the contrast between the fluidity of water and the stability of constructed landscapes. This ever-changing quality, reflected in varying qualities of light, images of motion, and sounds of waves and wind, makes waterfronts some of the most attractive landscapes in the modern American city. These are ideal places for a variety of recreational, touristic, and heritage-based activities, of literally countless types.

Telling river *Stories* speaks to some of the oldest ways humans communicate. "Tell me a story" we beg our parents, and, if we are from an indigenous culture such as Australian aboriginal people, or the native people of North America, that story contains a map of the surrounding landscapes. Stories establish relationships, traditions, and knowledge of what's important and holds us together.

The Mississippi riverfront in the Twin Cities was designated in 1988 as part of the Mississippi National River and Recreation Area (MNRRA), a unit of the National Park system. The waterfronts in St. Paul and Minneapolis therefore take their place among the "crown jewels" of urban waterfronts in the US. These places, and the public and private agencies that manage them for purposes of tourism, outdoor recreation, public interpretation, and heritage conservation, testify to the enduring importance of urban waterways as valued landscapes.

Conflict Management: Given the increasing pressure on urban water-based recreation resources, as well as diversifying use and place meanings, conflict is inevitable. Therefore, the need to acquire the expertise in conflict resolution is omnipresent among managers, planners, and administrators (Hammitt and Schneider 2000). While recreation conflict may appear as an antithesis of both managerial and visitor goals, it has a potential positive influence in that it can (1) indicate when an aspect of the current system needs attention, (2) lead to ideas and solutions of superior quality because of the multiple parties and perspectives involved, (3) keep an organization at a higher level of stimulation, and (4) at the very least, prevent stagnation. Conflict management can either be destructive or constructive in nature. Fortunately, conflict resolution potential is high among recreation groups compared to conflicts between other entities (Floyd et al. 1996). Conflict can develop among the variety of stakeholders involved in recreation management and has been modeled at nine levels (Little and Noe 1984; Fig. 7.2):

1. *Visitor to Visitor* – Emanating from a variety of sources including personal or social values as well as simply activity style, visitors conflict with one another either directly or indirectly. Motor boaters and canoes, walkers and joggers, or simply family groups and individual recreationists can conflict upon site or sound of one another. Evidence of others, such as litter, also induces visitor conflict. The percent of visitors experiencing conflict ranges from 10% to 40%.
2. *Visitor to Management* – Deviating from trails, ignoring regulations or interfering with other visitor's experiences can be sources of conflict for managers with regard to visitors.
3. *Visitor to Community* – Unexpected, unwelcome or unruly visitors can negatively impact the community and be a significant source of conflict between visitors and communities.
4. *Management to Visitor* – Changes in access, fee implementation, or other apparent infringements on the freedom associated with recreation are common sources of management conflict with visitors.
5. *Management to Management* – Given the multi-jurisdictional nature of water resources, conflicts among the many management and planning agencies are typical and may be due in part to organizational missions, values and operating procedures.

"Source" of Impact	"Recipient" of Impact		
	Visitors	Park	Community
Visitors	1 Visitors – Visitors	2 Visitors – Park	3 Visitors – Community
Park	4 Park – Visitors	5 Park – Park	6 Park – Community
Community	7 Community – Visitors	8 Community – Park	9 Community – Community

Fig. 7.2 Interaction model for various levels of recreation conflict (Little and Noe 1984)

6. *Management to Community* – Management plans for water recreation resources may not align with community plans and, as such, result in conflict between the agency and community.
7. *Community to Visitor* – The hustle and bustle of daily urban life is likely a stark contrast to the preferred recreation experiences. As such, the community itself may detract from and be a source of conflict for visitors.
8. *Community to Management* – A variety of stakeholder groups within the community—special interest groups, influential citizens—exert pressure on management with regards to operations and opportunities and, as a result, conflict may arise.
9. *Community to Community* – Like visitors or managing agencies, communities themselves have differing values, desired levels of visitors and preferences for development. As such, communities that work at odds to each other's missions or values will be at odds. Further, competition for visitors or market niches may drive conflict between communities as well.

Recreation management provides a variety of techniques to minimize visitor conflict. Typically, management tactics are categorized as: (1) direct, through regulations or rules; (2) indirect, through education and information; or (3) as building bridges and involving the public. Beyond visitor conflict, conflict resolution processes have involved three stages: (1) analysis, (2) confrontation, and (3) resolution. In the analysis phase, issue identification is the primary goal that provides the foundation for the second stage, confrontation. Confrontation involves further engagement to focus on the most contentious issues where major conflicts are defined, alternative courses of action are generated, and solutions are evaluated. Finally, conflict resolution is attempted. As resolution seeks to prevent further escalation, it works toward building sustainable relationships and structures that allow for equal identity among involved parties. However, this traditional approach is ineffective in that stakeholders neither are equally involved in the processes nor do singular answers to the conflict exist. As such, greater integration of all groups at the inception of any problem is highly recommended.

Though resource conflict may be inevitable, the opportunities and meanings afforded by urban water-based recreation areas are astounding. The high quality of life, environmental protection, and stimulated economic development reap benefits for both individuals and communities. These benefits provide only an outer layer of the importance of urban water environments, which ultimately create and possess meaning for urban dwellers and visitors alike. Understanding how the opportunities and meanings will change along with the US population and its increased interest in cultural/heritage tourism will be vital for continued progress in urban planning. Similarly essential for successful urban water-based areas is a comprehensive understanding and use of effective conflict management skills. Working with, and for, the diversifying public will ensure that urban water-based recreation areas develop in ways that effectively meet the needs of residents, visitors, and future generations.

7.5 Resources

Center for City Park Excellence: With the help of CCPE data, you can use the data provided at the link below to see how your city compares to others. http://www.tpl.org/tier3_cd.cfm?content_item_id=20531&folder_id=3208

References

Anderson, D., I.E. Schneider, S. Wilhelm, and J. Leahy. 2008. Proximate and distant visitors: Differences in importance ratings of beneficial experiences. Journal of Park & Recreation Administration 26(4):47–65.

Brown, P.J. 1984. Benefits of outdoor recreation and some ideas for valuing recreation opportunities. In: G.L. Peterson and A. Randall (eds.). Valuation of wildland resource benefits, pp. 209–220. Boulder, CO: Westview Press.

Budruk, M., S. Wilhelm, I.E. Schneider, and J. Heisey. 2008. Crowding and experience use history: A study of the moderating effect of place attachment among water-based recreationists. Environmental Management 41:528–537.

Caltabiano, M.L. 1994. Measuring the similarity among leisure activities based on a perceived stress-reduction benefit. Leisure Studies 13:17–31.

CDC. 2001. Physical activity trends: 1990–1998. Centers for Disease Control, Morbidity and Mortality Weekly Report 50:166–169.

Center for City Park Excellence. 2007. City park facts. Trust for Public Lands. http://www.tpl.org/tier3_cd.cfm?content_item_id=20531&folder_id=3208

Coleman, D., and S.E. Iso-Ahola. 1993. Leisure and health: The role of social support and self-determination. Journal of Leisure Research 25:111–128.

Cordell, H.K. 2000. National survey on recreation and the environment. Summary report #1. USDA Forest Service & NOAA, Washington, DC, http://www.srs.fs.usda.gov/trends/Nsre/summary1.pdf

Driver, B.L., D. Dustin, T. Baltic, G. Elsner, and G.L. Peterson. (eds.). 1996. Nature and the Human Spirit: Overview. Nature and the Human Spirit: Toward an Expanded Land Management Ethic. Venture Publishing, Inc., State College, PA.

Ewing, R., T. Schmid, R. Killingsworth, A. Zlot, and S. Raudenbush, S. 2003. Relationship between urban sprawl and physical activity, obesity, and morbidity. American Journal of Health Promotion 18:47–57.

Floyd, M. F., and L. Nicholas. 2008. Trends and research on race, ethnicity, and leisure: Implications for management. In M.T. Allison and I.E. Schneider (Eds). Diversity in the recreation profession: Organizational perspectives (revised ed.). pp. 189–209. State College, PA: Venture Publishing.

Floyd, D.W., R. Germain, and K. Ter Horst. 1996. A Model for assessing negotiations and mediation in forest resource conflicts. Journal of Forestry 29–33.

Frey, W.H. 2006. Diversity spreads out: metropolitan shifts in Hispanic, Asian, and black populations since 2000. Brookings Institution Metropolitan Policy Program, Washington DC.

Goines, L., and L. Hagler. 2007. Noise pollution: A modern plague. Southern Medical Journal 100:287–294.

Hammitt, W.E., and I.E. Schneider. 2000. Recreation conflict management, pp. 347–356. In: W.C. Gartner and D.W. Lime (eds.). Trends in Outdoor Recreation Leisure and Tourism. CABI Publishing, NewYork.

Hedley, A.A., C.L. Ogden, C.L. Johnson, M.D. Carroll, L.R.Curtin, and K.M. Flegal. 2004. Overweight and obesity among US children, adolescents, and adults, 1999–2002. Journal of the American Medical Association 291:2847–2850.

Humpel, N., N. Owen, and E. Leslie. 2002. Environmental factors associated with adults participation in physical activity: A review. American Journal of Preventive Medicine **22**: 188–199.
Iwasaki, Y., and R.C. Mannell. 2000. Hierarchical dimensions of leisure stress coping. Leisure Sciences **22**:163–181.
Iwasaki, Y., and B.J.A. Smale. 1998. Longitudinal analyses of the relationships among life transitions, chronic health problems, leisure, and psychological well-being. Leisure Sciences **20**: 25–52.
Lindsey, G., J. Man, S. Payton, and K. Dickson. 2004. Property values, recreation values and urban greenways. Journal of Park and Recreation Administration **22**:69–90.
Little, W. and F.P. Noe. 1984. A highly condensed description of the thought process used in developing visitor research for southeast parks. U.S. Department of the Interior, National Park Service, Southeast Regional Office, Atlanta, GA.
Manning, R.E. 1999. Studies in outdoor recreation: Search and research for satisfaction. Oregon State University Press, Corvallis, OR, 374 pp.
Moore, R.L. and B.L. Driver. 2005. Introduction to outdoor recreation: providing and managing natural resource based opportunities. Venture Publishing, Inc., State College, PA.
National Association of State Park Directors (2004) State Park Facts. http://www.naspd.org/ accessed November 24, 2007
NSRE. 2000–2002. National survey on recreation and the environment. The Interagency National Survey Consortium, Coordinated by the USDA Forest Service, Recreation, Wilderness, and Demographics Trends Research Group, Athens, GA and the Human Dimensions Research Laboratory, University of Tennessee, Knoxville, TN.
Nicholls, S., C. Vogt, and S.H. Jun. 2004. Heeding the call for heritage tourism. Parks and Recreation, 36–49.
Proshansky, H.M. 1978. The city and self-identity. Environment and Behavior **10**:147–169.
Relph, E. 1976. Place and placelessness. Pion Limited, London, 156 pp.
Roggenbuck, J.W., R.J. Loomis, and J.V. Dagostino. 1991. The learning benefits of leisure. In: B.L. Driver, P.J. Brown, and G.L. Peterson (eds.). Benefits of Leisure, Venture Publishing, Inc., State College, PA.
Roper ASW. 2004. Outdoor recreation in America 2003: Recreation's benefits to society challenged by trends. Prepared for: The Recreation Roundtable. Washington, DC.
Sallis, J.F., A. Bauman, and M. Pratt. 1998. Environmental and policy interventions to promote physical activity. American Journal of Preventive Medicine **15**:379–397.
Sasidharan, V., F. Willits, and G. Godbey. 2005. Cultural differences in urban recreation patterns: an examination of park usage and activity participation across six population sub-groups. Managing Leisure **10**:19–38.
Shafer, C.S., B.K. Lee, and S. Turner. 2000 A tale of three greenway trails: user perceptions related to quality of life. Landscape and Urban Planning **49**:63–78.
Stein, T.V., D.H. Anderson, and D. Thompson. 1999. Identifying and managing for community benefits in Minnesota State Parks. Journal of Park and Recreation Administration **17**:1–19.
Stokols, D., and S.A. Shumaker. 1981. People in places: A transactional view of settings. In: J. Harvey (ed.). Cognition, Social Behavior, and the Environment, pp. 441–488. Erlbaum, Hillsdale, NJ.
Travel Industry of America. 2003. The Historic/Cultural Traveler, 2003 ed. Washington, DC.
Travelers Digest. 2007. http://www.travelersdigest.com/visited'us'cities.htm Accessed November 24, 2007.
Ulrich, R.S., U. Dimberg, and B.L. Driver. 1990. Psychophysiological indicators of leisure consequences. Journal of Leisure Research **22**:154–166.
USFWS. 2007. 2006 National Survey of Fishing, Hunting, and Wildlife Associated Recreation. U.S. Fish & Wildlife Service, US Department of the Interior, Washington, DC.
US Census Bureau. 2006. Nation's population one-third minority. US Census Press Release. Retrieved January 21, 2007, from http://www.census.gov/Press-release/www/releases/archives/population/006808.html

Warzecha, C.A., and D.W. Lime. 2001. Place attachment in Canyonlands National Park: Visitors' Assessment of setting attributes on the Colorado and Green Rivers. Journal of Park and Recreation Administration **19**:59–78.

Williams, D.R., and J.J. Vaske. 2003. The measurement of place attachment: Validity and generalizability of a psychometric approach. Forest Science **49**:830–840.

Williams, D.R. 2008. Pluralities of place: A user's guide to place concepts, theories and philosophies in natural resource management. In: Kruger, L.E., Hall, T.E., Stiefel, M.C., tech. (eds.). Understanding Concepts of Place in Recreation Research and Management. Gen. Tech. Rep. PNW-GTR-744. Portland, OR: U.S. Department of Agriculture, Forest Service, Pacific Northwest Research Station, p. 204.

WHO. 2007. World Health Organization, About WHO, 2007. http://www.who.int/about/en/. Accessed May 15, 2007.

WTO. 1996. What tourism managers need to know: a practical guide to the development and use of indicators of sustainable tourism. World Tourism Organization.

Chapter 8
Urban Design and Urban Water Ecosystems

Kristina Hill

8.1 Background: Cities, Rain and Water Systems

This chapter addresses the role of urban design in the performance of urban water ecosystems, with an emphasis on urban rainwater runoff and future urban infrastructure systems. The main thesis is that new designs must be supported by an integrative framework for analysis and application in order to significantly change overall urban hydrologic performance. Designers, planners, and scientists do not currently share such a framework. A straightforward landscape-based heuristic is proposed here which uses simple categories of hydrological function to sort, map, and propose changes to diverse urban land uses within an urban drainage basin.

Urban stormwater drainage systems carry significant amounts of pollution to streams, rivers, lakes and marine shorelines across the world. For instance, the US National Research Council estimated that urban runoff in North America carries 1.4 million metric tonnes of oil and grease products to the sea each year, produced primarily by the consumption of oil in motorized vehicles (NRC 2003). That amounts to 44 times the oil released into the sea by the 1989 Exxon Valdez oil spill of 11 million gallons. In addition, urban runoff contains bacterial pathogens, drives sewer overflows full of untreated human wastes (EPA 1999), and carries significant loadings of nutrients and metals to aquatic environments (Gobel et al. 2007). These contaminants severely alter conditions for the survival of native plant and animal species over increasing miles of the North American coastline, and prevent millions of people from having access to safe swimming and fishing areas (Tallis et al. 2008).

In response to federal legislation such as the Clean Water Act and the Endangered Species Act, municipalities across the United States have begun to implement strategies to remove these pollutants from urban stormwater runoff (Novotny and Hill 2007). The management of water in and around urban areas that does not originate at an industrial site, but rather from rooftops, lawns, driveways, parking

---

K. Hill (✉)
University of Virginia, Charlottesville, VA, USA
e-mail: keh3u@virginia.edu

lots, and roads (known as non-point source pollutants), has become critical to the ecosystem health of rivers and oceans and to the industries that harvest fish and shellfish. In 2004, states reported that about 44% of assessed stream miles, 64% of assessed lake acres, and 30% of assessed bay and estuarine square miles were not clean enough to support uses such as fishing and swimming (US EPA 2007).

The problem of urban water pollution is significant, and appears to be expanding as the rapid urbanization of US coastal regions continues. Roofs are a source of metals in runoff such as zinc, which can be toxic to humans, or copper that makes its way to rivers and is often toxic to fish. As we expand urban areas, we typically also increase the miles of roadway and the average number of vehicle miles traveled to get around in those urban areas. The simple version of the problem is that the farther we spread out, the more we drive – making urban roads and parking spaces an ever-increasing source of pollution by metals worn off of brakes, leaking or spilled motor oil, de-icing salts, and nutrients (Beach 2002, Alberti et al. 2007).

Some writers have called for limits on the percentage of land that can be developed in any given watershed (see for example Beach 2002). But where these functions already occur, or cannot be limited in the future, planners and designers must approach them in new ways. Each roof, driveway, roadway, and parking lot can be designed to produce a much lower net change in the amount of runoff it produces, and can be designed to act as its own "kidney", filtering pollutants before they enter the larger water system in and around cities. Designers and planners have begun to experiment with these possibilities, and by monitoring some of their experiments, have developed a body of knowledge about how cities can be built and retrofitted to perform differently. The functional performance of buildings, parking lots, road rights-of-way, lawns, and open space can all be improved. In the process of addressing these functions, designers can reveal the flows of water and contaminants to the citizens of a democracy that must support these changes in order for them to be meaningful and widely implemented. Without broad implementation of new urban design standards, future generations of humans will have increasingly limited access to seafood and swimmable beaches.

8.1.1 Strategic Context

The high probability of climate change over the next 25–100 years will most likely produce new rainfall patterns and an increased rate of sea level rise that will make urban design interventions in water systems more difficult and/or more expensive (Ashley et al. 2005, Barnett and Hill 2008). Yet civic infrastructure must address these environmental trends, since transportation structures and piped water systems in particular are typically designed to have a useful lifetime of 25–75 years. It is surprisingly rare in the United States to find any recent infrastructure planning work or coastal development that considers the likely impacts of climate change and sea level rise. Meanwhile, the World Bank, global insurance companies, and planners in many other cities of the world have already begun this very serious work of assessing opportunities to adapt (Perkins et al. 2007, London Climate Change Partnership 2006).

Several regions of the United States have become international leaders in proposing and implementing new design approaches for urban water systems, with an emphasis on changing the way cities deal with rainwater runoff. Seattle, Washington; Portland, Oregon; and smaller communities in Prince George's County, Maryland, as well as the State of Maryland itself have set new standards in this area. Similarly ambitious efforts are being made more recently in Stockholm and Helsingborg, Sweden, and in Glasgow, Scotland. Many other communities in the United States and around the world have begun to allow or encourage experimentation with these design approaches, collectively referred to in the United States as LID (Low Impact Development) or NDS (Natural Drainage Systems), and in the United Kingdom as SUDS (Sustainable Urban Drainage Systems).

There is no longer any question that these approaches can function, although the extent of benefits may be debated. The large strategic questions are, first, what would it take to implement these designs broadly in cities in order to alter their overall hydrologic performance? Second, will these design approaches help cities adapt to climate change trends? And finally, are there barriers that prevent planners, engineers, urban designers (landscape architects and building architects), and policy makers from sharing a conceptual understanding of where the greatest benefits can be obtained from altering urban drainage systems?

8.2 Historical Questions and Examples

In the last third of the 19th century, the American designer Frederick Law Olmsted established urban park systems as a form of infrastructure that combined water systems, transportation systems, health concerns, biodiversity goals, and social goals. Particularly in his design for Boston's Emerald Necklace, Olmsted engaged in the activities of what is now considered the separate discipline of civil engineering, combining it with horticulture, public health, political economy, and design to propose a system that was uniquely American at the time. The Boston Fenway design was intended to control flooding that periodically stranded raw sewage on the grassy banks of a park where children and their care-givers were exposed to its bacterial pathogens. Tide waters from the Charles River estuary would block the flow of freshwater carrying sewage, causing it to back up onto the banks. Olmsted designed the entire corridor of the Muddy River through the Fenway, creating greater capacity within the channel and ending it with a floodgate that would prevent brackish water from entering the Muddy from the Charles. He experimented with using zones of native plants as well as zones of non-natives on the banks, and advocated for parks as social promenades where people from all economic classes could meet and, at least in theory, provide support to a shared system of political democracy. What we think of today as innovations in "natural drainage", or "green infrastructure", or in some circles referred to as "landscape urbanism", is actually one of the most successful and unique historical strategies of American urbanism dating from the 19th century (Meyer 1997, Hill 2002).

In early 20th century, Patrick Geddes of Scotland invented the geographical idea of an urban "region", a mappable entity connected by flows of people, water, goods, culture, and capital. Geddes went on to become an influential academic writer and an urban reformer, working to improve access to light, air, and education for working class people in Edinburgh and other cities. He participated in the planning and design of cities in British colonial India, and later in the founding and planning of Tel Aviv in Israel. Geddes was the first to abstract the analysis of flows in a way that crossed disciplinary boundaries and brought ecology, sociology, and economics together in looking at the city and its region as a system, and then apply those ideas to actual urban plans. Geddes rejected the idea that cities were necessarily unhealthful, which was the prevailing belief of his time. He was actively offended by the idea that disciplinary boundaries might restrict his ability to address the city and its region as a whole, and brought his work to the largest public audiences he could with lectures and exhibitions.

Chicago produced the next urbanist whose integration of form and function challenged the status quo of morphological thinking about cities. Kevin Lynch studied with the architect Frank Lloyd Wright at his Taliesin studio, then attended MIT and in 1948 began to teach in MIT's Urban Studies program. During the early 1960s, Lynch began to follow the work of psychologists who were exploring the idea of an ecological psychology – one that emphasized the role of environment on the development of cognition and self-awareness. Building on the Chicago School's approach to cities as sociological ecosystems, Lynch applied the methods of cognitive mapping to cities, asking his study participants to identify landmarks and paths they used to navigate urban districts and their city as a larger whole. This work linked the individual to the collective via the physical environment, leading Lynch to argue that the fundamental purpose of urban design was to increase the legibility of the city as a tool for healthy individual development and the parallel development of a collective civic life. His fundamental approach was to analyze movement through the city, beginning with those flows and the semiotics that informed them, and then evolving a sense of morphological structure only in relation to the flows. Lynch also anticipated that one day urban design would have to address the development of non-human species (Lynch et al. 1990).

A few years after World War II, a Scot arrived at Harvard to study landscape architecture in a department founded on Olmsted's ideas, and from which the discipline of urban planning had emerged only a few decades before. Ian McHarg worked to invent a way of planning and designing cities that would integrate them with flows of water and organisms, in order to support human health and the myriad links between humans and the ecosystems that support us. Inspired by Rachel Carson's writing about a connected web of species and the consequences of human actions in that web, and influenced by his own experiences as a tuberculosis patient, McHarg became an advocate for urban planning that integrated water systems and biodiversity into the infrastructure of cities and new towns at multiple scales. He persuaded a major client who was building a new town north of Houston to implement some of his ideas, and the first "natural drainage" system of the 20th century was constructed at The Woodlands in the late 1960s, early 1970s. The idea of such an infrastructure would have been familiar to McHarg from his studies in Boston, but the connection he made at The Woodlands contained a unique innovation. McHarg linked private parcels to public infrastructure with a shared responsibility to improve the hydrological and ecological performance of a city. He went on to re-map the city of Philadelphia using water systems and public health as themes, popularizing his ideas in an extremely successful book titled "Design with Nature" (McHarg 1969).

In Berlin during the same decades, a wetland ecologist named Herbert Sukopp began to realize that urban ecosystems contained surprising diversity, and seemed to have characteristic patterns that were unique to urban contexts. Sukopp was confined to a restricted geography by the Berlin Wall, but turned that restriction to his advantage when he conducted some of the earliest studies in the scientific field of urban ecology in the early 1960s. His detailed investigations of species diversity and local microclimatic patterns led to the establishment of endangered species lists in European conservation practice. Sukopp used isoline maps to study everything from humidity to precise flowering times for plant species, pioneering the extension of analytical ecological methods to urban space and inventing unique representations of the urban – rural gradient that unified knowledge about geology, climatology, hydrology, economics, and plant biology. These landscape-based sectional representations played a central role in Sukopp's conceptualization of urban spaces and the flows of energy, organisms and materials that occur in and through cities, not unlike the conceptual power of Geddes diagram of the Valley Section from 50 years earlier (for an overview of Sukopp's work, see Lachmund 2007).

Contemporary designers and theorists such as Michael Hough and Anne Whiston Spirn, both former students of Ian McHarg, developed and popularized the idea that cities can be designed to incorporate and mimic desired ecological functions. Spirn is noted for originating the idea of using "catalytic frameworks" in ecological urban design, which she described as structures that change intentionally over time in response to interactions with external processes. Similarly, designer and planner Joan Iverson Nassauer developed an intensive focus on the relationships between human perception and landscape ecology in the American Midwest. Her best-known work addressed the question of how some pre-development ecological functions can be returned to areas dominated by human activities, using strategies that make them visually appealing to the people whose decisions control the future of these landscapes (Nassauer 1995). Like Lynch, Nassauer identified visual perception as a critical influence on human choices that affect design and planning. Her early work studied the role of what she called "cues to care", referring to visible design elements that identify landscapes as intentionally maintained (fencing, signage, etc.). Although the ubiquity of these elements may make them appear mundane, Nassauer's work drew attention to their potential to allow changes in the way human-dominated landscapes are maintained and in the way they function. An urban application of Nassauer's observations can allow new design approaches to become more geographically widespread by making them acceptable to a broader public audience.

The ideas of 20th century designer-academics, from Geddes to Nassauer, have added important innovations to the original approach pursued by Olmsted in the 19th century. These include the recognition of the urban metropolitan region as a component of urban systems, the use of forms that are obviously constructed as well as naturalistic forms, the use of the scientific method to investigate both urban ecological relationships and the perception of those environments. The incorporation of recent scientific knowledge and concepts has also produced some significant changes in the original design approaches, such as the recognition of energy flows that support a continuum of river ecosystems. But the fundamental idea that cities can and should make infrastructure that serves multiple purposes as positive social spaces, as a vehicle for legibility, and in support of biodiversity, is an American invention that went largely unrecognized in its time and has been re-discovered in ours.

8.3 Establishing an Ecological Frame for Watershed Analysis in Urban Design

Many important strategic questions in design and policy are answered within narrow analytical frames. These narrow frames create crucial disadvantages when they exclude variables that will ultimately determine the success or failure of contemporary and future efforts to retrofit cities.

In contemporary urban design and planning, there are three distinct ways of trying to define and alter the urban water system. Each approach is practiced by a group that has (or could have) a significant influence on the future of urban water systems. These analytical frames are not mutually exclusive, but they are often applied as if they were. My goal in describing these different analytical frames is to establish a position from which we might integrate them, and thereby produce physical designs that improve the overall performance of urban water systems significantly. The key concept is that designers, scientists and planners may miss significant opportunities to alter urban performance because they lack a framework based on landscape function.

A landscape-based framework that creates substantial overlap among the analytical frames of scientists, designers, and planners will also position them to better persuade elected decision-makers that new or different investments should be made. If the conceptual approaches of experts in different fields could be hybridized within a simple spatial heuristic or analytical approach, that heuristic could become a useful vehicle for integrating science and design. My purpose in this section is to review the three dominant frames, and then propose a simple hybrid heuristic that would allow practitioners and academics from different fields to integrate their observations and proposals.

8.3.1 The Regulatory Frame

The first is a regulatory frame, in which social and legal structures such as parcel types, land uses, and development trends are mapped, described and assessed (qualitatively and quantitatively). It is focused on the present, and on assessing the political risks and practical benefits of changing present-day standards. If the analyses indicate acceptable risks and probably benefits, thresholds are set that trigger specific rules as development and re-development occur. These rules are implemented as private and public proponents of change seek required land development permits from a municipality, county, state, or federal agency, depending on whose administrative jurisdiction is engaged by a particular proposal. This is an unusually weak system of land use control compared to the relatively hierarchical regional planning that occurs in most other countries, but Americans often succeed in using the political system within particular jurisdictions to strengthen it. The leverage generated comes from the proponents' fear that citizen involvement will slow the permit process to such an extent that a private investor will not be able to afford to wait, and a public project will lose momentum and perhaps lose its proponent in a re-election campaign.

The statistically descriptive approach to land use that is employed by this regulatory concept of urban water systems allows policy makers to identify the degree of political risk inherent in choosing a particular threshold for proposed permit regulations, and also to identify the likely benefits (in hydrologic function, tax base, additional density, or other goals) that may accumulate as new rules are applied to a jurisdiction or district. Since multiple goals are involved and calculations of political risk can be conservative, it is likely that using this frame alone to evaluate and implement changes would fall short of moving a regional water system to a higher-functioning state, or even to conserve present-day functions. The EPA's Chesapeake Bay Program, as well as state-level efforts to conserve Puget Sound, are cases in point.

Since it begins with a statistical orientation, the regulatory frame is relatively likely to incorporate statistical observations from the environmental sciences, which in some cases become guiding elements of policy. Research in the 1990s, for example, showed a relationship between the amount of total impervious area within a watershed and the relative diversity of aquatic organisms that are taken as an indicator of system health. A threshold of 10% imperviousness was widely discussed among policy and planning professionals as a marker of a watershed that could sustain a high-quality aquatic environment. But this threshold represents a kind of hind-casting, since it includes measurements of existing development and does not consider that future development might perform differently than the designs of the past.

While this may be a reasonable argument for limiting the expansion of future development by placing land in private trusts, conservation easements, or public ownership, it does not address the need to improve performance in watersheds that have already exceeded the 10% threshold of total impervious surface. The millions of people who live in existing cities and suburbs would, in effect, be left out of the conversation about investing in improved performance; while all of the political and economic pressure would be placed on rural land owners who sometimes perceive their role in generating pollution as less than the urban contribution.

The regulatory frame has shifted recently in some areas, notably in Prince George's County, Maryland and the Pacific Northwest, to emphasize intervention in the runoff generated by small storms instead of demanding that development detain the runoff from ever larger storms. This represents a change in attitude about what really causes the changes in system state within streams that receive stormwater – very large flooding events that happen very seldom, or common, small flooding events that typically happen every year, or several times each year? Practitioners and regulators in some regions have begun to shift to the smaller storms in part because of evidence that they have a major impact on the morphology and the water quality of streams (Chapter 6). The emphasis on detention of larger and larger storms emerged from an analytical frame that assumes detention is good and more detention is better; the new emphasis on smaller storms matches a regulatory frame based on the concept of "eco-mimicry"– the idea that urban areas can and should perform more like pre-development landscapes (Hill 2002). In order to do this, cities can incorporate some of the disturbance events and dynamics of those non-urban conditions, such as unusually large floods, while limiting the everyday negative effects of urbanized conditions such as runoff from paved surfaces and pollutants generated by automobile use.

8.3.2 The Site-Based Frame

The second analytical frame is site-based, and seeks to demonstrate cumulative benefits at the site scale by combining as many "best practices" as possible. It is future-oriented, seeing each site in terms of its potential rather than its current state. Applying this frame is often a vehicle to establish new models of how to enhance human aesthetic experiences and the legibility of critical urban systems. There are clear strategic advantages in being able to show decision-makers and the public the built evidence of new approaches, when those projects are successful both functionally and aesthetically.

The practitioners of the site-based analytic frame often rely on metrics such as those that are incorporated into the US Green Building Council's rapidly-evolving LEED certification levels, in order to persuade owner/clients and permitting agencies that new models will succeed and will justify what is often an extra expense. Although most metrics are designed to evaluate building designs, a new standard for sites themselves is in the testing phase (known as the USGBC Sustainable Sites Initiative), as is a new standard for districts (LEED for Neighborhoods). These metrics can be seen as a badge of accomplishment for owners, and even for cities, as they incorporate them into their permitting incentives for property owners and begin to think of public LEED projects as a portfolio of investments in higher long-term performance.

The costs of these projects are seen within this frame as justified by the value of having demonstrable models that can be seen and touched, first of all. Once built, these projects can also be tested and used as a basis for improvement, which is the goal of most practitioners who use this analytical frame. They are aware that learning comes from building under real project conditions, if the limits to implementation are similar to other projects and if those experiences can be shared with other professionals. Since the number of "best practice" elements included tends to raise the LEED rating of a project, there may be a tendency to add components that may or may not be strategically useful in a given location or jurisdiction. And the ultimate value of a built prototype is in its ability to persuade visitors that it should become a new standard, or at least very common.

Practitioners who operate within the site-based frame may sometimes forget this, and enjoy the uniqueness of the design more than its ability to become common. Or they may forget that the more expensive elements of the prototype design are unlikely to ever become a new standard without a dramatic decrease in cost. The enduring advantage of this unique frame is that it pre-supposes that conventional development can become something very different from what it is today, and understands the value of tangible, sensate experience in persuading the majority of people that a particular change is positive.

8.3.3 The Geography-Based Analytical Frame

The third analytical frame is fundamentally geographical, approaching the water system as a network within a dynamic mosaic. The priorities and capabilities of a given part of that system may be defined as much, if not more, by the past than by the present. This can be observed in the influence of topography, geomorphology, and soils, but also in the legacy of existing infrastructure. This approach would consider the functional characteristics of pre-existing elements of the landscape, and pre-existing capital investments, before setting goals for the jurisdiction, for an individual parcel, or for a water body.

Starting with the physical geography of the landscape, a practitioner who uses this analytical frame to represent the urban water system might use measurable characteristics to ask questions about, for example, where streams might be expected to have once supported a particular species that is now of conservation concern, such as salmon in the cities of the Pacific Northwest. A stream with a steep longitudinal slope, for instance, may never have had salmon in it in the pre-urban era, and both a single reach and the larger watershed would probably make a poor choice for investments in present or future salmon habitat restoration. There may be other goals that justify those investments, such as downstream water quality benefits in the nearshore marine environment, but those should be clearly defined and examined in the light of other geographical concerns.

Similarly, the infrastructure history and current condition of a particular area have implications for the priorities that should be embodied in a jurisdiction's regulations, and in the goals set for site-scale designs. The most important distinctions are among urban drainage basins that drain directly to a large water body versus those that drain to a small stream versus those that drain to a sewage treatment plant. In the first two cases, the drainage pipe system is generally referred to as "separated", meaning that it uses different pipes than the sanitary pipe system that carries human wastes. A particular basin may also be referred to as partially separated, if roof runoff, for example, goes in to the sanitary sewer pipes but street runoff does not. Systems that use the same pipes for both human waste and stormwater runoff are known as "combined" systems.

The US EPA has stated that there are approximately 1000 communities in the United States with combined systems somewhere in their jurisdictions (EPA 1999). Many of these cities have been slowly separating their sewers since the 1960s and 1970s, but a recent EPA study of the gap in infrastructure performance to support clean water makes it clear that overflows from combined sewer systems are still a major cause of water quality problems in and around cities, and that the problem may be worsening as a result of deferred maintenance and other factors (EPA 2002). Urban drainage basins that have combined sewer systems need to make stormwater flow rate and volume control their first priority, in order to limit the number and magnitude of overflow events that dump untreated sewage into water bodies large and small. The faster rainwater moves into the sewer pipes, and the greater the volume of that rainwater that runs off of the surface instead of infiltrating the soil, evaporating, or being used by people and plants, the more raw sewage will be forced to overflow from pipes that are at full capacity. This is a very significant issue in urban water systems. EPA has estimated that if current trends continue, water quality will soon decline in urban areas to problematic levels not seen since the 1970s (EPA 2002).

Urban areas that use separated systems differ in important ways if those pipes drain to a large water body, such as a lake or marine bay, or to a small river or stream. In the first case, the volume of water discharged to the lake or bay is not likely to cause significant problems relative to the existing large volume of water. But water quality is an issue, since the location of the discharge point along the shoreline is not ideal to promote mixing and dilution of the contaminants that enter the lake, even if the lake ecosystem can theoretically absorb that pollutant load. The nearshore environment in lakes and marine bays has been found to be critical habitat for many deep as well as shallow-water species at some stage in their life history, and should be treated as the nursery environment of the lake or ocean it borders (Botsford et al. 2001). Animals such as Chinook salmon that spend their juvenile life stage in this shallow water zone may be even more vulnerable to the harmful effects of pollutants than they would be in an adult stage. For these reasons, regulations and designs that are evaluated for urban drainage basins that discharge to lakes and bays should address runoff water quality improvements as their primary priority.

Separated systems that discharge to small rivers and streams represent the most difficult situation, since water quality, water volume and the rate of discharge can all have severe negative effects on a stream or river ecosystem. The morphology of streams can be altered by high flow rates and volumes so that the stream bed does not provide habitat for species characteristic of the region (Walsh et al. 2005, Chapter 6). This has implications for the entire river and estuary system, since the absence of certain organisms that do the work of shredding leaves and other detritus in small streams can reduce the amount of energy that is transported downstream to the larger system, with consequences for fish and other animal populations. Water quality pollutants such as nitrogen or metals also have both local and downstream cumulative effects, as do physical changes such as increased water temperature or decreased groundwater inputs to streams (Booth 2005). Regulations and design interventions in urban systems that rely on separated storm drains that discharge to small rivers and streams must accomplish all of the goals of runoff management, improving water quality, detaining water volumes, and slowing the rate of water discharge.

This geographical frame of analysis can make effective use of social variables, such as income levels, home ownership rates, or ethnicity, that are associated with a lack of political influence and may help to justify public expenditures that could produce the benefits of additional recreational space, flood hazard mitigation, or water quality improvements in particular districts. Family size and the number of children in different age groups would also be useful information available through census data to display geographically, that is, using a high-resolution spatial map. Car ownership rates, traffic volume data, fixed transit stop locations, and the recent budget history of infrastructure investments would also be useful information to see in a spatial context, so that planners and designers can weigh the question of who pays for and who benefits from specific infrastructure expenditures as part of a social equity analysis.

Perhaps the most significant value of this geographical frame of analysis is in its ability to bring together the questions of ecological effectiveness and social equity, both of which are arguably essential to achieve successful human adaptations to environmental change.

8.4 Implications and Integration

Although it is rare for any agency, designer, or scientific researcher to take an approach that integrates all three of these analytical frames, it is necessary for each practitioner to consider them to some degree and include them in their analyses. Public agencies must contribute to requiring this integration as they make recommendations to elected officials, and professional societies can provide guidance to their members on how best to do this type of analytical work. If an integrated approach did become more standard, the necessary information could be assembled in places that are easy for practitioners of many types to access. It is not easy to erase more than a century of disciplinary specialization, and perhaps in some situations it is not even desirable. But in order for professionals to give the best possible advice under given resource conditions, it is nevertheless essential that we find efficient ways to bridge these frames.

8.4.1 Heuristics

Integration of these different analytical frames requires specific changes in approach; simply stating an intention to be holistic or comprehensive is not sufficient. Many disciplines, including physics and engineering as well as the fields of planning and design, have found it useful to develop heuristic diagrams in order to prompt changes in how practitioners approach their work. In this sense, "heuristic" refers to a tool for teaching oneself; a conceptual thinking tool that can lead to an integrative solution or insight that is not based on previous examples alone. Both Patrick Geddes and Herbert Sukopp used a section diagram to summarize their knowledge and ask new questions about processes that influence or are changed by urban dynamics. The type of heuristic diagram they both chose was a cross-section (Geddes' diagram dates from 1909, as described in Welter, 2003; Sukopp's section is presented in Sukopp 1973).

The "cross-section" represents a two-dimensional slice through a three-dimensional volume. Cross-sections emphasize vertical and horizontal relationships in space. Geddes used them to show the topographical relationships that link a waterfront town, where goods are shipped, to the mountains at the top of its watershed, where some key natural resources originate (wood and coal for fuel, building stone and minerals) (Fig. 8.1). It allowed him to speak about the controlling role of topography on both water flow and transportation, which was particularly significant in a regional economy that relied on boats and trains to transport people and goods. Sukopp used the cross-section to emphasize horizontal and vertical gradients in temperature, air quality, humidity, topography and landform changes created by humans, as well as depth to groundwater, all of which he saw as potential drivers of plant and animal population distributions (Fig. 8.2).

As a contemporary example of a diagram that was useful in highly integrated work with an urban water system, the staff involved in capital projects and planning at Seattle Public Utilities (SPU) in the early years of SPU's Natural Drainage program sometimes used a section diagram to represent the urban landscape in terms of water flows and functions, as they planned new components of their award-winning stormwater projects (Fig. 8.3, courtesy of M. Maupin, SPU). They used a section drawing to engage senior policymakers and technical staff in the effort to

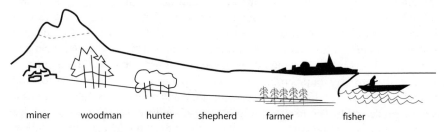

Fig. 8.1 Abstraction of Patrick Geddes' valley section, the first representation of an urban region

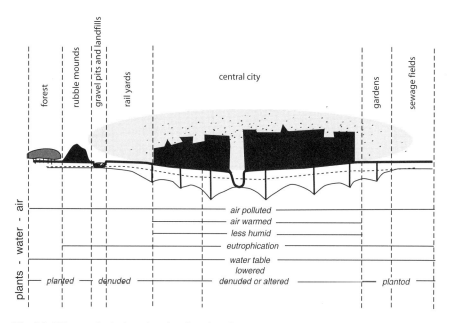

Fig. 8.2 Urban ecological section showing alterations caused by urban conditions, translated and simplified from Sukopp (1973)

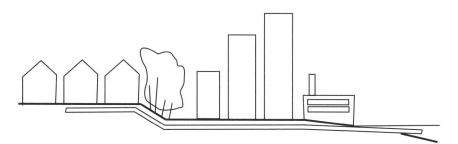

Fig. 8.3 A simplified section through an urban drainage system, showing sewered as well as unsewered areas, residential districts, parks, downtown, and shoreline industry/wastewater treatment plants, discharging to a large water body

mimic certain pre-development characteristics of landscape hydrology, while also acknowledging the infrastructure realities of the contemporary city.

8.4.2 Proposal for an Integrative Heuristic

In order to address the need for a simple but powerful integrative frame to re-examine urban water systems, I propose that planners and designers should approach the urban landscape as divisible into three categories that could match

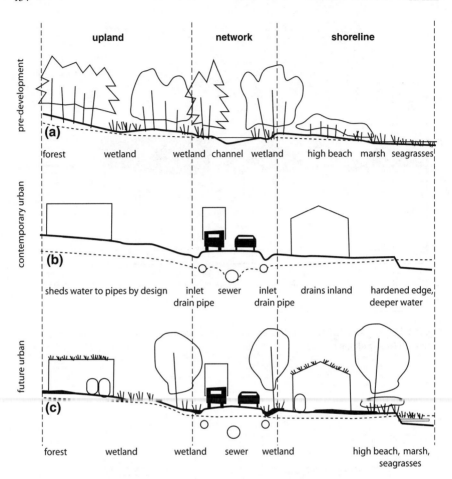

Fig. 8.4 Cross-sectional heuristic for water-based urban design, comparing pre-development patterns of water storage and flow to contemporary urban development and an improved future condition of urban development with cisterns, ROW rain gardens, private rain gardens, green roofs, and a restored shoreline edge. (**a**) indicates the pre-development water table height, (**b**) marks the lowered water table of traditional urban development, in which pipes siphon off groundwater, and (**c**) marks the raised water table of development with rain gardens, mulched soils, and pipes that have been internally sealed

portions of a cross-section diagram. Together with the cross-section, these categories are useful in providing a spatial organization of goals for hydrologic function within a fully constructed urban landscape (Fig. 8.4; Table 8.1).

The value of this heuristic for organizing urban land and water systems is that it is based on the pre-development hydrologic functions of landscapes and ecosystems, but can also incorporate urban landscapes quite readily within the same classification. It allows the user to immediately identify the primary functional goals for

Table 8.1 Categories for a function-based approach to urban water systems

Upland:

All locations where rainwater falls and can be collected, infiltrated, or dispersed as runoff, but where no significant runoff flows through from adjacent sites. This category includes building roofs, parking lots, driveways, and vegetated areas of private yards and public lands. Most urban land would fall in this category.

Network:

All locations and systems that convey significant amounts of stormwater runoff, including constructed pipe systems, roadway surfaces and roadside ditches, and streams.

Shoreline:

All locations that are adjacent to rivers, lakes, bays, and ocean beaches, or are subject to flooding from those waterbodies, would be included in this category.

a given area of land, and helps to remove conceptual barriers to thinking of constructed urban landscapes as ecosystems. It also allows the user to integrate the three analytical frames discussed above: regulatory, site-based, and geographical, in ways that lead to proposals for improvements in function as well as a recognition of political, economic, and geographic concerns. These other concerns can be brought into the classification as sub-categories, allowing the functional goals to remain as the primary concern for structuring the classification.

8.4.3 The Standard Approach Versus an Integrative Approach

Urban watersheds are typically dominated by privately-owned residential parcels when described in terms of percent area. For example, more than 50% of the largest watershed in Seattle, the Thornton Creek watershed, is made up of single-family residential lots. About 38% is made up of the public rights-of-way (ROW) for streets, which includes planting strips and sidewalks as well as paved street surfaces. The other 12% is a mix of multi-family residential parcels (<3%), private commercial parcels, and public land (including both public buildings and park lands). By taking a regulatory approach, we might identify the regulation of private single-family lots as the best "target" for changing the function of that urban watershed, since these are the typical land use by area. A political calculation might tilt that analysis to an emphasis on street rights-of-way, which are already controlled by the government sector, but which already have many other functional demands placed on them by transportation needs, cultural uses, and safety standards. At that point, voluntary action on residential parcels anywhere in the watershed might be seen as the initial path that bears the least political cost and has the potential for the greatest effect.

By taking a functional and geographical approach, we might target urban land that produces the largest volume of contaminants, such as roads and parking lots or in a combined sewer watershed, we might focus on parcels with permeable soils and relatively flat slopes that allow for high rates of safe infiltration (i.e., infiltration that is not likely to contribute to triggering a landslide on steep slopes nearby). We might also consider the socioeconomic patterns of ownership versus rental in order to determine where residents are more likely to control structural decisions about pollution-generating surfaces and drainage flows on their parcel, or we might use census information about economic status as a way to direct public investments toward neighborhoods where residents may be less able to make investments on their own. Income data might also be helpful to identify areas where residents are more likely to save money by changing their own engine oil in cars and trucks, making education about oil disposal a high priority.

From a site-scale design perspective, we might target sites that are visually accessible to a broad audience, hoping that changes on those sites would influence more people's decisions to change their own land or support new taxes and fees that would allow public lands to be transformed. Designers might also work opportunistically, using a willing client as an opportunity to experiment and build in as many different water management or design features as the client would be willing to purchase. The outcomes of these unique experiments often influence the choices made in future designs that can improve effectiveness or reduce costs, providing insights that are not available without full-scale applications.

Upland sites: If we used an integrative classification instead, sorting these parcels into upland, network, and shoreline, it would be possible to see not only space, but direction – the directionality of water flow. Without directionality, interventions tend to focus on the interior of developed land and not the shorelines – where all the pollutants and biological effects come home to roost. First, we would group all of the parcels where stormwater is first generated (public or private, commercial or residential) into the category of "upland" sites, before sorting them by use or by socioeconomic characteristics. In a Seattle urban watershed, the "upland" category might include 55–65% of the watershed by area. It would also include most of the land that generates certain kinds of pollutants, such as sediment or nutrients from lawn and garden fertilizers, pesticides and herbicides, pathogens from bird and pet wastes, or zinc from galvanized metal on roofs. Parking lots would also typically be included in this category, generating pollutants such as oil pan drippings and metals from car brakes. One common feature of all "upland" urban sites is that they are often the right place to encourage small-scale retention of stormwater, by using everything from rain barrels to mulch on landscape areas, and by turning parking lots into the hydrologic equivalent of wetlands by setting stringent storage and filtration goals for each one when re-development occurs.

In the Pacific Northwest, Portland and Seattle (and their surrounding county jurisdictions) have made successful efforts to use public sites as models for upland water management. Green roofs on public as well as private buildings are becoming relatively common, and valuable lessons about performance and soil specs are being learned and shared (BES 2006).

As one example, King County's Maple Valley Library outside Seattle was designed so that the building and parking lot would retain most of the second-growth woodland on site. The designers focused their efforts on keeping the footprint of the building small, distributing parking stalls into small groups set among the trees on a loop drive, and emphasized biological soil conservation. The building architect (James Cutler Architects) drained the roof toward the center of the U-shaped building, and into a round gravel-filled infiltration area that becomes a significant aesthetic feature of the design. The landscape architects (Swift and Company) stored and re-distributed soil that contained mycorrhizal fungi from the site, once construction was over. These fungi can play an important role in the trees' ability to survive changes in rainfall and soil nutrient levels. The designers also produced extremely constrained site access plans, so that a minimum of equipment traffic and materials storage would occur over tree roots close to the soil surface. The site engineers (SvR Design) distributed a series of stormwater "sumps" across the site to infiltrate additional runoff generated by the parking clusters, in order to mimic the pre-development capacity of native Douglas Fir (*Pseudotsuga menziesii*) forests to retain 6 centimeters or more of rainfall across an entire site.

The Maple Valley Library project represents a number of best practices for upland sites, such as forest retention and soil ecology that are often ignored by a single-minded focus on stormwater alone. Upland sites in regions that were forested before development should be designed and regulated so that, to the greatest extent possible, these landscapes mimic the ecological functions of a forested site that contains a gradient from dry to wet soil conditions. In a prairie region, upland sites should be designed and regulated to mimic the replicable components of that ecosystem type. The goal is to bring as many functions back to urban landscapes as possible, with the highest priority placed on those functions (such as stormwater transpiration by plants, infiltration, and detention) that affect ecosystems downstream. Categorizing urban lands as "upland" sites allows us to focus on their role in a cumulative set of processes first, and then expand our goals to include other processes and values.

Network sites: Surface water that runs off upland sites moves into channelized networks of flow. Those channels include everything from ruts or curbs along a roadway, grassy ditches, or underground pipes. In effect, extensive urban street networks have replaced what is often a widespread system of perennial streams and permanent streams in pre-urban landscapes (Chapter 2). The street is "the stream", although the flows are typically relocated underground. Depending on the basin's infrastructure history, the artificial streams created by urban streets may drain to a surface stream, a river, a lake, a marine bay, or a sewage treatment plant.

Categorizing street rights-of-way as network sites captures much of the opportunity to alter a public landscape to improve the hydrological function of a city and its urbanized region. Streets and their underground pipe systems are either the source or the conduit for most of the destructive pollutants that characterize non-point source stormwater runoff, such as petroleum byproducts, metals, biological pathogens, some nutrients, and sediment discharges. Studies have shown that with greater traffic flow, the volume of pollutants is higher (Patel 2005). Water-borne pollutants that escape a private parcel typically drain to this network as well, unless the private parcel drains directly to a waterbody or to the groundwater table. Network sites are places where flow concentrates, making them excellent strategic locations for intervening in those flows to reduce the downstream impact of many upstream acres of public and private land.

For example, Seattle's natural drainage program began with a street right-of-way project known as Viewlands Cascade (Fig. 8.5). Located on a cul-de-sac next to a public school in a relatively low-density residential neighborhood, the site provides detention, filtration and infiltration for stormwater that runs off of approximately 26 acres of land upstream. Its hydrologic performance has been monitored by a team of faculty and students from the University of Washington, who found that it was capable of reducing runoff and pollutant loading to nearby Piper's Creek by a factor of three, compared to a pre-existing ditch. The location of the Viewlands Cascade was strategic within the drainage network of Piper's Creek, allowing a cost-effective intervention in the sense that 26 acres of land were treated with a one-block vegetated swale system at a cost of about $225,000 USD (Horner et al. 2002). Subsequent SPU projects, such as the High Point community redevelopment, have

Fig. 8.5 Viewlands Cascade, Seattle's first natural drainage project. This vegetated swale receives runoff from approximately 26 acres of urban land, reducing the total runoff volume that enters a salmon stream below this location. Photo by K. Hill

Fig. 8.6 Network site design for detention and filtration at High Point, a community redevelopment project in Seattle, Washington with a density of approximately 16 residential units per acre. The planting strip within the street right-of-way contains an 8' deep installation of structural soil to hold and filter rainwater, but is designed as a lawn that children can play on. Photo by K. Hill

allowed the utility to experiment with designs in much denser residential districts (16 dwelling units per acre at High Point) (Fig. 8.6).

The Meadowbrook Pond project is another example of a strategically located network site. Between 1996 and 1998, SPU used the nine-acre site of a former sewage treatment plant to build a shallow pond and wetland complex that acts as an overflow area for an urban stream, along with planted areas that serve as a new park for the immediate neighborhood. The pond traps some of the excess sediment that would otherwise flow to Lake Washington, where shallow shoreline waters provide an important link in the passage of juvenile salmon through the region. The pond was built adjacent to a surface stream, a "network site" in this classification. It was intentionally designed to receive excess floodwaters and sediment that are produced by urban developments upstream, taking advantage of its strategic location where excess flows and pollutants are concentrated.

Many urban sites that may appear to be upland sites are actually network locations. Underground pipes convey not just stormwater and sewage water, but also often convey water that once flowed in surface streams. They can also convey groundwater inadvertently because of cracks in the pipes (known as inflow and infiltration, or "I and I"), and become critical influences on stream baseflows and the depth of a local water table. Interventions in urban water system design should recognize these hidden network sites and the important role they can play in improving the overall hydrological performance of cities.

One such site was recognized in Seattle, under a parking lot just south of the Northgate Mall. An underground pipe six feet in diameter conveyed a fairly constant stream of water measuring about one cubic foot per second (cfs), with storm-driven high flows that were predicted by watershed-scale modeling to reach up to 500 cfs. Neighborhood groups saw the re-development of the parking lot as an opportunity to "daylight" what they considered an upper branch of the local stream, Thornton Creek. The city did not consider this pipe a stream, and planned to treat it as a stormwater drainage pipe. Bringing the baseflow inside the pipe to the surface would be difficult, because the pipe was covered with up to 30 feet of fill that was used to level the parking lot when it was constructed. An eventual compromise was established to use new buildings on the site to step down to a lower grade, and allow the pipe's baseflow to be drawn out by a side pipe located at a weir inside the pipe. High storm flows in the pipe would still be allowed to flow above the weir and down the large pipe, as they had in the past.

The interesting thing about this design process was that the city's utility staff compared three alternatives in a quantitative analysis, intended to identify the strategy that would produce the greatest benefits in terms of the total amount of pollutants that could be prevented from entering Thornton Creek downstream of this site. The proposal for draining off the baseflow (referred to in public documents as the "water quality channel" design) was compared to daylighting the stream (building an exposed channel and removing the pipe altogether), or introducing a set of new surface bioswales in the area immediately around the parking lot that drains to the underground pipe. The conclusion of this comparison process was that the "water quality channel" would have significant benefits because the weir inside the pipe would trap a large quantity of fine sediment that would otherwise make its way into the streambed of Thornton Creek. The water quality channel itself would also provide some benefits, along the lines of a bioswale but with reduced efficiency because water would flow through without always achieving the ideal residence time for water quality improvements. Overall, trapping sediments behind the weir and periodically vacuuming them out was enough of a benefit that the water quality channel strategy was adopted.

By treating this large parking lot as a network site, design strategies were identified and benefits were achieved that an upland site could not accomplish. The special value of what seems to be an ordinary parking lot only becomes apparent when its role in a larger network is evaluated. This analysis is prompted using categories organized by hydrologic function (upland, network, and shoreline), not categories of human land use alone (parking, commercial, etc.). Categorizing urban sites by their present hydrological function, not just their historical role or their political significance to a community, is critical in altering the overall hydrological performance of cities.

Shoreline sites: Large lakes and marine areas are the ultimate recipients of urban stormwater pollutants, and of the ecological impacts of runoff on streams and wetlands. Yet most maps and datasets, such as topographic data, use waterbody boundaries as limits; terrestrial and aquatic systems are treated as if they were separate, when in fact they are directly connected by flows of water and materials. In spite of all the attention that stormwater has received, very little regulatory attention has been directed toward shoreline sites with regard to stormwater impacts and potential design strategies. The strategic question for urban design and planning is not whether "end-of-pipe" interventions are better than starting at the source, but instead, whether interventions at the shoreline can produce any additional benefits that interventions upstream do not provide. In other words, can design interventions at the end of the pipe add value to a broad set of interventions upstream?

In order to answer the question of what can be done at the shoreline itself, the problem must be re-framed to consider new scientific research that has emerged about ecological dynamics in shallow-water marine environments. A pattern is emerging in recent studies that links the spatial patterns of shallow sub-tidal and intertidal habitat, along with freshwater inputs and salinity gradients, to the successful reproduction and development of critical marine species (see Pineda et al. 2007, as one example). Research in the Pacific Northwest has shown that juvenile salmon not only travel along shallow waters at the edges of lakes and marine bays, but they also congregate at the mouths of freshwater streams where they discharge into larger water bodies (Tabor et al. 2006). Marine researchers have known for decades that crabs and other animals orient their movement in part by sensing salinity gradients in salt water that are created by freshwater inputs (Johnson 1960). Larval dispersal of marine organisms can be affected by nearshore habitat patterns as well (Gerlach et al. 2007). Shoreline development must begin to support and enhance this set of relationships as it expands to affect more and more of what is now undeveloped coastline, and as we re-evaluate the importance of older urban shorelines in regional dispersal patterns. Human benefits can also result, both in terms of enhanced tourism and a renewed focus on swimmable and fishable waters around cities.

In a few short decades, the broad environmental trends of rising sea levels, disrupted storm patterns, and changes in temperature or water supply that affect these biodiversity issues will link them to urban water management as a whole. Sea level changes will inundate shallow intertidal habitat areas if they do not accumulate sediment with a parallel increase in rate. Similarly, the number of tidally-influenced drainage outfall pipes will increase, as will the number of coastal water supply intakes that are affected by saltwater contamination. As cities and other jurisdictions begin to address flooding, drainage system changes, and water supply challenges, the physical structures associated with shoreline designs will become critical components of adaptation. Sea walls, pipe outfalls and intakes, piers, wetland conservation areas, artificial islands, and even storm surge barriers will be manipulated as coastal developments adjust to new relationships with storms and tidal processes. Urban areas that can adapt their shorelines to support the biological as well as physical needs of coastal systems will be in the best positions to maintain the special resources associated with water-based economies, from residential property values to tourism and fishing industries.

For example, urban seawalls and fixed storm surge barriers typically eliminate sub-tidal, inter-tidal, and supra-tidal ecosystems. Recent scientific work has observed that these gaps in what were once relatively continuous corridors of shoreline or submerged vegetation can be very significant to the dispersal and reproductive success of aquatic organisms. Coastal areas that will experience increased rates of sea level rise or storm surges are likely to expand these structures. It is possible that such structures could, in the future, include systems of buoyant modular planting beds, designed to float at the surface or at key depths to support specific animals and plants. This strategy could also overcome some of the effects of increased algal densities in marine water that have prevented successful restoration of submerged aquatic vegetation in some coastal regions, by allowing sub-tidal plants to become established at ideal depths.

It is also likely that key transportation structures may be raised or re-located as an adaptation to storm surge and sea level changes, creating opportunities to redesign coastal landscapes to include these key ecosystems. Although state Departments of Transportation have not yet released adaptation plans, a US Transportation Research Board report identified the need for them (TRB 2008). European countries and World Bank researchers are actively studying alternative coastal designs. The San Francisco-based Bay Area Conservation and Development Council has begun planning for adaptations that will enhance aquatic systems while protecting two major airports and associated transportation lines. In non-governmental organizations across the United States, citizen activists have begun to link infrastructure and ecology along shorelines. In New York City, citizen groups are advocating for the removal of a 1960s-era highway segment along a tidal section of the Bronx River, creating opportunities for new tidal wetlands. Activists in Seattle are working for the removal of a highway viaduct along a marine bay, proposing that it should be replaced with a series of shallow inter-tidal beaches. The key point here is that investments will be made in infrastructure adaptation in the future, and as this occurs the health of marine species and ecosystems can also benefit.

8.4.4 Strategic Implications for Science and Design

The three basic objectives of urban design that supports the health of aquatic systems are (1) to mimic the pre-development hydrology of an urbanized landscape, (2) to limit the movement of pollutants into aquatic systems that would undermine the health of humans and other species, and (3) to re-establish or sustain a nearshore environment that supports the biodiversity of river, lake and ocean ecosystems and human fisheries. The main strategic question is, how can we accomplish those goals as quickly as possible with a limited budget of money and political resources?

Physical design is a sub-category of urban strategies, in which each instance embodies specific objectives and proposes social, material, and spatial tactics. Scientific rigor should be employed to help identify these specific objectives based on the historical and contemporary patterns and processes of an urban ecosystem, including its human community. Designers need scientists to help them know what processes have changed significantly, what the future trends might be, and how these changes are linked to other important processes and patterns. The rigor of design as a cultural action, seeking to achieve aesthetic performance as well as functional and ethical goals, should be employed to produce proposals of material, social and spatial tactics that represent a cohesive design strategy. The likelihood of success for any given design strategy can be evaluated by reflecting on a record of past attempts, or using knowledge of present interactions. The design strategy itself can be treated as an experiment as it is first tested and then implemented, ideally in a staged process through which initial lessons are incorporated in future expansions. This partnership between science and design has historically been the basis for engineering, and can be a much more vital basis for urban design and planning.

The key here is that uncertainty alone cannot justify inaction. Larger environmental trends such as increased flooding inland and sea level rise on the coasts will force that action eventually. Imperfect but reversible design strategies must be tested in actual cities in order for better solutions to be developed incrementally. Once identified, affordable solutions that contribute to achieving significant goals must be implemented in as many relevant locations as possible, or overall urban performance will remain unchanged and may even decline. The use of a shared conceptual and analytical frame among planners, designers, and scientists should be helpful in generating the political will to make these investments as quickly as possible.

8.4.5 Value of Visibility/Public Awareness

Interventions in the physical design of cities, whether in uplands, on network sites, or at shoreline locations, must be primarily designed to achieve functional performance benefits. As noted above, they must also be designed for replication, in order to have significant cumulative benefits. The need for replication creates two additional criteria: First, these designs must be both cost-effective and inexpensive enough to fit the current and future maintenance and capital improvement budgets of municipalities and other public authorities which are typically funded by development or user fees. Second, they must be supported by the public in order for elected officials to approve the expense and tolerate the inevitable construction disturbances (Hill 2003).

Prototypical design interventions can help to establish the widespread public support that is needed to achieve replication, if they are designed with that explicit goal. Seattle's SEA-Street project has become a classic case in point, in which a one-block demonstration project helped to establish a multi-million-dollar public program of investments in roadway runoff improvements. Neighbors and elected officials who visited the demonstration project expressed satisfaction with its aesthetic and functional characteristics, allowing the city's public utilities and transportation departments to continue a successful partnership on similar projects. Similarly, Portland Oregon has experimented with urban street designs that have become quite popular (Fig. 8.7). Ecologists, planners, engineers and designers who advocate for interventions that would not receive similar approval are likely to face difficult implementation battles that become more challenging with each instance of implementation, and can lead to an eventual abandonment of strategies that may be functionally

Fig. 8.7 Downtown street design in Portland, Oregon, that filters runoff water from traffic and parking lanes. Grates cover the channels that bring water from the street into the planting areas. Photo by K. Hill

successful, but are not aesthetically or politically acceptable (for additional examples and solutions to this dynamic, see Nassauer 1995).

8.5 Current Drivers of Innovation

Although larger political and economic trends, as well as new scientific insights, provide specific openings for change, individuals must drive the initial phases of change in the way cities are designed and built. Innovation in urban design related to urban water systems has been lead by a few creative and dedicated practitioners over the past twenty years, who mastered the factual arguments and have persuaded their jurisdictions to make and monitor changes. In particular, practitioners in Seattle, Washington; Portland, Oregon; Prince George's County, Maryland; and Los Angeles, California deserve recognition for their successful work to establish demonstration projects. Elected officials in a few states have become advocates as well, and the work of public agency staff in these states is now supported by larger professional societies and other non-governmental organizations that disseminate knowledge of successful case studies.

These individuals and organizations were supported by federal and state legislation that sought to protect aquatic systems, such as the federal Clean Water Act (CWA), Endangered Species Act (ESA), and to a lesser extent, the Coastal Zone Management Act (CZMA). Much of the water-related urban design innovation in the Pacific Northwest was driven by a need to respond to the listing of Chinook salmon as a threatened species under the federal ESA in 1998. Advocates linked this goal to others in order to build successful demonstration projects. For example, Seattle's SEA-Street project linked ESA goals with the city's political commitment to provide sidewalks in neighborhoods that had been annexed to become part of the city in the 1950s. Another award-winning Seattle project, the High Point housing re-development effort of 2003-6, linked the ESA goals for stormwater that drained to a salmon stream and a marine bay to federal goals for public housing that were part of the Hope VI funding program.

Other major trends that might be described as driving change in the relationships between urban design and water systems include increasing public awareness of climate change and aquatic ecosystem degradation, increasing advocacy or actual proposals by elected officials, evidence from European countries and cities that are taking action to adapt, and an increase in the number of cases available that can be used to argue by example. Changes within the design and planning disciplines are also happening, in which civil engineers are learning more about decentralized system design, building architects are learning more about keeping water on the roof instead of shedding it, landscape architects are bringing more ecological rigor to expanding their recent role in urban design, and planners are re-considering the value of physical planning using spatially-explicit tools such as geographic information systems. Legislation in some states is promoting the use of physical design as a new tool in addressing cumulative environmental problems, such as Maryland's revision of its stormwater act to include innovative site design as a requirement, not merely an option (see the State of Maryland stormwater website for further information on this program, cited as Maryland Department of Environment, 2007).

Future trends that will drive innovation are likely to include an intensified awareness of the need for adaptation to climate change, the need to invest in infrastructure that has been subjected to deferred maintenance for decades, and the need to expand transportation options within urban regions. If a new national administration instigates an era of greater international cooperation, the approach is likely to accelerate these trends, as social and political norms from outside the United States will most likely reinforce the need for rational planning and design related to climate change and water resources.

8.6 Conclusions

The most important general lesson discussed in this chapter is that, while similar broad trends drive the need for innovation in the relationships between urban design and water systems, practitioners and academics from different fields may conceptualize the problem very differently. While there is no single best approach, the hybridization of several analytical frames is likely to provide more valuable insights than the choice of any one frame alone. In particular, urban designers and planners would benefit from using a spatial analytical approach that reminds them to look for intervention opportunities in network sites and shoreline sites, as well as the more typical upland locations.

The history of these ideas extends to at least the late 19th century, when American city design and planning was re-shaped by Olmsted's efforts to piggy-back social and ecological functions onto the spatial patterns of water systems. It has evolved significantly since that time, particularly through the establishment of more specific knowledge of the patterns and mechanisms of ecology and of the psychology of perception. But the simple, direct strategy of constructing a land-based infrastructure to manage water flow and water quality which also serves as a recreational infrastructure dates back to a critical 19th century urban innovation.

In practical terms, this chapter has proposed that a very simple set of three categories (upland, network, and shoreline) can be an effective basis for an integrative approach to urban design for water systems, especially in combination with a cross-sectional representation of larger urban patterns. Examples are provided of recent experience with urban design that explicitly addresses water systems, and that demonstrate the value of the proposed landscape-based analytical frame. Taken as a whole, these examples provide a vision of how the hydrological performance of cities and urbanizing landscapes can be altered on a large scale – from upland to shoreline.

The insight gained from reviewing these cases of successful design interventions, or in the case of shorelines, a set of proposed strategies, is that implementing demonstration projects can create many benefits. The first is to test whether the demonstration can achieve its functional goals. The second is a test of whether the design meets the aesthetic and political needs of its social context, the latter of which includes measures or estimates of its costs and benefits.

In terms of current trends and future drivers of innovation, increases in both public awareness and political courage to address adaptations to climate disruptions, and to channel development pressures, are most likely to influence the future of urban design related to water systems. Several regions of the United States have established themselves as leaders in the development of voluntary and mandatory design and planning standards, including the Pacific Northwest, southern and central California, and the state of Maryland. While these models have generated new initiatives in other regions, national leadership and international partnerships are needed to move to broader implementation – within those regions as well as nationally. Leadership is also needed to inspire regional leaders to continue to expand those initiatives, especially with regard to shoreline design and development.

8.7 Research

As Winston Churchill is reported to have said, "However beautiful the strategy, you should occasionally look at the results." The most important need for urban runoff systems research is in the area of monitoring. If new urban approaches either fall short of their goals or surpass them, this is critical information for future design strategies. Yet very few jurisdictions or practitioners engage in monitoring. There are obvious disincentives – for instance, few jurisdictions are willing to create public documents that identify design failures that lead to the continued pollution of streams or other water bodies by runoff from public streets. Previous case law has established the vulnerability of governments to lawsuits for such pollution. In practice, while most practitioners make the argument that past experience establishes their firm's expertise, evidence of a major design failure would undermine that argument. And if the client doesn't pay for the monitoring, the firm would be paying for evidence that undermines its credibility. Evidence of success would be very valuable, but very few firms see themselves as able to dedicate the financial or staff resources to demonstrate that success.

In addition, there are several areas in which better information or more innovation is needed, including:

- assessing benefits to property values and operating costs (this would provide support to the idea of using local improvement districts as a financing tool for water system improvements);
- developing better methods for oil/grease removal;
- studying urban retrofit benefits in different regions (such as the recent Washington DC study of green roof retrofits and tree planting; Deutsch 2007);
- testing proposals for shorelines designs, especially in relation to habitat support that can be adapted as sea level rises or storm patterns change;
- detailed regional assessments of sea level rise and storm surge impacts on drainage and other infrastructure along the coasts, and rain/snowfall changes inland;
- comparative studies of alternative implementation plans for replication (financing, cost, performance, maintenance, etc.).

Linked to these research needs, there is also a need for expanded education within the urban design professions (civil engineering, planning, landscape architecture, and building architecture). Students need to enter their professional careers with greater knowledge of landscape-based strategies for improving hydrologic performance. This can be taught using case studies, and by framing discussions of the performance of those cases in a bioregional context. Students also need an enhanced focus on shorelines as the ecosystem at the end of the pipe, or protecting systems like the Chesapeake Bay, the Great Lakes, or Puget Sound will not be understood as the fundamental reason for (and test of) efforts to improve urban performance. Finally, students need to participate in developing innovations related to "positive impact re-development" – in order for urban design to move beyond the current focus on the development of new urban areas, and re-center professional attention on urban infill and brownfield re-development. The market of design clients will continue to push for the development of new land, but new professionals need to see this type of development as inextricably linked to central cities and their performance (especially in terms of fiscal performance and infrastructure systems).

8.7.1 Implementation

In order for any of the ideas or approaches mentioned in this chapter to make any difference to urban performance as a whole, broad implementation is necessary. While a discussion of municipal public finance and infrastructure budgets is beyond the scope of this chapter, implementation requires a new approach in both these areas that will, like all major urban changes, sometimes involve political and financial risks as well as social and environmental benefits. For example, most large jurisdictions have manuals that set standards for street design, and have code language that sets standards for parking lot design on public or private sites. These two areas of public administration are the "low hanging fruit" of an urban revolution in water quality and aquatic biodiversity. But barring a dramatic change in federal contributions to local infrastructure budgets, any plan for investing in improved streets and parking lots must be paid for through a combination of local public and private dollars. As different jurisdictions try experiments in funding, anti-tax organizations or groups that seek to represent the narrow financial interests of utility rate-payers often challenge these innovations in court as well as in a political arena. In other cases, higher levels of government – such as the state versus the city – may block financing instruments as a political maneuver designed to create or demonstrate leverage, not as a way of achieving environmental goals.

Non-governmental organizations can exert a significant and important influence on these political outcomes, as can changes in the incentives established by federal funding programs, or implementation of federal laws like the CWA and the ESA. A sustained alignment of interests must be established to achieve the creation of new standards, as well as the financing and enforcement of those standards, if cities are to improve their cumulative hydrological and ecological performance. Benefits to human health issues, transportation, recreation, the fishing industry, and tourism will have to be aligned with efforts to improve aquatic health more generally.

A combination of inspiration, regulation and common sense will be needed to change any particular city. Individuals as well as non-governmental groups and public agencies will be responsible to create that combination of characteristics. In the end, it comes down to how many miles of roadway, how many roofs, how many parking lots, and how many miles of shoreline can be held to a different standard as communities develop, re-develop, and adapt to changes in climate and the rates of sea level rise.

Acknowledgments The author wishes to thank the editor, Larry Baker, and two reviewers, Doug Johnston and Lance Neckar, for suggestions that improved this chapter significantly.

References

Alberti, M., D. Booth, K. Hill, C. Avolio, B. Coburn, S. Coe, and D. Spirandelli. 2007. The impact of urban patterns on aquatic ecosystems: An empirical analysis in Puget lowland sub-basins. Landscape and Urban Planning **80**(4):345–361.

Ashley, R.M., D.J. Balmforth, A.J. Saul, and J.D. Blanskby. 2005. Flooding in the future: predicting climate change, risks and responses in urban areas. Water Science and Technology **52**:265–273. IWA Publishing.

Barnett, J. and K. Hil. 2008. Design for Rising Sea Levels. Harvard Design Magazine, Number 27, Winter 2008.

Beach, D. 2002. Coastal Sprawl: The Effects of Urban Design on Aquatic Ecosystems in the United States. Pew Oceans Commission, Arlington, Virginia.

Booth, D. 2005. Challenges and prospects for restoring urban streams: a perspective from the Pacific Northwest of North America. North American Benthological Society **24**:724–737.

Botsford, L.W., A. Hastings, and S.D. Gaines. 2001. Dependence of sustainability on the configuration of marine reserves and larval dispersal distance. Ecology Letters **4**:144–150, 2001.

Bureau of Environmental Services (BES). 2006. 2006 Stormwater Management Facility Monitoring Report Summary. Sustainable Stormwater Management Program, Portland, Oregon, September 2006.

EPA. 1999. Combined Sewer Overflow Management Fact Sheet. EPA 832-F-99-041. US Environmental Protection Agency, Washington, D.C., September 1999.

Deutsch, B. 2007. The green build-out model: quantifying the stormwater management benefits of trees and green roofs in Washington, DC. Casey Trees Foundation and LimnoTech, Washington, DC.

Gerlach, G., J. Atema, M.K. Kingsford, K.P. Black, and V. Miller-Sims. 2007. Smelling home can prevent dispersal of reef fish larvae. Proceedings of the National Academy of Sciences of the United States of America **104**:858–863.

Gobel, P., C. Dierkes, and W.G. Coldewey. 2007. Storm water runoff concentration matrix for urban areas. Journal of Contaminant Hydrology **91**: 26–42.

Hill, K. 2002. Urban Design and Ecology. In CASE: Downsview Park Toronto, ed. by Julia Czerniak, Prestel Publishers.

Hill, K. 2003. Green Good, Better, Best: Effective ecological design in cities," Harvard Design Magazine, Number 18, Summer 2003.

Horner, R., H. Lim, S., and Burgess. 2002. Hydrologic monitoring of the Seattle ultra-rrban stormwater management projects. Water Resources Series, Technical Report No. 170, University of Washington, Seattle, WA, September 2002.

Johnson, M.W. 1960. The offshore drift of larvae of the California spiny lobster *Panulirus interruptus*. California Cooperative Oceanic Fisheries Investigations Reports **7**:147–161.

Lachmund, J. 2007. Ecology in a walled city: researching urban wildlife in post-war Berlin. Endeavor **31**:78–82.

London Climate Change Partnership. 2006.Seattle: Managing stormwater, pp. 34–45. In: Adapting to Climate Change: Lessons for London. Greater London Authority, London.

Lynch, K., M. Southworth, and T. Banerjee. 1990. City sense and city design: writings and projects of Kevin Lynch, MIT Press, Cambridge, MA.

Maryland Department of the Environment. "Maryland Stormwater Act of 2007," http://www.mde.state.md.us/Programs/WaterPrograms/SedimentandStormwater/swm2007.asp, accessed June 2008.

McHarg, I. 1969. Design with Nature, Natural History Press, Garden City, NewYork.

Meyer, E. 1997. Ecological Design and Planning. Wiley, New York.

Nassauer, J.I. 1995. Messy ecosystems, orderly frames. Landscape Journal **14**:161–170.

National Research Council. 2003 Oil in the Sea III: Inputs, Fates, and Effects. The National Academies Press, Washington, DC.

Novotny, V., and Hill, K. 2007. Diffuse pollution abatement – a key component in the integrated effort towards sustainable urban basins. IWA Publishing, London, UK.

Patel, J. 2005. Briefing: review of CIRIA Report 142 on highway pollutants. Proceedings of the Institution of Civil Engineers, Transport **158**(TR3):137–138.

Perkins, B., D. Ojima, and R. Correll. 2007. A Survey of Climate Change Adaptation Planning. The Heinz Center, Washington, DC.

Pineda, J., J.A. Hare, and S. Sponaugle. 2007. Larval transport and dispersal in the coastal ocean and consequences for population connectivity. Special Issue on Marine Population Connectivity, Oceanography, **20**(3):22–39.

Sukopp, H. 1973. Die Großstadt als Gegenstand Oekologischer Forschung. Schriften des Vereins zur Verbreitung Naturwissenschaftlicher Kenntnis zu Wien. Bericht ueber das 113. Vereinsjahr, 90–139.

Tabor, R., H. Gearns, C. McCoy, and S. Camacho. 2006. Nearshore habitat use by juvenile Chinook salmon in lentic systems of the Lake Washington Basin. Annual Report, 2003 and 2004, U.S. Fish and Wildlife Service, Western Washington Fish & Wildlife Office, Lacey, Washington, DC.

Tallis, H., Z. Ferdana, and E. Gray. 2008. Linking terrestrial and marine conservation planning and threats analysis. Conservation Biology **22**:120–130.

Transportation Research Board (TRB). 2008. Special Report 290: Potential Impacts of Climate Change on U.S. Transportation. The National Academies, Washington, DC, March 2008.

US EPA. 2007. National Water Quality Inventory: Report to Congress, 2002 Reporting Cycle, EPA 841-R-07-001, October 2007.

Walsh, R., A. Roy, J. Feminella, P. Cottingham, P. Groffman, and R. Morgan. 2005. Urban stream syndrome: current knowledge and the search for a cure. Journal of the North American Benthological Society **24**:690–705.

Welter, V. 2003. Biopolis: Patrick Geddes and the City of Life. MIT Press, Cambridge, MA.

Chapter 9
Legal Framework for the Urban Water Environment

Robert W. Adler

9.1 Introduction

The complexity of the legal framework for the urban water environment approaches the complexity of the scientific and technical aspects of urban water management addressed in other chapters, although for different reasons and with different possible solutions. Urban water resources are addressed, to varying and sometimes overlapping degrees, by private, local, state, regional, federal, and sometimes even international law. Some aspects of urban water resources management are governed by *common law* (legal principles derived from a series of decisions reached by judges in individual cases), while other requirements are dictated by statutes passed by federal, state, or local legislative bodies, or regulations issued by administrative agencies. Separate (although sometimes linked) legal regimes address aspects of water supply, water treatment and distribution, and the environmental and human health and safety aspects of wastewater, storm water, and drainage or flood control.

This chapter will outline the major legal doctrines and sources of law that govern or affect urban water management most directly. (Other legal principles, such as those governing contracts, affect urban water use and management more tangentially and largely in the same way as they affect other public and private activities.) Even this brief summary, however, covers a wide range of statutory, regulatory, common law and other legal aspects of the urban water environment. The chapter concludes with a critique of the manner in which the fragmentation of that law and policy impedes efforts to promote more sustainable and efficient uses of water in urban areas.

9.2 Governing Legal Principles and Doctrines

The legal regime governing the urban water environment could be organized in a number of ways. This analysis focuses on four primary governmental functions

---

R.W. Adler (✉)
University of Utah, S.J. Quinney College of Law, Salt Lake City, Utah
e-mail: adlerr@law.utah.edu

related to urban water use and management. First, especially in areas where demand approaches or exceeds supply, cities must secure adequate supplies of fresh water for their citizens (the "front end" of the urban water environment). Second, they must treat and convey water of acceptable quality to all necessary points of use. Third, they must deal with the "back end" of the urban water cycle by making sure that urban wastewater, in the form of sewage and storm water runoff, does not pose threats to human health and safety, property, and aquatic ecosystems. Fourth, they can take advantage of urban waterways as environmental and recreational assets rather than simply as resources to exploit (Chapters 6 and 8).

9.2.1 Water Supply

Municipalities cannot simply stick a pipe in the ground or in a body of surface water and extract water for use by their citizens. All states (and most other countries) have legal systems governing allocation of water rights among competing users (Sax et al., 2000). Water law in most U.S. states has deep origins in common law, but all states now have statutes, regulations, and other administrative mechanisms to implement those principles. Although most states confer some degree of legal preference on municipalities to meet the basic, health-related needs of domestic users, attention to water law is increasingly important for growing cities to secure adequate water supplies. As outlined below, however, the nature of the legal systems governing water rights varies in different parts of the United States. Eastern states use modified versions of the riparian rights doctrine inherited from England. Arid western states adopted the prior appropriation doctrine in response to very different hydrological and geographic conditions.[1] Water law in some states in the intermediate zone along the 100th meridian, and along the west coast, reflects a hybrid of riparian and appropriative rights.

Riparian rights doctrine: The original riparian rights doctrine developed in England is a system of property rights in which only riparian (waterside) land owners had the right to withdraw and use water from a stream or other water body (Sax et al., 2000, pp. 20–97). The doctrine was based on a concept of "no harm", meaning that riparian landowners could use water so long as they did not substantially impair either the quantity or quality of water for downstream users. Although designed to protect the rights of downstream landowners rather than as a system for environmental protection, the system was inherently protective of aquatic ecosystems and ensured that neither upstream owners nor the earliest users could dominate water resources at the expense of others. This rights-based system also made sense in a country where most landowners had access to some supply of water, and at a time when water supply far exceeded demand. Courts reconsidered the pure riparian rights doctrine during early American history when mill users and others increased demand and competition for scarce water. As a result, American courts modified

---

[1] A few western states that were part of Mexico before the Treaty of Guadalupe Hidalgo still observe some remnants of Spanish water law.

the doctrine to allow more significant stream depletions when justified to promote industry, agriculture, and other development (*Snow v. Parsons*).

Under traditional riparian doctrine, domestic uses for culinary purposes, to cultivate gardens and other subsistence uses, enjoyed an absolute preference, but those uses typically were not large enough to cause significant depletions or harm. Other, more intensive uses, for example to run mills or other economic uses, are subject to the *reasonable use doctrine*, in which uses are permissible if reasonable relative to the rights of other riparian landowners for other reasonable uses. The reasonable use doctrine requires courts to balance the rights of competing users based on factors such as the purpose of the use, the suitability of the use to the water body, the economic and social value of the use, the harm caused, ways to avoid the harm, etc. (American Law Institute, 1979, Restatement (Second) of Torts, §850A).

Under the balancing principles of riparian rights, courts seek to allocate shortages fairly among all legitimate users, so no single user is likely to be shut off completely. However, this doctrine generate uncertainty for municipal and other users because no fixed quantity of water is assured, and some courts have limited the ability of cities to withdraw water even for general public uses in violation of strict riparian doctrine (Sax, et al., 2000, pp. 55–58). To address this problem, state legislatures have often intervened by enacting special legislative preferences or authorities for municipal water supply. For example, New York State's water supply law provides: "The acquisition, storage, diversion and use of water for domestic and municipal purposes shall have priority over all other purposes" (Sax et al., 2000, p. 58, *quoting* New York Environmental Conservation Law Title 15 – Water Supply). Many riparian doctrine states also now have statutes and regulations that establish somewhat clearer rules governing municipal water rights, often modeled after the Regulated Riparian Model Water Code (American Society of Civil Engineers, 1997). The exact status of municipalities with regard to riparian water rights, then, varies from state to state. The inherent uncertainty of riparian rights law—along with practical issues of storage and distribution—also prompted cities to build reservoirs to store water during wet periods as a hedge against later shortages.

Riparian rights also presented challenges for growing U.S. cities because of the original limitation that only riparian landowners could use water from a stream, and only on riparian lands. The riparian land limitation was environmentally protective because it kept water within watersheds, and worked reasonably well in the context of non-municipal water users. However, this limitation restricted the ability of cities to provide sufficient common water supplies for large bodies of citizens over a larger area. Courts first modified the limitation on place of use, so that riparian landowners could use their water supplies on non-riparian parcels, or sell it to non-riparian owners (*Connecticut v. Massachusetts*). With that change, a city could purchase—or acquire through eminent domain[2]—riparian parcels necessary to assure municipal

---

[2]Eminent domain, or "condemnation", is a legal process by which a government can obtain land from non-willing sellers for legitimate public purposes. Under both federal and state constitutions, the government must pay "just compensation," or fair market value, for the land, and provide a fair set of judicial or other procedures ("due process") to justify the public use, if challenged, and to determine fair compensation.

water supply. Later, as cities expanded and as demand for water for industrial, commercial, and other purposes on non-riparian lands grew, courts further loosened the requirement that water be used within the watershed of origin, allowing cities to import water from other watersheds. For example, New York City began to import water from the Croton River in Westchester County (*Hudson River Fisherman's Association v. Williams*), and later from more distant supplies in the Catskill and Adirondack Mountains.

Prior appropriation doctrine: By the mid-nineteenth century, riparian rights principles did not serve the needs of water users in the arid west. Water was needed for mining, irrigation and other uses on non-riparian lands; greater certainty of water supply was needed to justify investments; and demand often exceeded supply, especially during droughts. To address those needs, western states developed the prior appropriation doctrine (Sax et al., 2000, pp. 98–279). Under prior appropriation law, water rights are quantified specifically (x cubic feet per second (cfs) or y acre-feet (af)) and priority in times of shortage is determined in order of seniority ("first in time, first in right"). Priority dates are determined by the time at which water is first diverted and put to a legally "beneficial use", such as irrigation or municipal water supply, and during times of shortage senior water rights are honored in full before junior rights-holders receive any water at all. Prior appropriation law does not limit the place of use, meaning that large amounts of water can be—and are—transported out of the watershed of origin to distant locations where it is needed.

Unlike the riparian rights doctrine, in which water rights are attached to ownership in land and continue whether or not water is used, appropriative rights are "usufructory" in nature. The public, through the state, owns the water but individuals are given the right to use it, at certain times and for certain purposes, and subject to various conditions. Thus, under the "use it or lose it" tenet of prior appropriation law, rights to use water can be forfeited if not exercised. This ostensibly ensures that water is not "wasted" or that water rights are not held purely for speculation. Although in theory water must be used efficiently to prevent waste, the incentive is to use one's full water right so as not to lose it, and rules against inefficient waste are rarely enforced. Moreover, until relatively recently, traditional prior appropriation law has recognized as "beneficial uses" only off-stream uses for human economic purposes, at the expense of in-stream and other environmental "uses".

As was true in riparian rights states, western legislatures recognized that domestic uses of water to sustain basic human needs warranted some priority over other uses, and that concept translated to some degree of preference for municipal water supply as well. Nevertheless, the fact that so much western water is held for agricultural and other non-municipal purposes, with very early priority dates, has created problems for rapidly growing western cities, especially in areas with inadequate proximate supplies of fresh water. That led to infamous "water grabs" such as Southern California's raids on the water resources of Owen's Valley and Mono Lake (Reisner, 1986).

Growing western cities governed by prior appropriation law have tried to address water shortages in various ways (Adler, 2007; Chapter 12). Cities in urban Southern California have facilitated water transfers from agricultural areas with superior water rights (such as the Imperial Valley) by paying those areas to implement more efficient water conveyance and use measures (Hadad, 2000), and are now beginning to use expensive desalination of ocean water. Las Vegas adopted very aggressive water conservation measures, and is trying to augment existing supplies through groundwater from nearby basins. Denver and other cities along Colorado's Front Range import Colorado River water through tunnels beneath the high peaks of the Rockies. Many cities are reclaiming urban waste water to re-use for irrigation water (Furumai, 2007). Despite all of those measures, each region faces shortages under the pressure of impending growth, and both legal and technological solutions will be required in response.

Inter-jurisdictional conflicts: Water supply needs for expanding cities in all parts of the country have led to inter-basin, interstate, and even international conflicts over supplies from major river basins. The stakes are high, as water supply can be one key factor in determining which cities will grow and which will face limits. One key example is the ongoing tension between Colorado's Front Range, Southern California, Phoenix, and Salt Lake City over water from the Colorado River Basin, a problem that is likely to exacerbate as those and other cities in the region continue to expand, and as global warming potentially reduces runoff in the basin (Adler, 2007). But cities have fought over rivers in the east as well, including the Connecticut, the Hudson, the Delaware, and most recently the Appalachacola-Chatahoochee-Flint basin in the southeast.

There are three main legal responses to interstate water conflicts. First, on a number of occasions the U.S. Supreme Court has issued decrees allocating water among states under a doctrine known as "equitable apportionment," in which the court balances a number of factors such as need, priority, and fairness to determine how to apportion scarce water resources among states (e.g., *Wyoming v. Colorado*). Second, under the Compact Clause of the U.S. Constitution (Art. I, §7, cl.3), states can negotiate settlements through interstate compacts, such as the Colorado River Compact, which require congressional ratification. Third, Congress can step in independently and pass laws allocating interstate water resources among states. The United States has also negotiated international treaties over allocation of water from international waters such as the Colorado and Rio Grande Rivers (Meyers and Noble, 1967).

Groundwater allocation: The law governing groundwater allocation is even more perplexing than it is for surface water for two reasons. First, five separate doctrines (and variations within those doctrines) apply to groundwater in various U.S. states, so water allocation law varies even more widely among states with respect to groundwater than it does for surface water (Sax et al., 2000, pp. 359–385). Groundwater law evolved initially during an era when groundwater was a mystery—when people knew nothing about where it came from and how much was there. Especially where groundwater seemed essentially unlimited, that led to doctrines such as the "rule of capture," in which landowners had the right to extract as much water as they needed from wells drilled on their own property, thus providing no protection to those whose wells may have been sucked dry as a result.

Few jurisdictions continue that simple doctrine now that groundwater demand often exceeds supply (Glennon, 2002). Other doctrines, therefore, modify the rule of capture idea to varying degrees (Sax et al., 2000, pp. 364–365). The American Reasonable Use Doctrine, for example, recognizes a right of capture so long as the water is put to a reasonable use on the land from which it is withdrawn. The Correlative Rights doctrine is similar to the riparian rights doctrine for surface water, and requires a balancing of competing uses based on a series of equitable factors. Other states similarly modify the reasonable use principle borrowed from surface water law, and many western states apply the prior appropriation doctrine to groundwater as well as to surface water.

Second, applying separate legal regimes to surface and groundwater supplies, which again hales from a time when little was know about the source of groundwater supplies, makes little sense where those resources are interconnected. This problem has generated legal conflicts in which it is not entirely clear, for example, how a groundwater withdrawal relates to water rights from an adjacent surface water even when it is clear that the two are hydrologically connected, and that the groundwater pumping will reduce surface water supplies (or vice versa) (*City of Albuquerque v. Reynolds*). The result depends on variations in state law, including the precise relationship between a state's common law and statutory treatment of the two water sources, as well as variations in the relevant hydrogeology. As the modern science of hydrogeology evolves, and given our understanding of hydrological cycles that connect surface water, groundwater, and atmospheric water, it would make more sense to merge the doctrines and to treat water as a single resource.

Regulatory overlays: Cities planning projects to expand or to improve water supplies also face a maze of federal, state, and local regulatory requirements which are mixed blessings from the municipal perspective. Although those laws and regulations serve important roles in protecting water, aquatic, and other environmental resources, they also can complicate and delay water project planning and development. Only the most prevalent federal law examples are outlined below.

Any water project that involves a major federal action with potentially significant impacts on the human environment triggers compliance with the National Environmental Policy Act (NEPA). In many instances that requires an environmental impact statement to evaluate and disclose the environmental impacts of the proposed project, to identify feasible project alternatives and mitigation measures, and to solicit and integrate comments on those issues from other interested or affected agencies and the public (NEPA, § 102(2)(C)). A "major federal action" includes federal funding or other direct involvement in project planning, construction, or implementation, or a range of federal licenses or permits. Moreover, courts have interpreted the kinds of environmental impacts to be addressed quite broadly (Driesen and Adler, 2007, p. 330). However, NEPA is primarily an environmental full disclosure law that demands only analysis and public airing of impacts, alternatives, and mitigation measures. It does not dictate particular decisions or results once those procedural requirements for analysis, disclosure, and public discourse are met (*Robertson v. Methow Valley Citizen's Council*).

Other federal regulatory laws impose more direct substantive requirements on decisions regarding urban water supply. Projects that involve the discharge of fill material into the "waters of the United States", for example, require permits under

9 Legal Framework for the Urban Water Environment

Section 404 of the federal Clean Water Act (CWA). Permits may not be issued if they would have unacceptable adverse impacts on the aquatic environment, or if there are less damaging practicable alternatives that would meet project goals without causing the same level of impacts. Similarly, any project involving major federal actions that would jeopardize the continued existence of any species listed as threatened and endangered under the federal Endangered Species Act (ESA) requires consultation with the U.S. Fish and Wildlife Service (for inland species) or National Marine Fisheries Service (for marine mammal, anadromous fish, and marine species). Approval of such projects must prevent jeopardy to the species, and require implementation of all reasonable and prudent measures to protect the species.

9.2.2 Water Treatment and Distribution

Securing adequate water supplies and building the dams, well-fields, conveyances, and other projects necessary to transport water to cities is only the first step in providing water that is fit for various urban end uses. Cities also must treat water to appropriate standards for distribution and sale to public and other users, and administer a system for the sale of water in ways that balance affordability, equity and efficiency.

Water treatment: Public water supply systems are regulated by the federal Safe Drinking Water Act (SDWA), under which the U.S. Environmental Protection Agency (EPA) establishes drinking water standards known as maximum contaminant levels (MCLs) and maximum contaminant level goals (MCLGs) (U.S. EPA, 2003). MCLs are mandatory and enforceable standards that balance public health and safety against treatment costs, while MCLGs are stricter but unenforceable goals for optimum water safety independent of cost. The appropriate balance to strike between treatment costs and health benefits is the subject of considerable controversy between EPA and cities, for example, when EPA established new treatment standards for arsenic and for disinfection byproducts (U.S. EPA, 2001). Congress provided states with primary enforcement authority for public water systems, so long as EPA approves the state program as complying with federal law and regulations. Specific requirements can vary depending on the size of the public water supply system, the source of water used, and other factors.

The SDWA also authorizes cities to adopt and implement measures to protect "sole source aquifers". This allows communities to establish a "critical aquifer protection area" and to adopt comprehensive plans to protect a particularly important water supply at its source ("wellhead protection") rather than simply providing treatment after water is collected for use. This provision is notable for its focus on the relationship between land use and drinking water quality, as opposed to a purely treatment-based approach.

Distribution and sales: As a purveyor of water to industrial, commercial, and residential customers, cities also stand in a position similar to that of a public utility selling a good (water) and service (distribution and delivery) to those end users

(Tarlock, 2005). In addition to the full range of business transactions necessary to run this kind of complex operation, municipal water entities also may be subject to legislative or regulatory requirements governing rate structures and other aspects of serving residential, commercial, and industrial end users.

Regulations governing prices and sales implicate important policy considerations regarding the pricing of public urban water supplies. Because clean water is so fundamental to basic public health and welfare, cities have a responsibility to provide water at affordable prices to all urban residents, who range widely in income and prosperity. Excessively cheap water, however, which was the mainstay in U.S. cities for generations, can encourage profligate use. In the absence of economic incentives to use water efficiently, many consumers will waste it. Public water can be subsidized in a number of ways, meaning that even the full direct costs of providing public water are not passed on to consumers, leading to inefficient use. Nor do most cities include in consumer water rates the external costs of water supply, such as the environmental damage caused by dams or dewatered rivers and streams.

Cities can balance the competing goals of providing adequate, safe water to all urban users while encouraging more efficient use in several ways. They can adopt a "least cost first" approach to water supply and demand, under which cities may purchase water-saving devices for end users (such as water-saving toilets or shower heads) if it is cheaper to do so per unit of water supplied than to build and operate new water supplies. Water efficiency can also be encouraged through increasing block prices, in which all consumers purchase up to a fixed amount of water deemed appropriate for basic purposes (indoor drinking, bathing, etc.) at very affordable prices, but additional "blocks" are incrementally more expensive. Those who wish to use more water to irrigate lawns, fill swimming pools, or for other less essential uses must pay for that privilege. Cities can also adopt building code regulations designed to encourage or to require water efficiency in new homes and other structures. In the Energy Policy and Conservation Act of 1992, Congress adopted national efficiency standards for plumbing fixtures and fittings.

Growing cities face additional policy choices, again guided by various legal principles, regarding the extension of water distribution and supply systems to newly-developing areas (Tarlock, 2005). New developments and increasing sources of water demand can impose significant capital costs on already strained institutions, or stretch limited water supplies to the point where reliability for all consumers is jeopardized. Under one view, municipalities or the related legal institutions responsible for water supplies have an obligation to provide safe water supplies to all of the public, whether they live or operate businesses in existing or new areas. This might require cities to implement stricter conservation measures on existing users, to use pricing structures to allocate limited supplies efficiently, or to take extraordinary steps to import water from beyond their traditional supply areas.

Another perspective, however, is that sustainable communities should not grow beyond their natural resource limits with respect to water supply (and other resources) (Nolan, 2001). This could result in effective or overt growth limits, and legal conflicts between developers and local governments who either impose bans on new development or simply decline to expand service areas (effectively preventing new hookups) (Thompson, 2005; Tarlock and Van de Weterling, 2006). Courts have sustained such moratoria based on limited water supplies, but political pressure can cause cities to abandon those policies to accommodate growth (Arnold, 2005). A related issue is whether new developers should bear the full marginal costs of providing new water infrastructure and supplies, or whether those costs should be spread among existing and new users.

9.2.3 Wastewater, Stormwater, and Drainage

The "back end" of the urban water environment is governed by a separate (but in some respects overlapping) legal regime designed to protect aquatic and other resources from the impacts of water pollution. That body of law is driven primarily by the CWA and complementary state and local laws and regulations governing water pollution control. It is also affected by state common law rules and local ordinances and regulations regarding drainage, as well as state and local land use and planning laws and regulations that affect the location, timing, intensity, and other attributes of development in ways that can affect water quality and aquatic ecosystem health dramatically, as described in other chapters.

The Clean Water Act: Since 1972, when Congress adopted major amendments to the Federal Water Pollution Control Act (known more typically by its short title as the CWA), water pollution control has been governed under a system of cooperative federalism involving some combination of the federal, state, and local governments. In the CWA, Congress established minimum principles and requirements that apply nationwide, supported by the federal government's authority over navigable waters under the commerce clause of the U.S. Constitution (U.S. Const. Art I, Section 8, cl. 3). The U.S. Supreme Court identified control over navigability as part of the federal government's commerce clause powers in an early case (*Gibbons v. Ogden*). Those minimum federal requirements also preempt non-complying state and local laws under the Supremacy Clause of the Constitution (U.S. Const. Art. VI, Section 2). However, Congress left considerable latitude to state and local governments to implement the CWA according to local conditions and priorities, so long as those minimum requirements are met. (CWA §101).

The 1972 CWA transformed U.S. water pollution control law by flatly prohibiting discharges of pollutants without permits that assure application of minimum treatment requirements (end-of-pipe obligations of individual dischargers) and compliance with ambient water quality standards (standards that establish goals for whole bodies of water in the face of pollution from multiple sources) (Adler et al. 1993). Previously, discharges were allowed presumptively so long as no harm could be proven. Beyond that simple principle, however, the confines of the law become more complex. More precisely, the CWA prohibits any person from discharging any pollutant into any navigable water from any point source (CWA §301). Efforts by dischargers to narrow the scope of the statute have led to series of legal cases regarding the meaning of the terms "discharge of a pollutant", "point source", and "navigable waters" (*Miccosukee Tribe v. South Florida Water Management District*; *Concerned Area Residents for the Environment v. Southview Farm; Rapanos v. United States*). As explained further below, the battle over the latter term ("navigable waters") involves at a minimum the nature of water bodies that Congress intended to cover in the federal law, but also suggests questions about the kinds of waters over which the federal government has authority under the Commerce Clause.

Dischargers covered by the CWA must obtain a National Pollutant Discharge Elimination System (NPDES) permit from either EPA or from a state water quality agency with an EPA-approved program (CWA §402). Those permits impose effluent limitations that reflect the stricter of two kinds of controls. First, all dischargers must meet *technology-based standards* at least as stringent as those that can be met using what EPA has determined reflects the best technology available to treat that kind of waste from that category of facility. The best technology findings vary with the type of discharger, kinds of pollutants, and other factors, but aspire to a statutory goal of zero discharge, that is, the complete elimination of pollutant discharges into the navigable waters.

Second, dischargers must meet stricter "water quality-based" limits where necessary to assure attainment of in-stream water quality standards (CWA §303). Those standards consist of designated uses for all waters (such as contact recreation, drinking water, or protection of various kinds of fish and aquatic life), and water quality criteria deemed necessary and sufficient to protect those uses. Water quality standards are established by individual states, but require EPA review and approval, and EPA must adopt federal standards where the state standards are not sufficient.

Sewage treatment regulation: The CWA imposes several legal responsibilities on municipalities. Most notably, much of the fresh water that cities supply during the front end of the municipal water process becomes polluted sewage when it exits homes and businesses. Because nearly all U.S. communities do not separate "gray water"—mildly contaminated water from sinks—from the wastes generated in toilets and other more heavily contaminated waste, this generates millions of gallons a day of contaminated sewage that must be conveyed to treatment plants and properly treated and discharged (Tang et al., 2007). That requires the construction, operation and maintenance of both an extensive system of sewers and treatment facilities to meet applicable discharge requirements (Heany, 2007).

Through the CWA, beginning in 1948 and then expanded significantly in 1972, Congress provided large amounts of federal funding to assist states and cities in designing and building the sewerage infrastructure necessary to comply with those collection and treatment obligations (Adler, et al. 1993). Along with that funding, however, came legal responsibilities to adopt comprehensive wastewater treatment plans, in large part to prevent communities from growing without adequate sewerage capacity to treat the resulting wastewater to acceptable levels as prescribed in the CWA. The program was controversial in part because critics claimed that the resulting infusion of sewerage infrastructure did as much to fuel urban sprawl and related environmental problems as it did to control water pollution. In 1987, Congress replaced this system of outright federal grants with state revolving loan funds to continue to finance municipal sewerage systems.

Along with all other dischargers of pollutants into navigable waters, municipal sewage treatment plants must meet the minimum requirements of the CWA. They must obtain and comply with an NPDES permit for each treatment plant and outfall into navigable waters, and those permits must ensure compliance with both technology-based and water quality-based effluent limitations. The minimum technology-based requirement for municipal sewage is known as secondary treatment, the numeric limits for which are set forth in EPA regulations (40 C.F.R. Part 133). Strict limits to meet state water quality standards vary depending on the stringency of the applicable standards and the size of the receiving water relative to discharge volume (i.e., dilution or assimilative capacity), but often result in imposition of tertiary or other advanced treatment requirements.

Another way to comply with the CWA's presumptive discharge prohibition is not to discharge sewage waste into waters at all, but to reuse it beneficially to irrigate city parks and other green spaces, so that both the water and nutrients contained in properly treated sewage are not wasted (Furumai, 2007). Arid western cities increasingly are relying on beneficial reuse of sewage to augment their water supplies (or more precisely, to make better use of existing supplies by using available water multiple times). Doing so, however, still requires appropriate standards to ensure that contaminants do not interfere with beneficial reuse or create health or other hazards in the process. Beneficial reuse also involves environmental tradeoffs for western cities, because that effluent no longer returns to downstream waters which often have been significantly dewatered due to human diversions for agricultural, municipal, and other uses.

Municipalities also must comply with at least two other significant federal regulatory requirements in connection with sewage treatment plant operation. First, they must protect their own facilities (sewers and treatment plants) and receiving waters from toxics and other pollutants discharged into the sewer system at homes, commercial businesses, and industries within their service areas (CWA §307(b)). Most modern sewage treatment plants use a process of biological treatment in which live bacteria are critical to break down organic pollutants. Toxic pollutants can kill those "bugs" and thus impair treatment plant effectiveness. Some pollutants (such as heavy metals) are not effectively removed by sewage treatment processes designed to deal mainly with municipal sewage. Those substances can pass through the plant into receiving waters (and the plant operators are legally responsible for those releases), contaminate the sludge (also known as biosolids) produced as a byproduct of treatment, and even endanger the health or lives of treatment plant workers.

The CWA and EPA regulations require discharge limitations on these commercial and industrial "indirect dischargers" into sewers (to distinguish them from direct dischargers of the same pollutants into surface waters) through the "pretreatment" program. Indirect dischargers must pre-treat their sewage discharges so that the combination of pretreatment and treatment at the sewage treatment plant is at least as effective as requirements that would be imposed on the same facility discharging directly into a water body. Pretreatment also must be sufficient to prevent pass-through of toxic pollutants in unacceptable amounts, to prevent sludge contamination, and to protect sewage treatment plant workers and infrastructure. Most notably for municipal sewage treatment program managers, the primary responsibility for ensuring that indirect dischargers comply with pretreatment requirements falls on the municipal sewage treatment plant, rather than EPA or the state water quality agency. Thus, although there is considerable state and federal oversight, the municipality itself must develop, implement, and enforce a pretreatment program that complies with federal statutory and regulatory requirements. Somewhat unusually, then, municipal sewerage programs are both regulators and regulated entities in the CWA process.

Second, a related requirement is to comply with EPA's sludge contamination regulations issued under Section 405 of the CWA (40 C.F.R. Part 503). The solids generated during the sewage treatment process can be viewed either as a valuable resource for recycling and reuse ("bio-solids"), or as a problematic, high-volume waste ("sludge") which must be disposed of properly. For either purpose, EPA regulations ensure that excess contamination by heavy metals and toxic organic contaminants do not either increase waste disposal hazards or interfere with valuable reuses of that material as fertilizer. For example, by minimizing contaminants and maximizing the nutrient content of its sewage by-products, the City of Milwaukee markets a soil conditioner for home gardening called Milorganite.

Stormwater runoff ("nonpoint source") pollution control, and urban land use: Although flooding occurs as a natural process even in undeveloped environments, those areas consist largely of permeable surface in which precipitation can infiltrate into soils and discharge more slowly into aquatic systems. Natural features such as wetlands and vegetated riparian flood plains further buffer the impacts of storms on streams and other aquatic systems. Urbanization changes regional hydrology in ways that have significant implications for city environments, and that must be addressed by legal rules (Brown et al., 2005). The increase in impervious surface areas (roads, buildings, parking lots) and the decrease in wetlands, vegetated flood plain habitats, and other natural features dramatically increases peak flows following storm events, causing erosion and damage to local stream morphology and aquatic habitats (Bledsoe and Booth, Chapter 6 of this volume). The resulting runoff water can also be contaminated badly by a range of chemical and other pollutants in the urban environment (U.S. Department. of the Interior, 2002).

Common law principles govern drainage problems between private landowners, with the usual variations among jurisdictions (Sax et al., 2000, pp. 92–93). In fact, two competing traditional common law rules of drainage produced entirely opposite presumptions. In one, the so-called "civil law" approach, landowners are liable for damages from any diversion of surface water from its natural flow, thus limiting the ability of property owners to protect development on their own properties from flooding or other drainage problems without potentially compensating other affected owners. Some states modified this approach to allow small diversions for which there is no reasonable alternative and where damage to others is minor. The opposite "common enemy" doctrine allows landowners to alter surface water flow on their properties in any way, regardless of harm to others, so long as they do not harm others through negligence. Many states, however, now adopt a "reasonable use" approach in which a range of factors are considered in weighing the rights of landowners affected by drainage problems.

More wide-reaching drainage-related issues, such as runoff pollution in urban environments, however, must be addressed by public law. In addition to the point source control system discussed above, the CWA regulates pollution from diffuse sources, including runoff of contaminated precipitation from various land disturbances, artificial modification of stream channels and banks, and similar impairments of physical and biological characteristics of water bodies (Adler et al., 1993). Outside of the municipal context, those forms of pollution are not subject to the same kind of mandatory permitting, treatment, and control requirements as are point sources. Instead, the CWA requires states to adopt statewide nonpoint source pollution control plans with significant discretion to the states to determine how to address runoff pollution from agriculture, other land uses, and hydrological modifications (CWA §319).

Because of flooding and related hazards to property, human health and safety, municipalities cannot simply allow precipitation water to course through city streets. Cities construct and maintain networks of storm sewers to channel runoff water away from property (or roads) and into nearby rivers and streams during storm events. This water is contaminated by a range of pollutants from motor vehicles, building materials, lawn chemicals, organic matter, and other sources. Those discharges of pollutants into navigable waters from point sources (municipal storm sewers and outfalls), therefore, require NPDES permits like any other point source discharge. However, because of the multiple and diverse sources of pollutants in storm water discharges, the potentially massive volumes of storm water releases during heavy storm events, and the existence of large numbers of storm water outfalls in large municipal areas, use of concentrated treatment plant strategies similar to that used for municipal sewage is not viable. In essence, municipal storm water can be viewed as a hybrid because contaminants from nonpoint sources are channeled into storm sewers and then treated as a point source for legal purposes.

To address this hybrid nature of municipal storm water, in 1987 Congress enacted a separate storm water control provision within the NPDES program (CWA §402(p)). Because of delays in implementing that program, only recently have most cities been required to obtain discharge permits from EPA or delegated states (with deadlines for those permits dictated by city size). By regulation, EPA established requirements for cities to mitigate the impacts of storm water pollution through pollution prevention, land use and other control efforts (40 C.F.R. Part 122). As with the pretreatment program for sanitary sewer systems, this places cities in the position of implementing regulatory or quasi-regulatory programs to reduce storm water contamination from other property owners, while simultaneously operating the regulated discharge system and being responsible to reduce contamination from road systems and other public sources.

Many older cities initially built combined sewer systems, in which sanitary and storm sewers are combined (Chapter 1). Those cities face particularly serious water pollution episodes when storm intensity causes flows that exceed the capacity of the storm sewers, and when the combined flows from the storm and regular sewage exceed sewage treatment plant capacity, causing combined sewer overflows (CSOs). After years of controversy regarding the extreme cost of addressing this set of issues, EPA adopted a CSO permitting strategy based on a combination of sewer system retrofitting and maintenance, treatment, and storm water management strategies as appropriate to individual cities (U.S. EPA, 1994).

Water pollution caused by land development is also addressed to some degree by a separate CWA provision governing discharges of dredge and fill material into waters covered by the Act, including most notably wetlands (CWA §404). Under this statutory program, such discharges require permits from the U.S. Army Corps of Engineers, which may be issued only after a finding that there is no less damaging practicable alternative, no unacceptable adverse impacts to the aquatic environment, and other requirements (40 C.F.R. Part 230; 33 C.F.R. Part 330). Wetlands and undeveloped floodplains are critical resources in urban and other environments, because they can help to buffer the hydrologic impacts of storms, filter pollutants, and provide open space and important habitats (National Research Council, 1995). Thus, regulatory and other strategies to preserve those areas can help cities to meet storm water management and other regulatory requirements discussed above.

At the same time, many wetlands are on private property, and denying applications to fill (and therefore to develop) those areas often generate claims by those landowners that the government has taken their property without due process or without just compensation (Sax, 1993). Although those claims are likely to fail in most circumstances (as discussed below), they generate pressure to minimize impacts of the program on private property, for example by narrowing the scope of waters covered by the CWA Section 404 program (*Rapanos v. United States*).

While water pollution and other forms of aquatic ecosystem impairment caused by land use and development are governed by all of these federal laws and regulations, state and local planning, zoning, and other land use controls also play significant roles in minimizing the effects of development on urban aquatic environments (Arnold, 2005; Nolan, 2001). For example, cities may impose setback requirements that prohibit or limit development within prescribed distances from streams and other aquatic resources. They can zone sensitive areas for lower densities, mandate open space and riparian area protection within large developments, or prohibit developments from exceeding a specific percentage of impervious surface.

Just as the federal government must take care to avoid unconstitutional takings of property in administration of the CWA and other federal statutes, states and cities must negotiate the appropriate balance between regulations that serve legitimate public purposes and those that arguably result in takings of private property without due process or just compensation. Although the law of "takings" is complex and sometimes confused, however, municipal land use programs and regulations will probably pass constitutional scrutiny if they are reasonably proportionate to the public purpose to be protected, do not result in a complete diminution of private property value, and are adopted and implemented with notice and opportunity for affected parties to participate (Martinez, 2006). Under these principles, the U.S. Supreme Court has invalidated some local requirements designed to protect water resources where they were not shown to be proportionate to the goals to be served, but has also upheld broad government regulations designed to protect water quality and aquatic ecosystem health (*Nollan v. California Coastal Comm'n*; *Palazzolo v. Rhode Island*; *Tahoe-Sierra Preservation Council, Inc. v. Tahoe Regional Planning Commission*).

Some cities and regions are using comprehensive plans and taking other more comprehensive steps to protect their aquatic resources using state and local legal authority, alone or in combination with relevant federal programs (Arnold, 2005). As discussed in Chapter 11, a large number of collaborative watershed management programs around the country are considering more comprehensive approaches to protect urban ecosystems, especially in rapidly developing areas.

Groundwater pollution: Municipal programs and activities can also affect ground water quality in ways that are addressed both by the common law of nuisance and by a wide range of federal statutes and regulations, in addition to the SDWA wellhead protection programs discussed above. Cities can use some of these tools to protect the quality of their ground water resources (quantity issues are addressed above) and the health and welfare of their citizens, but they are also subject to those requirements with respect to municipal activities such as solid waste disposal. Groundwater pollution can pose particular challenges because once an aquifer is contaminated, it is not likely to have the same flushing capacity as a river or other surface water. If polluted seriously, the resource might be lost for human consumption without expensive and lengthy remediation.

Urban areas generate tremendous volumes of solid waste. This waste burden can be reduced through aggressive recycling and reuse efforts, but significant amounts of waste are unavoidable. When disposed of improperly, solid waste disposal can cause serious groundwater and surface water pollution as well as public health risks from disease vectors and other problems. Under the federal Resource Conservation and Recovery Act (RCRA, an amendment to the SWDA), individual states operate non-hazardous solid waste programs according to general EPA regulations and standards, while EPA promulgates more specific waste treatment, transportation and disposal requirements for more dangerous hazardous wastes. Municipal groundwater and other resources can be protected through compliance with landfill siting, permitting, design and operation standards adopted under RCRA, and cities can affirmatively use this statute, either directly or through EPA and a state environmental agency, to require other parties to protect groundwater supplies or to clean up contaminated groundwater.

The federal "Superfund" statute (more formally the Comprehensive Environmental Response, Compensation and Liability Act, or CERCLA) also serves as a two-edged sword for municipalities. Under Superfund, several broad categories of "responsible parties" (current property owners, past owners at the time of disposal, some kinds of transporters, and persons who arrange for hazardous substance disposal) are liable for releases of hazardous substances into the environment. Superfund is a particularly potent legal tool because any responsible party can be held liable for releases regardless of fault (e.g., merely by virtue of current property ownership even if they were not responsible for the wastes), and because individual parties can be held "jointly and severally liable" for releases, meaning that one party can bear full liability for an entire cleanup even if many parties contributed to the problem. These seemingly unfair provisions, however, are tempered by several somewhat complicated exceptions, and the fact that private landowners can sue other potentially responsible parties to clean up contaminated sites, or to seek contribution from other responsible parties to offset joint and several liability.

Municipalities can be responsible parties under Superfund, although they enjoy some narrow exceptions in addition to those available to private landowners. For example, cities are not liable for hazardous substance releases on properties acquired through tax sales, unless the municipality is responsible or partially responsible for the release. Thus, municipal governments must take the same care with hazardous substance disposal and management as do other parties, in part to avoid potentially significant Superfund liability and, more importantly, to protect their groundwater and other resources. However, like other parties, municipalities can also use Superfund affirmatively to require other property owners to clean up contaminated sites, or to bear their fair chare of the costs of doing so.

9.2.4 Benefits of Urban Aquatic Ecosystems

Thus far, we have discussed laws and regulations designed to address *problems* in the urban water environment. Although a full discussion of legal principles governing the use of public land as an amenity is beyond the scope of this text, it is a mistake to ignore the fact that urban aquatic environments can and should also be viewed as tremendous resources to enhance quality of life in a community. Many cities are taking steps to restore urban rivers and streams, to promote recreational and environmental "greenways" using riparian corridors as assets rather than liabilities, and to preserve and protect green spaces in and around riparian zones to enhance and protect water resources and habitats for fish, wildlife, and other ecological communities (Chapter 7). Those efforts can benefit landowners by increasing property values, and aquatic ecosystem restoration can be part of urban renewal efforts in previously undesirable areas.

9.3 Legal Barriers to a Sustainable Urban Water Environment

This brief (and only partially complete) survey of the law relevant to the urban water environment suggests that a very wide range of laws, regulations, judicial decisions and other sources of law can influence the ways in which cities manage their water resources and aquatic environments. Some of the law provides useful tools that cities can use to achieve the objectives of providing safe and sufficient water supplies, and protecting urban waterways and aquatic ecosystems for the benefit of their citizens. While serving those needs, the same set of legal doctrines present a maze of compliance challenges for city water managers, sewerage officials, planners, zoning officials, and others.

9.3.1 Common Law Versus Statutory Approaches

One famous, early twentieth century case in the U.S. Supreme Court (*Missouri v. Illinois*) illustrates a number of these issues simultaneously: the interactions between the "front end" and the "back end" of the urban water environment, resulting conflicts among jurisdictions, and the need for innovative, holistic legal approaches to those problems. At the end of the nineteenth century, the City of Chicago realized that its water supply was being polluted by its own sewage discharges to Lake Michigan, which were close to the city's water intake structure, thus causing serious public health epidemics. To solve this problem, Chicago literally reversed the course of the Chicago River by means of an artificial channel that diverted the city's sewage into the Desplaines River, which empties into the Illinois River and then the Mississippi River upstream of St. Louis. When St. Louis experienced an increase in the incidence of typhoid fever, it alleged that Chicago had eliminated a public nuisance that affected Chicago's citizens at the expense of other communities downstream. The Supreme Court rejected this famous public nuisance lawsuit in the face of conflicting scientific evidence about the presence and residence time of the typhoid bacillus in the river, other possible sources of contamination from cities much closer to St. Louis, and ambiguous epidemiological data.

The *Missouri v. Illinois* saga illustrates the difficulties that lawyers and judges face when dealing with new, rapidly evolving and conflicting science and technology. After all, the case was brought only a matter of decades after Louis Pasteur demonstrated the role of bacteria in human disease. At the same time, however, it highlighted the need for legal solutions designed to prevent unsound urban water management practices rather than relying on the uncertainties of proof in isolated common law nuisance lawsuits. Although Justice Oliver Wendell Holmes in *Missouri v. Illinois* suggested that might come in the form of filters on the St. Louis public water supply to protect against pollution from multiple sources, in the CWA Congress ultimately intervened to require all cities to treat sewage to appropriate standards at the source, so that public health impacts would not turn on the location and fate of any particular discharge.

Although discrete legal challenges are pervasive, as they are in most aspects of municipal affairs, this survey suggests that two significant kinds of legal barriers may impede efforts to attain a more sustainable urban water environment. Unfortunately, neither will be easy to "solve" within the confines of the current legal regime governing urban water issues.

9.3.2 Fragmentation in Water Law

The first major barrier is fragmentation. As described above, urban water management is governed to varying degrees by common law, statutory law, and administrative regulation; and by federal, state, and local law. In part to summarize some of the contents of this chapter, Table 9.1 illustrates in a highly simplified fashion the

Table 9.1 The legal landscape for the urban water environment (simplified)

	Supply	Storage and Conveyance (dams and distribution systems)	Water treatment	Distribution and sale	Waste treatment (sewage and stormwater); drainage control
Goals	Water rights; water source protection	Adequate supply; minimize environmental impacts	Safe community drinking water	Affordability, efficiency, equity	Water quality, aquatic ecosystem protection, flood control
Problems	Competing users; growing demand; pollution and land use	Environmental conflicts	Contaminants	Increasing costs of acquisition, storage, treatment, distribution	Pollution sources; precipitation intensity; urbanization
Legal tools	State water law (riparian rights, prior appropriation, groundwater); planning and watershed protection; tort law; pollution laws (RCRA, CERCLA, CWA, SDWA); local building codes, plumbing and efficiency standards, water pricing	Federal, state and local environmental laws and regulations (NEPA, ESA, CWA)	Source protection (planning and zoning, land use, pollution prevention); SDWA standards	Public utility and rate regulation, pricing methods	Water quality laws (CWA and state); wetland and floodplain protection; land use planning and zoning; flood control laws; drainage law

complex relationship between legal sources and water sources and uses in the United States. Taken together, the legal regime distinguishes in various ways between water quantity and water quality; between water supply, distribution, treatment and discharge; and between surface water and groundwater (from both supply and pollution perspectives). Likewise, distinct sets of laws and regulations designed to provide water to direct land use through planning and zoning and to minimize water quality impacts from urbanization serve different functions which are not always well coordinated (Arnold, 2005). Urban water managers and other officials must navigate a maze of different laws, regulations, procedures, and agencies to address various parts of their missions, and those different sources of authority are not necessarily consistent. No wonder the urban water environment itself is fragmented and often poorly coordinated, as discussed in Chapters 11 and 12.

It is highly unrealistic to think that this problem of fragmentation will be solved with any magic bullet, that is, with some kind of "superlaw" governing all aspects of the urban water environment. Indeed, there are some very good reasons for dealing with various aspects of water law and management in different ways. For example, local governments may be best suited to making land use decisions that affect water quality and quantity based on a range of local conditions, preferences, and other factors. However, establishing minimum national requirements for water quality (for example, through the CWA), ensures that all citizens receive certain basic protections against water pollution, and prevents some communities from simply exporting their wastes to others downstream.

It is realistic, and probably essential if we are to achieve greater sustainability in urban water resources and management, to take incremental steps to better coordinate various components of the laws and regulations that apply to these issues, even at the cost of some short-term disruptions and conflicts. For example, some administrative disruption would occur if states merged their systems for allocating surface water and groundwater rights, and some water rights would likely be affected during the transition. In the long run, however, addressing all water sources within hydrologically-connected basins and aquifers would make more sense than the current separate regimes.[3] Likewise, a formal merging of water quality and water quantity law—the absence of which Supreme Court Justice Sandra Day O'Connor referred to as "an artificial distinction"(PUD No. 1 of *Jefferson County v. Washington Department Ecology*)—would likely create a significant number of difficult transitional issues. It would be easier for municipalities and others to engage in comprehensive, integrated water resource planning and management, if the two systems were merged or at least better integrated. Finally, the linkages between water, land use, and growth are far too profound to continue to deal with them through entirely disconnected legal regimes. Sustainable urban water use and healthy urban aquatic environments require a more holistic consideration of the relationships between land use, water use and disposal, and aquatic ecosystem health.

---

[3] Not all aquifers are geographically coextensive with surface water basins, but in most cases hydrological connections predominate over discontinuities.

9.3.3 Public Versus Private Rights

The second major legal barrier to a more sustainable urban water environment, and one that may be even more challenging to address, is the traditional conflict between public welfare and private property rights, especially at the land–water interface ("the water's edge") (Adler, 2005). As noted above, the Supreme Court has accepted a range of legitimate government regulatory and other programs designed to protect water and aquatic resources in the face of challenges that unlawful takings have occurred. (*Palazzolo v. Rhode Island*; *Tahoe-Sierra Preservation Council, Inc. v. Tahoe Regional Planning Commission*). In others, it has rejected those controls as insufficiently proportionate to the public objectives sought (*Nollan v. California Coastal Commission*), or has ruled that the regulation would constitute an unlawful taking without just compensation (*Lucas v. South Carolina Coastal Council*). Even if most formal takings challenges to governmental regulation fail, the prospect of those lawsuits and the very effective political advocacy to protect private property rights against perceived or real governmental abuse can have a chilling effect on public measures to ensure sustainable water resources and healthy aquatic ecosystems in urban areas and elsewhere.

9.3.4 The Public Trust Doctrine

The common law public trust doctrine, however, illustrates that private property rights are not always paramount, and that single legal doctrines cannot necessarily be viewed in isolation (Adler, 2005). The public trust doctrine has ancient origins in Roman law, was adopted in many European countries during the middle ages and later was embedded in modified form in English common law. As adopted in England, the trust doctrine entailed ownership by the sovereign on behalf of the people in common, and restricted the ability of the Crown to alienate those trust resources in favor of private individuals. As such, the trust concept imposed on the government a duty to manage and protect those resources for the common purposes of commerce, navigation, and fishing.

In the United States, the colonies and then the states inherited both public trust ownership and responsibility. In a seminal public trust doctrine case, the U.S. Supreme Court held that the Illinois legislature and other government trustees have only limited discretion to dispose of public trust resources, and may not make a disposition that is fundamentally inconsistent with the purposes of the trust (*Illinois Central R.R. Co. v. Illinois*). However, the doctrine was limited to the common law triad of commerce, navigation, and fisheries. Born in a time when ecological awareness and understanding was virtually nonexistent, and when population pressures had not yet generated the magnitude of environmental harm that is occurring and understood today, ecological values were not included in the original doctrine.

Beginning in the early 1970s, in parallel with the Nation's growing interest in and understanding of water pollution, loss of species and habitat, and other forms of environmental harm, the public trust doctrine was revitalized in an effort to provide a common law basis for broader protection. In a now-famous article, Professor Joseph

Sax argued for the renovation and later expansion of the doctrine as a means of providing a legal right, vested in the public, and enforceable against the government, to vindicate commonly-held expectations in environmental values (Sax, 1970). The courts soon took up the banner, most notably in the famous "Mono Lake" decision, and expanded the doctrine in terms of both geographic reach and the scope of common values to be protected (*National Audubon Society v. Superior Court*; *Marks v. Whitney*; *Just v. Marinette County*).

While some courts have viewed the trust duty expansively, one key critic argues that on a national scale, the doctrine remains limited to issues of public access and navigability, rather than broader issues of environmental protection (Lazarus, 1986). Moreover, as with the statutory applications and solutions discussed above, those proposed expansions will be met with heavy opposition from those who argue that expanding the doctrine beyond its traditional reach violates private property rights and other constitutional limitations (Huffman, 1987). These and more pragmatic barriers to case-by-case litigation brought on behalf of public resources historically has rendered common law approaches to environmental protection potentially effective for specific cases, but less so on a national scale.

9.3.5 Beyond the Public Trust?

More fundamentally, despite its recent expansion in some jurisdictions to address environmental as well as commercial resources, the public trust doctrine remains rooted in anthropocentric notions of property law in which the trust assets are held by the government for the common benefit of human users. This foundation, along with the legal nature of the trust analogy itself, poses a serious impediment to the doctrine's effectiveness as a means of environmental protection.

Just as we distinguish artificially between water quality and water quantity, in some ways we draw an artificial boundary between private property rights on riparian and other waterside lands, and public rights in the water itself and in the beds and banks of navigable water bodies. In reality, also as recognized by the Supreme Court, the dividing line between land and water is often far from clear (*United States v. Riverside Bayview Homes*). A more fine-tuned concept of public and private property rights (and responsibilities) in these transition zones, in which the government's ability to protect public aquatic resources increases with the aquatic nature of public resources, would more realistically account for the shifting nature and benefits of those resources. Likewise, it would allow federal, state, and local governments to implement legitimate programs to manage and protect public water and aquatic resources, and to do a better job of promoting sustainability in the urban water environment.

References

Adler, R.W. (2007). *Restoring Colorado River Ecosystems: A Troubled Sense of Immensity*, Washington, D.C.: Island Press.

Adler, R.W. (2005). The law at the water's edge: Limits to "ownership" of aquatic ecosystems. In Arnold, C.A., ed. *Wet Growth: Should Water Law Control Land Use?* Washington, D.C.: Environmental Law Institute.

Adler, R.W., Landman, J.C., and Cameron, D.M. (1993). *The Clean Water Act 20 Years Later*, Washington, D.C.: Island Press.

Arnold, C.A. (2005). Introduction: Integrating water controls and land use controls: New ideas and old obstacles. In Arnold, C.A., ed. *Wet Growth: Should Water Law Control Land Use?* Washington, D.C.: Environmental Law Institute.

Brown, L.R., Gray, R.H., Hughes, R.M., and Meador, M.R. (2005). Introduction to effects of urbanization on stream ecosystems, 47 *American Fisheries Society Symposium* 1.

American Law Institute. (1979). *Restatement (Second) of Torts*.

American Society of Civil Engineers, Water Laws Committee, Water Resources Planning and Management Division. (1997). *Regulated Riparian Model Water Code*.

City of Albuquerque v. Reynolds, 379 P.2d 73 (N.M. 1962).

Clean Water Act, 33 U.S.C. §1251 et seq.

Comprehensive Environmental Response, Compensation and Liability Act, 42 U.S.C. §9601 et seq.

Concerned are Residents for the Environment v. Southview Farm, 34 F.3d 114 (2d Cir. 1994).

Connecticut v. Massachusetts, 282 U.S. 660 (1931).

Driesen, D.M., and Adler, R.W. (2007). *Environmental Law: A Conceptual and Pragmatic Approach*, New York: Aspen Publishers.

Endangered Species Act, 16 U.S.C. §1531 et seq.

Energy Policy and Conservation Act of 1992. Public Law 102–486, October 24, 1992. 102nd Congress. Washington, D.C.

Furumai, H. (2007). Reclaimed stormwater and wastewater and factors affecting their reuse. In Novotny, V. and Brown, P. eds. *Cities of the Future: Towards Integrated Sustainable Water and Landscape Management*, London: IWA Publishing.

Gibbons v. Ogden, 22 U.S. (9 Wheat) 1 (1824).

Glennon, R.L. (2002). *Water Follies: Groundwater Pumping and the Fate of America's Fresh Waters*, Washington, D.C.: Island Press.

Hadad, B.M. (2000). *Rivers of Gold: Designing markets to Allocate Water in California*, Washington, D.C.: Island Press.

Heany, J.P. (2007). Centralized and decentralized urban water, wastewater & storm water systems. In Novotny, V. and Brown, P. eds. *Cities of the Future: Towards Integrated Sustainable Water and Landscape Management*, London: IWA Publishing.

Hudson River Fisherman's Association v. Williams, 139 A.D.2d 234 (N.Y. S.Ct App. Div. 1988).

Huffman, J.L. (1987). Avoiding the takings clause through the Myth of public rights, 3 *Journal of Land Use and Environmental Law* 171.

Illinois Central RR. Co, v. Illinois, 146 U.S. 387 (1892).

Just v. Marinette County, 201 N.W.2d 761 (Wis. 1972).

Lazarus, R.J. (1986). Changing Conceptions of Property and Sovereignty in Natural Resources: Questioning the Public Trust Doctrine, 71 *Iowa Law Review* 631.

Lucas v. South Carolina Coastal Council, 505 U.S. 1003 (1992).

Marks v. Whitney, 491 P.2d 374 (Cal. 1971).

Martinez, J. (2006). *Government Takings*, St. Paul, Minn: Thompson-West Publishing.

Meyers, C. and Noble, R. (1967). The Colorado River: The Treaty with Mexico, 19 *Stanford Law Review* 367.

Missouri v. Illinois, 200 U.S. 496 (1906).

National Audubon Society v. Superior Court, 658 P.2d 709 (Cal. 1983).

National Environmental Policy Act, 42 U.S.C. §4321 et seq.

National Research Council. (1995). *Wetlands: Characteristics and Boundaries*, Washington, D.C.: National Academy Press.

Nolan, J.R. (2001). *Well Grounded: Using Local Land Use Authority to Achieve Smart Growth*, Washington, D.C.: Environmental Law Institute.

Nollan v. California Coastal Comm'n, 483 U.S. 825 (1987).
Palazzolo v. Rhode Island, 553 U.S. 606 (2001).
Phillips Petroleum Co. v. Mississippi, 484 U.S. 469, 475–476 (1988).
PUD No. 1 of Jefferson County v. Washington Department Ecology, 511 U.S. 700 (1994).
Rapanos v. United States, 126 S.Ct. 2208 (2006).
Reisner, M. (1986). *Cadillac Desert: The American West and its Disappearing Water*, 2d ed., New York: Viking.
Resource Conservation and Recovery Act, 42 U.S.C. §6901 et seq.
Robertson v. Methow Valley Citizen's Council, 490 U.S. 332 (1989).
Safe Drinking Water Act, 42 U.S.C. §300f et seq.
Sax, J.L., Thompson, B.H., Jr., Leshy, J.D., and Abrams, R.H. (2000). *Legal Control of Water Resources*, 3d ed. St. Paul, Minn.: West Group.
Sax, J.L. (1993). Property Rights and the Economy of Nature: Understanding *Lucas v. South Carolina Conservation Council*, 45 *Stanford Law Review* 1433.
Sax, J.L. (1970). The public trust doctrine in natural resources law: Effective judicial intervention, 68 *Michigan Law Review* 471.
Snow v. Parsons, 28 Vt. 459 (Vermont, 1856).
South Florida Water Management District v. Miccosukee Tribe, 541 U.S. 95 (2005).
Tahoe-Sierra Preservation Council, Inc. v. Tahoe Regional Planning Comm'n, 122 S. Ct. 1465 (2002).
Tang, S.L., Yue, D.P.T., and Ku, D.C.C. (2007). *Engineering and Costs of Dual Water Supply Systems*, London: IWA Publishing.
Tarlock, A.D. (2005). We are all water lawyers now: Water law's potential but limited impact on urban growth management. In Arnold, C.A., ed. *Wet Growth: Should Water Law Control Land Use*? Washington, D.C.: Environmental Law Institute.
Tarlock, A.D. and Van de Weterling, S.B. (2006). Western growth and sustainable water use: If there are no "natural limits," should we worry about water supplies? 27 *Public Land and Resources Law Review* 33.
Thompson, B.H., Jr. (2005). Water management and land use planning: Is it time for close coordination? In Arnold, C.A., ed. *Wet Growth: Should Water Law Control Land Use*? Washington, D.C.: Environmental Law Institute.
United States v. Riverside Bayview Homes, 474 U.S. 121 (1985).
U.S. Department of the Interior, U.S. Geological Survey (2002). Fact Sheet FS-042-02, Effects of Urbanization on Stream Ecosystems.
U.S. Environmental Protection Agency, Office of Water (2001). National Primary Drinking Water Standards, EPA 816-F-03-016, available at http://www.epa.gov/safewater.
U.S. Environmental Protection Agency (2001). National Primary Drinking Water Regulations; Arsenic and Clarifications to Compliance and New Source Monitoring, 66 Fed. Reg. 6975 et seq. (January 22, 2001) (codified at 40 C.F.R. Parts 9, 141, and 142).
U.S. Environmental Protection Agency (1994). Combined Sewer Overflow (CSO) Control Policy, EPA 830Z94001.
Wyoming v. Colorado, 259 U.S. 419 (1922).

Chapter 10
Institutions Affecting the Urban Water Environment

Robert W. Adler

10.1 Introduction

The complexity of the legal framework discussed in Chapter 9 is mirrored by an equally intricate mosaic of legal and political institutions that govern, manage, and otherwise affect the urban water environment. Those include legislative bodies that pass the statutes, ordinances, and other enactments that affect urban water use and management; courts that interpret and enforce those legal rules and obligations; administrative agencies that implement and often further interpret applicable legislation; and governmental, quasi-governmental, and private entities that provide water and related services to end users of water and other beneficiaries of aquatic resources. As is true for the relevant sources of law, institutions affecting urban water resources operate at the local, state, regional, federal, and sometimes international levels. Water institutions can also consist of collaborative mixtures of organizations of various kinds and at various levels of government, working together to address problems of mutual interest, as well as a wide range of private entities not specifically discussed in this chapter.

Although it is not possible to identify every kind of legal and political institution affecting the urban water environment in a single chapter, this chapter will survey the most common and most significant institutions within each of the above categories. It also comments on the manner in which the sheer number and fragmentation of those institutions, and their frequently overlapping or even inconsistent jurisdictions and principles and methods of governance, can impede efforts to promote more sustainable and efficient uses of water in urban areas.

As was true for the legal regime, discussion of the wide array of institutions affecting urban water could be organized in several ways. For some purposes, it is useful to distinguish between institutions that serve largely regulatory functions and those that provide more direct customer services, such as water supply and sewerage. Institutions serving largely regulatory functions operate at all levels of government (local, state, regional, national, and international). Institutions with

---

R.W. Adler (✉)
University of Utah, S.J. Quinney College of Law, Salt Lake City, Utah
e-mail: adlerr@law.utah.edu

largely service functions more typically operate at the local level. However, as with any organizational structure for complex and variable systems, this distinction between regulatory and service functions can be misleading and overly simplistic, because some entities serve both types of functions, and some kinds of functions are difficult to label one way or the other. Water institutions also serve all four of the major functions discussed in Chapter 9, that is, water supply, water treatment and distribution, wastewater and drainage (or storm water) management and treatment, and recreational and environmental services and amenities.

Because institutions often function and can best be understood in terms of their sources of authority and constituent bases, then, for purposes of this chapter it makes most sense to describe water institutions by level of government. The ensuing sections, therefore, discuss the various kinds of institutions affecting the urban water environment at the federal, state and local, and regional and international levels (including, to some degree, multi-agency and multi-interest watershed or basin management institutions, although those kinds of efforts are also addressed in Chapters 11 and 12).

10.2 Federal Institutions and Agencies

The federal government exercises pervasive but often indirect authority and other influence over urban water issues. In some cases that entails federal regulation of water supply and management issues, to protect public health and safety, environmental integrity, and other values that Congress has deemed to implicate sufficiently important national interests to justify federal intervention. In other cases, the federal government influences urban water issues and environments more through capital investments and related physical infrastructure, and any accompanying conditions of those expenditures and services. Although virtually every federal agency or other institution can affect the urban water environment in one way or another, and one source identified 37 agencies concerned with water issues in some way (Fiero, 2007), and other sources organize federal and other water agency responsibilities in different ways (Cech, 2003; Sax et al., 2000), the following discussion tries to identify and describe briefly federal agencies that address water issues as one or more of their *primary* missions.

10.2.1 Environmental Protection Agency (U.S. Environmental Protection Agency Website)

The U.S. Environmental Protection Agency (EPA) administers and enforces most of the major federal environmental pollution statutes.[1] The most important of

---

[1] Many federal environmental laws, including those implemented by agencies other than EPA, involve environmental pollution. However, the major statutes implemented and enforced by EPA are those designed specifically to prevent or regulate the release of pollutants into various environmental media, such as air, land, surface water and ground water.

those laws for purposes of urban water systems are the Clean Water Act (CWA) and the Safe Drinking Water Act (SDWA), both described more extensively in Chapter 9. Briefly, the CWA is designed to protect the chemical, physical, and biological integrity of the nation's surface waters, and to regulate the discharge of pollutants into those waters from a wide range of sources and activities. The SDWA governs the quality of water provided to consumers by public drinking water suppliers, and various efforts to protect their water sources. However, other statutes implemented and enforced by EPA also help to protect water sources. Those laws include the Resource Conservation and Recovery Act (governing hazardous and solid waste management and disposal), Comprehensive Environmental Response, Compensation, and Liability Act (CERCLA, more commonly known as "Superfund") (facilitating and regulating the cleanup of hazardous substances, and allocating liability and responsibility for those clean-ups), the Toxic Substances Control Act (TSCA) (governing the manufacture and use of toxic chemicals), the Federal Insecticide, Fungicide, and Rodenticide Act (FIFRA) (governing the manufacture and use of those materials), and even the Clean Air Act (CAA) (governing the regulation and control of air pollutants, including those that affect water bodies through atmospheric deposition).

Institutionally, EPA is principally in charge of implementation and enforcement of only some of those laws, including most notably TSCA and FIFRA, because of their national scope and applicability. (It would be difficult and confusing, e.g., for manufacture and use of the same toxic chemical to be governed by 50 different states.) Congress directed EPA to manage many of the environmental pollution laws, however, under a system of *cooperative federalism* in which EPA oversees statutory implementation in conjunction with various state environmental agencies. In the CWA, for example, states have the primary responsibility to adopt ambient water quality standards (WQS) for all surface waters, but EPA reviews and approves those standards, and adopts federal WQS if a state fails to do so or issues standards that EPA deems insufficient. Federal standards are rare, because EPA typically works with states more cooperatively to revise any deficient standards until they can be approved. Conversely, but with the same general result, EPA is charged statutorily with implementation and enforcement of the National Pollutant Discharge Elimination System (NPDES), the program under which all point source discharges to the "waters of the United States" must obtain permits limiting their releases in various ways. However, states have the opportunity to implement their own permitting programs instead, with EPA oversight and review authority, and most states have chosen to do so. Similar arrangements apply under the SDWA, RCRA and CAA. (A more detailed description of the cooperative federalism approach to water pollution control is included later in the chapter.)

Cooperative federalism suggests both benefits and challenges for urban water managers. The system is designed to ensure that minimum national environmental standards and procedures are observed, while allowing individual states the flexibility to implement and attain those requirements and standards as appropriate to more localized and regional conditions and circumstances. This allows those involved in urban water policy to work with state officials who are more likely to be familiar with local conditions and problems than federal officials might be. However, it also means that both federal and state agencies may be involved in many issues and decisions, which potentially complicates a range of urban water decisions and activities (e.g., CWA permitting of a new sewage treatment plant).

Municipalities and other local water institutions are both beneficiaries of EPA's activities and regulated parties under many of the laws and regulations EPA oversees or implements. Effective pollution control and other environmental programs can safeguard urban water supplies from both a quality and a quantity perspective. They also help to restore and protect healthy aquatic ecosystems, which provide critical ecosystem services and valuable aesthetic and recreational amenities for urban areas. Of course, municipal water entities are also subject to those same laws and regulations when they build and operate urban water infrastructure, such as water storage, treatment and distribution systems, and sanitary and storm sewers and treatment facilities.

10.2.2 Army Corps of Engineers (ACE Website)

The U.S. Army Corps of Engineers (ACE), a branch of the Department of Defense, also has responsibilities for implementing some components of the CWA, as well as other functions that affect urban water issues and policy. Although it may seem counter-intuitive for a component of the military to be involved in water policy, this involvement stems from nineteenth century legislation—particularly the Rivers and Harbors Act—designed to promote national defense as well as interstate and international commerce. ACE was authorized to take steps to ensure that navigable waterways remained unimpaired by physical structures and other alterations built by other parties, and were improved for those purposes with canals, navigation channels, and other structural changes. Under section 10 of the Rivers and Harbors Act (often known as the "Refuse Act"), ACE continues to review and issue permits for projects that might obstruct those waters.

In a series of decisions issued in the 1960s, before Congress enacted any of the modern federal pollution control laws, the U.S. Supreme Court interpreted the Rivers and Harbors Act broadly to apply to discharges of pollutants into *navigable waters* (Rodgers, 1971). ACE adopted a permitting program in response to those cases. When Congress amended the Federal Water Pollution Control Act in 1972 (now commonly known as the CWA), it retained ACE's permitting authority with respect to the discharge of dredge and fill material into the "waters of the United States" embodied in section 404 of the CWA, while transferring to EPA the authority to issue permits for other kinds of pollutants (the NPDES program described above). However, EPA also retains an important role in the section 404 permitting process, because EPA issues guidelines that ACE must follow to protect aquatic ecosystems. EPA writes regulations governing ACE permits issued under this section, and has the authority to veto permits it believes will have an unacceptable adverse effect on the aquatic environment.

ACE has also built, and continues to exercise control over, a large number of water storage, flood control, channelization, and other projects designed, among other purposes, to support urban water needs around the country. Although the primary purpose of ACE water projects is to promote and protect navigation on the nation's waters, they often were designed to provide, and justified economically on the basis of, multiple purposes. Controversy remains, however, about whether many of those projects did more harm than good to aquatic ecosystems and the resources they provide, and whether ACE levees and other flood control programs are properly designed and maintained to serve their intended functions. Failure of ACE levees, for example, is cited as one major contributing cause of the Hurricane Katrina disaster (van Heerden et al., 2007). In addition, especially along major navigational waterways such as the Missouri and Mississippi Rivers, ACE and other agencies often encounter conflicts about how much water should remain in the river to support navigation, and how much may be diverted for irrigation, municipal, and other uses.

10.2.3 Fish and Wildlife Service (U.S. Fish and Wildlife Service Website)

The U.S. Fish and Wildlife Service (FWS) manages a large series of fish and wildlife refuges around the United States, some of which are in or near urban areas. Because one principal focus of the refuge system has been to provide breeding, rearing, and staging areas for waterfowl and other aquatic-dependent bird species, many of those refuges include significant wetlands and other water bodies. Key examples of wildlife refuges in urban areas include the Oyster Bay Refuge on Long Island, the San Francisco Bay National Wildlife Refuge, the San Diego Bay National Wildlife Refuge, and the Bayou Sauvage National Wildlife Refuge in New Orleans.

FWS is also responsible for implementation and enforcement of the federal Endangered Species Act (ESA) with respect to terrestrial and fresh water species.[2] The ESA is one of the most potent federal environmental statutes, with the power to halt entirely any project involving a federal action (including project funding or approval) that might jeopardize the continued existence of any federally-listed threatened or endangered species or impair designated critical habitat for those species, unless "reasonable and prudent alternatives" are available that will allow the project to proceed without resulting in such jeopardy. The statute protects species that might be important components of the urban water environment, and at times can act as a check on urban and suburban development in some areas. The land use implications of those constraints might affect urban water planning. The ESA is also part of the regulatory process that urban water managers may face when planning,

---

[2]The National Marine Fisheries Service (NMFS) exercises that authority with respect to marine and anadromous species. Technically, Congress assigned that authority to the Secretary of the Interior and the Secretary of Commerce, who delegated that authority to FWS and NMS, respectively.

constructing, and operating various kinds of urban water infrastructure, especially new dams or other water supply projects.

10.2.4 *Natural Resources Conservation Service (Natural Resources Conservation Service Website)*

Formerly the Soil Conservation Service, the Natural Resources Conservation Service (NRCS) is a branch within the U.S. Department of Agriculture. Formed in 1935 during the dust bowl era to help reduce the massive erosion that plagued American farmers but also impaired waterways throughout the nation, the agency implements a wide range of programs to provide farmers with technical and financial assistance. However, NRCS also built and manages a large number of water projects, largely but not entirely in the Midwest, designed to serve irrigation and other water supply needs as well as flood control purposes. Although agriculture is the primary intended beneficiary of NRCS programs and projects, urban areas can also benefit from water storage in multiple use projects, and be affected (beneficially or detrimentally) by the manner in which the flood control functions of those projects are managed.

NRCS also implements erosion control, wetland protection, and other environmental programs adopted as part of federal Farm Bill legislation. Examples include the so called "swampbuster" and "sodbuster" programs, as well as the more recent Environmental Quality Improvement Program (EQIP) program, in which farmers are given incentives to protect wetlands, steep slopes and other highly erodible areas, and to reduce loadings of nutrients, pesticides, and other pollutants. Where those programs protect water quality upstream from urban regions, they can also benefit urban water quantity and quality, and potentially reduce advanced treatment or other burdens on urban water supply and wastewater treatment systems.

10.2.5 *Federal Emergency Management Agency (Federal Emergency Management Agency Website)*

The Federal Emergency Management Agency (FEMA) oversees a wide range of emergency management and preparedness programs, to anticipate and respond to events ranging from natural disasters to terrorism. Urban water systems are affected most notably by FEMA's flood control programs, including delineation of floodplain zones within which building limitations might apply. Although the National Flood Insurance Program is based more on incentives than on prescriptive prohibitions on construction within floodplains and riparian areas, cities that wish to protect those areas from excess development can use that authority to limit impervious surfaces and other hydrologic and environmental impacts of development within those areas, as well as to protect property from flooding.

10.2.6 Federal Land Management Agencies

In addition to FWS, which as mentioned above manages the National Wildlife Refuge System—a relatively small percentage of federal land holdings in terms of acreage[3]—several federal agencies manage federal lands in ways that may affect urban water resources. The National Park Service (NPS) and the Bureau of Land Management (BLM) are agencies within the Department of the Interior. NPS manages the specially protected lands within the NPS (national parks, national monuments, national recreation areas, and national historic sites) for ecological, scenic, historic, geological, recreational, and other public values. Like national wildlife refuges, some national parks are in or near urban centers, and many of those have important aquatic ecosystem values. Examples include Golden Gate National Recreation Area and Boston Harbor Islands National Recreation Area. Many other national parks include the headwaters of important watersheds whose downstream drainages affect urban regions and their water supplies, such as Yosemite National Park and the Great Smoky Mountains National Park.

Particularly noteworthy in this category of federal land management agencies, however, is the U.S. Forest Service (USFS) within the U.S. Department of Agriculture, which manages the vast National Forest System throughout the country. Although national forests are now used heavily for both timber supply and recreational uses, when Congress first established the system in the late nineteenth century, it identified preservation of watersheds and downstream water flows as one of the two primary functions of national forest lands. In many parts of the country, the manner in which these critical headwater regions are managed can have significant implications for urban water quantity and quality. As one group of researchers noted, "[t]he watersheds of 3,400 community water systems (CWSs) serving 60 million people in 900 cities are located within National Forest lands" (Sedell et al. 2000).

10.2.7 U.S. Geological Survey (U.S. Geological Survey Water Resources Website)

The U.S. Geological Survey (USGS) is relatively unique among government agencies because its mission is predominantly scientific. Its mission is to conduct scientific research and analysis of many of the nation's natural resources, including water as well as extraction of resources such as minerals, oil and gas. Because of this role, USGS is the source of invaluable data on water resources for urban water managers and other users. For example, USGS maintains an extensive system of stream gages around the country, real-time data from which can be accessed readily on its website. USGS also conducts extensive monitoring of surface and ground water quality, and identifies water quality problems from a range of sources, including urbanization.

---

[3]This is not true in Alaska, where some National Wildlife Refuges are extremely large.

10.2.8 Council on Environmental Quality (Council on Environmental Quality Website)

The Council on Environmental Quality (CEQ) is within the Office of the President rather than an independent agency like EPA or a branch of a cabinet-level department, such as the BLM within the Department of the Interior or USFS within the Department of Agriculture. As such, it has no independent responsibility for implementing programs, constructing projects, or issuing permits or approvals. However, it does serve an important role in overseeing implementation of the National Environmental Policy Act (NEPA) by every other federal agency. As explained in Chapter 9, every federal agency or office is required to comply with NEPA for every federal action that may have a significant impact on the human environment. While all agencies have their own NEPA-implementing regulations and procedures, they also must comply with minimum requirements adopted by CEQ (40 C.F.R. Part 1500 et seq.). Environmental Impact Statements and other procedures required by the CEQ regulations can have significant impacts on urban water projects and resources.

10.2.9 Bureau of Reclamation (U.S. Bureau of Reclamation Website)

The U.S. Bureau of Reclamation (BOR), another agency within the Department of the Interior, is a regionally-significant force in water resources and politics in the western United States. Under the Reclamation Act of 1902, BOR was authorized to construct water projects in the western states designed to help western settlers "reclaim" arid lands, primarily for agricultural uses. Users of reclamation water were supposed to repay the federal costs of those projects through water payments, but Congress frequently extended payment schedules and lowered project interest rates in ways that rendered reclamation water heavily subsidized. Moreover, initially small, local projects gave way to massive dams and conveyance projects such as the Hoover Dam and the All-American Canal to California's Imperial Valley (Adler, 2007). Those projects were supported in part by revenues from the hydroelectric facilities built into the dams. Electricity and water from those projects also helped to fuel urban growth in Southern California, Southern Arizona, the Colorado Front Range and other regions. Although irrigated agriculture remains the predominant user of BOR water, those resources form an increasingly important component of urban water supply in the west, especially as thirsty cities purchase additional reclamation water from existing agricultural users. Metropolitan areas that use water from the Colorado River Basin, for example, compete for scarce reclamation project water needed to support additional growth, and also negotiate deals to transfer water supplies from agricultural regions.

10.2.10 Native American Tribes and the Bureau of Indian Affairs (Bureau of Indian Affairs Website)

Also primarily in the west, federally-recognized Native American Tribes can play an important role in the allocation of scarce water resources. Tribes are not part of the federal government, but rather independent sovereigns that co-exist with the national government in a relationship analogous to states. However, through the Bureau of Indian Affairs (BIA), the federal government owes a trust responsibility to ensure that Indian needs and interests are protected (*Seminole Nation v. United States*). More important, the Supreme Court ruled early in the twentieth century that when Congress set aside Indian reservations around the west, it reserved sufficient water rights to allow for tribal agricultural and other economic development (*Winters v. United States*). Those "federal reserved water rights", which the Supreme Court later ruled also apply to other federal lands reserved for special purposes (such as national parks, wildlife refuges, and military reservations), are particularly significant in the context of the prior appropriation doctrine of western water law, because they bear a "priority date" (see explanation of the prior appropriation doctrine in Chapter 9) defined by the time in which Congress reserved those lands, which often makes them senior to many urban and agricultural water rights in the system. Quantification and settlement of reserved water rights claims can affect the amount of water remaining for urban uses. On the other hand, cities also might be able to purchase latent reserved rights from tribes, providing needed water for cities and cash for tribal economic development.

10.3 State Institutions

State water management institutions vary greatly with regional climate, hydrology, and other factors (Fiero, 2007). The biggest divide is between the generally wet regions in the eastern parts of the country, and the much more arid states to the west of the 100th Meridian. Intermediate states include moderately or regionally arid states in the Great Plains, or areas with both extremes along the west coast. Eastern states often combine water allocation and water quality within a single, statewide environmental agency (although those functions may be handled by separate offices or other entities within that umbrella agency). Western states employ a range of separate institutions to oversee and administer the prior appropriation doctrine and other laws and programs with respect to surface water and ground water allocation and use (Ayotte et al., 1993).

10.3.1 State Water Quantity Institutions

Both riparian states and prior appropriation states relied initially on self-implementation aided by judicial enforcement of legal rights to implement and enforce those doctrines (Sax et al., 2000). As the number of users increased and competition for scarce resources intensified, however, most states have moved to some formal institutional system involving permits, recordkeeping, judicial decrees of water rights, or some other administrative means of tracking and enforcing water rights.

Variations in institutional arrangements among states make generalizations difficult, but there are two basic patterns that parallel hydrologic conditions. As noted above, and following the pattern of water rights practices discussed in Chapter 9, in most humid states those functions are usually combined within a single agency. In some of those states, however, responsibilities for water-related issues are divided among state environmental agencies (primarily water quality issues) and state natural resource agencies (primarily water resource allocation issues). In arid states, prior appropriation systems are usually managed by a department of water resources, a state engineer's office, or sometimes by multiple entities serving information, planning, and administrative or regulatory and enforcement functions, respectively. In Utah, for example, the Division of Water Resources within the Department of Natural Resources serves information and planning functions, while water rights determinations are made by the State Engineer. In some states, such as Colorado, water rights determinations are made through a series of basin-specific water courts, and in other states general stream adjudications to determine comparative rights among users in the same system are made by courts of general jurisdiction (courts with the authority to address a wide variety of legal claims).

10.3.2 *State Water Quality and Environmental Agencies*

Issues of water quality, aquatic habitat and ecosystem restoration and protection are usually addressed by state environmental, natural resource, and fish and wildlife agencies. Although the exact organization of those agencies again vary widely, they often parallel in some way their respective federal counterparts, described above, and all of them may affect urban water resources and aquatic ecosystems in some way. For example, the state water quality agency may be responsible for such varied water-related issues as drinking *water quality standards*, testing sport fish for contaminants and issuing fishing advisories for urban streams, and similarly testing recreational waters for pathogens and issuing appropriate advisories or closings for urban beaches. A state department of fish and wildlife may be involved in urban stream restoration efforts, as well as the regulatory task of administering and enforcing recreational fishing laws.

Probably most important for urban water managers, state water quality and other environmental agencies are usually the first step in complying with CWA and other federal statutory and regulatory requirements involving water (including the SDWA, RCRA, CERCLA, and other programs). Because EPA has delegated administration of the NPDES program to most (but not all) states, most municipalities obtain discharge permits for their sewage treatment, storm water, and other discharges from the state water quality agency, and are responsible for submitting discharge monitoring reports and other compliance information to both the state and to EPA for compliance and enforcement purposes. Likewise, most states have primacy over the SDWA program that regulates public water supply systems, and implement the wellhead protection and sole source aquifer programs under that law. All states issue permits and otherwise administer regulatory programs governing municipal landfills, and most states do for hazardous waste landfills as well. Compliance with and enforcement of all of those regulatory programs (or lack thereof) can affect the quality of urban water supplies.

10.4 Local and Regional Water Institutions

In most ways, local institutions are at the front line of urban water issues. Local entities usually procure, transport, treat, and distribute water to end users in urban areas (Sax et al., 2000; Arnold, 2005). They collect, treat, and manage sewage and contaminated storm water runoff, and control drainage patterns to protect public and private property, as well as aquatic ecosystems, through land use planning, street design, and other means. Local planning and zoning officials also affect both water use and the hydrological and ecological impacts of development through a range of land use decisions and planning efforts. The specific nature and organization of local water institutions and entities varies greatly, making broad generalizations even more difficult than is true for state institutions. For example, in some cities all water issues from supply, treatment and distribution to wastewater collection, treatment, and discharge are governed by a single water department, whereas other areas separate those functions among different entities. The degree to which those functions are integrated or divided can affect the region's approach to water management in significant ways. Indeed, in some areas the diversity and fragmentation of local water institutions, sometimes with conflicting or overlapping jurisdictional boundaries, can create inefficiencies and other problems for urban water planning and management.

10.4.1 Local and Regional Water Suppliers

As described in Chapter 9, municipalities face increasing challenges in procuring adequate water supplies, especially in areas of rapid growth and short supply (Tarlock and Van de Wetering, 2006). There is also an increasing consensus, based on both actual data trends and modeling results, that global warming will exacerbate water supply shortages (while also increasing the volume and intensity of precipitation in other areas, leading to equally difficult issues of flood control and storm water management).

In many cities, water is procured and supplied by private water companies acting effectively as regulated public utilities much like a local gas, electric, or telephone company. State law may impose universal service obligations, or regulate the water rates that may be charged to end users in ways that affect water supply and demand. Some states integrate water supply planning with land use planning and zoning in various ways, such as assured supply laws that require developers to demonstrate sufficient water supplies to serve a new development before the necessary permits and approvals may be granted (Davies, 2007; Arnold, 2005; Chapter 12). In other urban areas, water is supplied directly by government agencies such as city water departments, or other institutions such as municipal utility districts. Those institutions may be governed by city employees or officials reporting to the mayor or city council, or they may be managed independently by elected or appointed boards.

Under some state laws, special water districts (and sewer districts described below, or joint water and sewer districts) are necessary to issue public bond offerings to raise sufficient capital to design, build and operate urban water infrastructure. Depending on the particular institutional arrangement, those offerings might be in the form of general obligation bonds that can be repaid out of general tax revenues, or revenue bonds in which investors are paid out of the operating revenues from the specific project. Because general obligation bonds are typically viewed as less risky, they might require lower interest rates, but they also encumber the general local tax base. Revenue bonds may bear higher interest rates, and because they must be paid in full out of water agency revenues, they can affect water rates accordingly. Those arrangements, of course, arguably eliminate general tax subsidies to water users, and may serve as some incentive to increase efficiency of water use as a result. (Because water rates are subsidized in so many different ways, it cannot be said that this factor alone will serve to impose the full cost of water on end users in urban areas or elsewhere.)

Especially in arid western states where urban water supplies often compete with irrigation uses, and where supplies can be concentrated in large storage facilities built and operated by the BOR or other federal or state entities, other water supply institutions can also affect urban water supplies. Irrigation districts or mutual water companies may be formed as the primary holder of water rights in those storage facilities, and then the purveyor of water to various end users. Although designed primarily to serve irrigation users, where a project is designed to serve municipal and industrial end users as well, urban water managers may need to procure water or water rights from those entities, especially as agricultural to urban water transfers become one of the viable remaining sources of new water for growing cities.

As core cities expanded into a mass of sprawling, separately incorporated suburbs, and other smaller cities grew together, many urban areas have faced an increasingly diverse, complex and fragmented array of water supply institutions, often competing for common but limited supplies. This can lead to an inefficient and potentially overlapping management system and infrastructure, sometimes literally including duplicative and criss-crossing conveyance systems, reservoirs and other storage facilities. Some regions either replaced that mélange or combined the existing entities into larger, more efficient regional water authorities designed to deal more efficiently with water acquisition, storage, and distribution on a regional basis.

Examples of this kind of regional authority include the Southern Nevada Water Authority and the Metropolitan Water District of Southern California.

10.4.2 Local Sewerage, Water Pollution Control, Storm Water Management and Flood Control Agencies

The manner in which "back end" water issues are addressed also varies significantly among different states and localities. As noted above, sewerage and storm water management are often handled by the same municipal water department or water and sewer district as is responsible for water supply. This arrangement might promote a more integrated approach to urban water use and management, and might facilitate implementation of some of the potential reforms suggested elsewhere in this book. For example, combined water institutions may be better suited to separate potable from urban irrigation water, because financial savings realized in the wastewater treatment process might help to offset costs of separating urban water systems.

Likewise, some municipalities have separate flood control districts or agencies designed to finance and to operate levee systems, storage facilities, and other flood control infrastructure. Although such entities provide important flood control and property protection functions, it may be that such special purpose entities interfere with more holistic efforts to reduce flooding risks through integrated land use planning, storm water management, and other methods, in favor of traditional structural approaches. Local flood control management districts necessarily operate in coordination with ACE and other federal entities that are responsible for construction and operation of so much of the nation's physical flood control infrastructure.

10.4.3 Local Planning and Zoning Institutions

Finally, decisions made by an equally variable array of local and regional land use planning and zoning institutions can have profound effects on the urban water environment (Thompson, 2005; Waterman, 2004). The density of development, including the percentage impervious surface in developed areas, can affect the hydrology of urban systems dramatically, and intensify storm water pollution and other impacts to urban aquatic ecosystems and water supplies. Cities can choose either to allow development right up to the water's edge, with accompanying loss and degradation of riparian and aquatic habitats and deterioration of water quality, or they can plan to protect riparian areas and flood plains, and to limit the density and nature of development to protect those areas for water supply, environmental, recreational, and aesthetic purposes. In making those choices, urban planners must address issues of private property and development rights discussed in Chapter 9, but also balance them against increased private property values and increased public values of healthy urban aquatic areas that serve important ecosystem functions and provide valuable amenities to local communities. Development decisions made by planning and zoning officials also affect water supply and demand issues, especially in areas where existing supplies are inadequate to support new urban growth, or where shifts from other water uses or increased efficiency is needed to support that growth.

All of those interactions between land use decisions and urban water systems suggest that land and water institutions either should be integrated structurally, or that procedures should be implemented to ensure that decisions made by both categories of institution are coordinated adequately. Some states, such as Oregon, have adopted statewide planning requirements designed to integrate land use and water decisions in some way, although the manner and degree of implementation of those requirements can vary greatly around the state. Other areas have adopted more comprehensive and integrated land use and water planning at the country or regional levels. In many areas, however, land and water decisions are made with little or no coordination, resulting in unnecessary and often unforeseen impacts on urban water systems.

10.5 Institutional Fragmentation as a Barrier to a Sustainable Urban Water Environment

Fragmentation in water resources programs and institutions has long been identified as a barrier to effective and efficient water use and management, and to the protection and sustainability of aquatic ecosystems (Adler, 1994; Reuss, 1993). One scholar described the "system" as "similar to a marbled cake, with several levels of government intermingled in an irregular pattern" (Whipple, 1989). Another summarized this fragmentation as falling into three broad categories: "(1) political fragmentation—the overlapping and conflicting division of responsibilities among multiple levels of government and agencies; (2) issue fragmentation—the artificial division of related water issues into separate programs (such as water quality and quantity, land and water use, and surface and groundwater); and (3) gaps in program design and implementation" (Adler, 1994). A third source identified "vertical disconnects" (fragmentation among the federal, state, and local levels of government), "horizontal disconnects" (fragmentation within a level of government with respect to water and related land use issues), and "internal disconnects" (conflicting or mutually inconsistent policy goals affecting water resources (Arnold, 2005).

As illustrated above and in Chapter 9, fragmentation is no less evident in urban water programs than in water resources programs more generally. Table 10.1 summarizes—in a necessarily simplified fashion—just the most important institutions (or categories of institutions) designed to address a range of urban water issues:

In addition, because so many entities are, or can be, involved in urban water issues, and because political, hydrological, geographic, and other conditions vary so much around the country, different cities divide responsibility for water resources planning, management, and service provision in different ways. The following

10 Institutions Affecting the Urban Water Environment

Table 10.1 Examples of institutions involved in urban water issues

	Land Management	Water Supply	Waste Water	Environmental Protection	Dispute Resolution
Federal	Bureau of Land Management; Forest Service; National Park Service; Fish and Wildlife Service;	Bureau of Reclamation; Army Corps of Engineers; Natural Resources Conservation Service	EPA; Army Corps of Engineers	EPA; Army Corps of Engineers; Council on Environmental Quality; Federal Energy Regulatory Commission	Federal courts and administrative agencies
State	State land agencies; state land use planning agencies	Water Resources Departments; State Engineers; Public Utility Commissions	State water quality, natural resource, and fish and wildlife agencies	State water quality, natural resource, and fish and wildlife agencies	State courts and administrative agencies
Local	Land use planning and zoning agencies, boards and commissions	Municipal Water Districts and Companies; Public Utility Districts	Publicly Owned Treatment Works; Water Districts; Public Utility Districts; Stormwater Management Agencies	Local health and environmental agencies	Local courts, boards, and commissions
Private and quasi-public	Landowners and developers	Private water suppliers; Irrigation Districts; Mutual Water Companies; Private Users	Private wastewater treatment entities; Private Customers (residential, commercial and industrial)	Regulated parties (landowners; industry; agriculture)	Arbitration; mediation; facilitation

examples of institutional arrangements from one large Midwestern city (Chicago), and one medium-sized city in the Intermountain West (Salt Lake City), illustrate a trend toward efforts to integrate water resources management within cities in a more coordinated way. However, those institutions vary greatly within and among different regions of the country, due to physical, political, legal, and sometimes purely historical factors.

10.5.1 Example: Integrated Water Institutions in Chicago (Chicago Water Website)

For decades, the "front end" (water department) and the "back end" (wastewater department) of the Chicago urban water cycle were managed separately. Chicago combined its water and sewer departments in 2003 to create the Department of Water Management. Within that combined department, a water department obtains water from Lake Michigan, and handles public water treatment and distribution, while the sewer department handles sewage and stormwater collection and treatment, as well as flood control. In 2004, Cook County ceded authority to a regional entity, the Metropolitan Water Reclamation District of Greater Chicago (MWRDGC), an independent government agency run by a board of elected commissioners, to manage stormwater (and combined sewer overflow (CSO)) policy for the entire county (MWRDGC website).

10.5.2 Example: Integrated Water Institutions in Salt Lake City (Salt Lake Water Website)

In Salt Lake City, a unified Department of Public Utilities manages drinking water, stormwater, and wastewater in Salt Lake City. A drinking water branch obtains water supplies from artesian wells, streams and associated reservoirs in the adjacent Wasatch Mountains, and through trans-basin diversions from the Colorado River watershed, and handles public water treatment and distribution. A separate stormwater division was created in 1991 to address stormwater collection and treatment, and flood control. A wastewater division handles sanitary sewage collection, treatment and disposal, through both traditional treatment plants and wetlands treatment. In cooperation with the U.S. Forest Service, the department also administers a comprehensive watershed protection program designed to protect the significant portion of the city's water supply that comes from canyons streams and reservoirs.

10.5.3 Competing Policy Factors in Allocating Responsibility for Urban Water Quality

Finding optimal solutions to water resources fragmentation, and especially solutions that work universally throughout the country, has been a difficult quest, and involves difficult tradeoffs with no presumptively "correct" answers. The issue of water pollution control illustrates the competing policy factors involved in deciding how to allocate regulatory and other governmental authority. What entities and levels of government are best suited to address problems of surface water quality? In 1972, Congress passed comprehensive amendments to the Federal Water Pollution Control Act (now known by its popular name the Clean Water Act (CWA)), in response to the widespread failure of individual states to reduce serious water pollution around the country (Adler et al., 1993). In allocating pollution control authority and responsibility among the federal, state and local governments, Congress employed a strategy known as "cooperative federalism". Through this approach, it is fair to say that Congress expressly addressed several key issues of political fragmentation (whether or not one agrees with the manner in which that was accomplished), but was far more ambiguous with respect to issue fragmentation.

10.5.4 Addressing Issues of Political Fragmentation

For some purposes, Congress centralized water pollution control in the federal government (EPA and the U.S. Army Corps of Engineers (ACE)). Congress gave EPA the task of adopting relatively uniform discharge requirements for particular classes and categories of discrete (or "point source") dischargers (e.g., certain kinds of steel mills or POTWs), tied to best available control technologies rather than variable water quality conditions (CWA §§301, 304). The decision was designed to level the economic playing field among dischargers across the country, to establish minimum applicable control requirements, and to prevent states from competing for jobs by lowering treatment obligations. Similarly, Congress assigned to EPA and ACE the presumptive authority to issue permits to discharge pollutants into the "waters of the United States" (CWA §§402, 404). However, states have the option to assume permitting responsibility for municipal and industrial discharges (which most states have exercised), and for discharges of dredge and fill material into wetlands and other waters (which most states have not exercised), subject to federal government approval, review, and ongoing oversight (including potential veto power over individual permits or, in the extreme, withdrawal of state permit program delegation). This system is designed to provide flexibility to states while ensuring that national water quality and economic equity goals are met.

At the same time, Congress directed states to adopt ambient surface water quality standards tailored to variable uses and conditions around the country (CWA §303). EPA must review and approve those standards as sufficient to meet the broadly-defined water quality goals set by Congress, but states have considerable flexibility under the statute and EPA regulations (40 C.F.R. Part 131) to account for different uses of water bodies, and different ecological, hydrological, and other physical conditions. However, EPA retains the responsibility to issue water quality standards for any state that fails to do so, or to do so adequately. Through this approach, Congress required states again to meet minimum federal goals and requirements, but allowed them to do so in different ways, and to exceed federal requirements if they choose to do so.

Likewise, Congress chose a more deferential strategy vis-à-vis the states with respect to more diffuse, "nonpoint source pollution" (or "polluted runoff") from agriculture, development and other land disturbance activities (CWA §§208, 319). States are required to develop and implement nonpoint source pollution control plans, subject to EPA review and approval. However, EPA's only remedy in the case of inadequate state programs is withdrawal of relatively small amounts of federal funding. Unlike the point source permitting or water quality standards programs, EPA has no authority to adopt a federal nonpoint source pollution control plan or program in the face of inadequate state programs. Here, despite the significant nature and scope of the problem, Congress deferred to state and local preferences because nonpoint source pollution is far more tied to land use planning and economic policies traditionally reserved to states and localities. Ultimately, there is some accountability because states are required in theory to ensure that the combination of point source and nonpoint source controls are sufficient to meet their ambient water quality standards through a process known as *total maximum daily loads* (TMDLs; CWA §303(d)). However, implementing the TMDL process has been both scientifically complex and fraught with tension between EPA, the states, end environmental groups that have tried to force that process along through litigation (Houck, 2002).

Allocating responsibility for pollution control becomes even more challenging when one adds the municipal dimension. Cities are responsible in some way for both compliance with CWA requirements and for regulating other sources of municipal water pollution, in compliance with the CWA and EPA rules. For many years, Congress has provided either direct federal grant funding or loans to help cities to meet those significant financial burdens. However, whether or not they have received federal financial assistance, municipal POTWs must acquire and comply with discharge permits issued by either EPA or the state. And, when the POTW receives potentially toxic wastes from commercial or industrial customers, they must implement a regulatory program designed to protect the POTW itself, the sewer system, workers, and the quality of both the resulting treatment plant effluent and solid waste material (sewage sludge or "biosolids"). Likewise, municipal stormwater discharges require permits from EPA or the state, but compliance with those permits requires the city to implement its own education and regulatory programs to prevent or to clean up pollutants that might contaminate those discharges into urban waterways.

10.5.5 Remaining Problems of Issue Fragmentation

Congress' main focus in enacting the 1972 CWA was surface water pollution. As a result, in that legislation and in subsequent amendments it has painted with a much broader brush with respect to three key areas of institutional issue fragmentation in water resources law and policy: surface water versus ground water; water quality versus water quantity; and water quality versus land use. In defining the "waters of the United States", Congress did not expressly discuss the issue of groundwater, but the CWA is generally interpreted as applying largely to surface water. Thus, for the most part groundwater protection is left largely to other federal statutes (discussed in Chapter 9), or to the states.

As discussed above, Congress adopted a deferential policy toward the states with respect to the land use implications of nonpoint source pollution control (and to a large degree, municipal stormwater control programs). Likewise, Congress has been reluctant to infringe on traditional state authority over water quantity (water supply and allocation decisions), even in the guise of protecting water quality and aquatic ecosystem health. Thus, in the 1972 legislation Congress provided: "It is the policy of the Congress to recognize, preserve, and protect the primary responsibilities of States to prevent, reduce, and eliminate pollution, [and] to plan the development and use ... of land and water resources ..." (CWA §101(b)). In 1977 amendments, Congress added the following policy statement regarding the relationship between water quality and water quantity:

> It is the policy of Congress that the authority of each State to allocate quantities of water within its jurisdiction shall not be superseded, abrogated, or otherwise impaired by the [Act]. It is the further policy of Congress that nothing in this [Act] shall be construed to supersede or abrogate rights to quantities of water which have been established by any State. Federal agencies shall co-operate with State and local agencies to develop comprehensive solutions to prevent, reduce and eliminate pollution in concert with programs for managing water resources. (CWA §101(g)).

Despite those general policies, as discussed in Chapter 9, the Supreme Court has ruled that *states* may use their own water quality standards to require minimum water flows to protect salmon and other resources and uses (PUD No. 1 of *Jefferson County v. Washington Department Ecology*). And water volumes clearly affect water pollution control efforts, for example, where low flows result in higher concentrations of pollutants, higher temperatures, and similar problems. In general, however, the CWA does not establish clear and effective linkages between water quality and water quantity, a policy that has been defended by some (Hobbs and Raley, 1989) and criticized by others (Benson, 2005).

10.5.6 *Working Toward Solutions: Interstate, International, Watershed Management and Other Collaborative Institutions*

Various inter-jurisdictional and other watershed-based initiatives have also evolved as one potential way to address fragmentation in water resources planning and management, and to address urban and other water issues in a more holistic, coordinated way. Some of those initiatives are described briefly below, but are the main subject of Chapter 11.

Few urban water systems are in isolated watersheds that cross no geopolitical boundaries. Geographic linkages between urban water uses and water users can span multiple municipalities, counties, states, and even cross international borders. Moreover, in addition to the remarkably wide range of governmental and quasi-governmental institutions involved in urban water policy, a large number of private businesses, landowners, nongovernmental organizations, and other entities either play a significant role in urban water issues, or are affected by urban water decisions and policies. This more complex level of interactions can be addressed at a wide range of levels, both geographically and institutionally.

Geographically, a large and diverse set of interstate and international entities have been formed to address trans-boundary water pollution, water supply, and other water issues. Examples of interstate institutions include the Chesapeake Bay Commission, the Delaware River Basin Commission, the Ohio River Basin Sanitary Commission, and the Colorado River Basin Salinity Control Forum. International institutions include the International Joint Commission between the United States and Canada, and the International Water and Boundary Commission between the United States and Mexico. Interstate and international water disputes can also be addressed through interstate compacts, such as the Colorado River Compact, and international treaties, such as the U.S-Mexico Water Treaty.

There is also a long but checkered history of efforts to implement comprehensive basin-wide or watershed-based planning and management in the United States. A nationwide program of basin-wide planning established by Congress in 1965 was abandoned in the 1980s, but has been replaced by a more ad hoc collection of watershed-based programs around the country, ranging geographically from small urban stream systems to interstate and international basins such as the Great Lakes and the Chesapeake Bay. Although those institutions again vary very widely in composition, organization, operational structure, authority, and implementation, they share the common idea of bringing together multiple stakeholders to work collaboratively to restore, protect, and manage urban and other water resources and aquatic ecosystems. Those kinds of institutions are highlighted in the next chapter.

References

Adler, R.W. (2007). *Restoring Colorado River Ecosystems: A Troubled Sense of Immensity*, Washington, D.C.: Island Press.
Adler, R.W. (1994). Addressing Barriers to Watershed Protection, 25 *Environmental Law* 973.
Adler, R.W., Landman, J.C., and Cameron, D.M. (1993). *The Clean Water Act 20 Years Later*, Washington, D.C.: Island Press.
Arnold, C.A. (2005). Introduction: Integrating Water Controls and Land Use Controls: New Ideas and Old Obstacles. In: Arnold, C.A., ed. *Wet Growth: Should Water Law Control Land Use?* Washington, D.C.: Environmental Law Institute.
Ayotte, K.A., Dougherty, T.G.Z., and Tener, D.M. (1993). State Coordination of Water Allocation Management and Water Pollution Regulation, 4 *Villanova Law Review* 129.
Benson, R.D. (2005). Pollution Without Solution: Flow Impairment Problems Under Clean Water Act Section 303, 24 *Stanford Envrionmental Law Review* 199.
Bureau of Indian Affairs Website, http://www.doi.gov/bureau-indian-affairs.html
Cech, T.V. (2003). *Principles of Water Resources: History, Development, Management, and Policy*, New York: John Wiley & Sons, Inc.
Chicago Water Website. http://egov.cityofchicago.org/city/webportal/home.do
Clean Air Act, 42 U.S.C. §7401 et seq.
Clean Water Act, 33 U.S.C. §1251 et seq.
Comprehensive Environmental Response, Compensation and Liability Act, 42 U.S.C. §9601 et seq.

Council on Environmental Quality Website, http://www.whitehouse.gov/ceq/
Davies, L.L. (2007). Just a Big, 'Hot Fuss'? Assessing the Value of Connecting Suburban Sprawl, Land Use, and Water Rights Through Assured Supply Laws, 34 *Ecology Law Quarterly* 1217.
Endangered Species Act, 16 U.S.C. §1531 et seq.
Federal Emergency Management Agency Website, http://www.fema.gov/
Federal Insecticide, Fungicide, and Rodenticide Act, 7 U.S.C. §136 et seq
Fiero, P., Jr. (2007). Agencies and Organizations. In: Fiero, P., Jr. and Nyer, E.K. eds., 3rd edn. *The Water Encyclopedia, Hydrologic Data and Internet Resources*. Boca Raton, FL: CRC Press.
Hobbs, G.R. and Raley, B.W., (1989). Water Rights Protection in Water Quality Law, 60 *University of Colorado Law Review* 841.
Houck, O.A. (2002). *The Clean Water Act TMDL Program: Law, Policy and Implementation* (2d ed.), Washington, D.C.: Environmental Law Institute.
MWRDGC Website. http://www.mwrdgc.dst.il.us/
National Environmental Policy Act, 42 U.S.C. §4321 et seq.
Natural Resources Conservation Service Website, http://www.nrcs.usda.gov/
PUD No. 1 of Jefferson County v. Washington Department Ecology, 511 U.S. 700 (1994).
Reclamation Act of 1902, Pub.L. no. 57-161, Chapter 1093, 32 Stat. 390 (codified as amended at various sections of 43 U.S.C.).
Resource Conservation and Recovery Act, 42 U.S.C. §6901 et seq.
Reuss, M. (1993) *Water Resources Administration in the United States: Policy, Practice, and Emerging Issues*, E. Lansing, MI: Michigan State University Press.
Rivers and Harbors Appropriates Act of 1899, 33 U.S.C. §§403, 407, and 411.
Rodgers, W.H., Jr. (1971). Industrial Water Pollution and the Refuse Act: A Second Chance for Water Pollution Control. 119 *University of Pennsylvania Law Review* 761.
Safe Drinking Water Act, 42 U.S.C. §300f et seq.
Salt Lake Water Website. http://www.slcgov.com/utilities/ud˙home.htm
Sax, J.L., Thompson, B.H., Jr., Leshy, J.D., and Abrams, R.H. (2000). *Legal Control of Water Resources*, 3d ed. St. Paul, Minn.: West Group.
Sedell, J., Sharpe, M., Dravenieks, D., Copenhagen, M., and Furness, M. (2000). *Water and the Forest Service*. U.S. Forest Service, U.S. Department of Agriculture, Washington, D.C. http://www.fs.us/publications/policy-analysis/water.pdf
Seminole Nation v. United States, 316 U.S. 286, 297 (1942).
Tarlock, A.D. and Van de Weterling, S.B. (2006). Western Growth and Sustainable Water Use: If There Are No "Natural Limits," Should We Worry About Water Supplies? 27 *Public Land and Resources Law Review* 33.
Thompson, B.H., Jr. (2005). Water Management and Land Use Planning: Is it Time for Close Coordination? In Arnold, C.A., ed. *Wet Growth: Should Water Law Control Land Use*? Washington, D.C.: Environmental Law Institute.
Toxic Substances Control Act, 15 U.S.C. §2601 et seq.
U.S. Army Corps of Engineers Website, http://www.usace.army.mil/
U.S. Bureau of Reclamation Website, http://www.usbr.gov/
U.S. Environmental Protection Agency Website, http://www.epa.gov/
U.S. Fish and Wildlife Service Website, http://www.fws.gov/
U.S. Geological Survey Water Resources Website, http://www.water.usgs.gov/
Van Heerden, I.L., Kemp, G.P., and Mashriqui, H. (2007). Huricane realities, models, levees and wetlands. In Novotny, V. and Brown, P., eds. *Cities of the Future: Towards Integrated Sustainable Water and Landscape Management*, London: IWA Publishing.
Waterman, R. (2004). Addressing California's Uncertain Water Future by Coordinating Long-term Land Use and Water Planning: Is a Water Element in the General Plan the Next Step? 31 *Ecology L. Q.*117.
Whipple, W., Jr. (1989). Future Directions for Water Resources. In: Johnson, A.I. and Viessman, W., Jr., eds. *Water Management for the 21st Century*. Middleburg, VA: American Water Resources Association.
Winters v. United States. 207 U.S. 564, 576-77 (1908).

Chapter 11
Institutional Structures for Water Management in the Eastern United States

Cliff Aichinger

11.1 Introduction

Cities are faced with numerous and difficult tasks related to the management, treatment and disposal of water. Institutional structures (agencies, organizations, utilities, and departments) are needed to address these water issues. The choice of institutional arrangement is varied across the country and affected by local factors such as politics, history, geography, resources, water management issues and individuals involved in water issues.

Historically, large cities have organized their departments to address stormwater management, water supply, and, sewage disposal and treatment. In many of these metropolitan areas, the services provided by the major cities have been expanded beyond the borders to adjacent suburbs. In other cases it has become more efficient to centralize and combine services like sewage treatment and water supply into regional agencies. This has often resulted in higher quality sewage treatment and improved drinking water quality at reduced costs. However, the management of stormwater has typically been left to individual cities. Cities have historically collected their sewage and stormwater and discharged these flows into large rivers that transport water and waste out of the city. There has been little if any incentive for cities to have water leave their community in as good or better quality than when it entered.

As our population has grown, so have the magnitude of resource management problems and the recognition of these problems by the general population. As people recognize and understand the problems, they also recognize the impacts of one city or geographic area on another. This brings public demands for solutions and action. Even though many of these issues have been recognized, like downstream pollution caused by inadequate sewage treatment, action to resolve the problems have been slow. In some cases it has taken lawsuits under the authority of the Clean Water Act (CWA) to get one area to respond to the impacts on another.

C. Aichinger (✉)
Ramsey-Washington Metro Watershed District, St. Paul, Minnesota, MN, USA
e-mail: cliff@rwmwd.org

Water management tends to focus around resource issues where planning and implementation involves multiple actors at different levels of government who may vary in their individual and collective capacity to solve the problems. Who is involved greatly impacts the success at resolving issues. In particular, issues like flooding, water quality, CSO, and water supply are rarely issues impacting only one city and totally within its political boundaries to address and manage.

11.2 Legal Framework for Water Management

Understanding the complexity of federal and state laws addressing water management is essential to develop a viable local approach to water management that addresses local needs, conditions and problems. Chapters 9 and 10 identified the legal principles, doctrines, federal laws and programs that impact city water management. As discussed in those chapters, these laws have a tremendous impact on the water management approach used in cities across the country. These laws, in combination with state laws and programs shape the methods and approaches used as well as influence the scope and scale of the program.

A recent analysis conducted in Minnesota by the Minnesota Stormwater Steering Committee[1] identified the 19 state laws and programs related to water management in its analysis of the potential of watershed-based stormwater permitting in Minnesota's MS4 (Municipal Separate Storm Sewer System) program. The laws were listed and analyzed to illustrate the multiple laws and rules that are already in place to influence and shape how water is managed and regulated. In combination with the federal laws identified earlier, this illustrates the complexity of water law and the difficulty in developing coordinated water management programs.

The challenge for local government is how to comply with state and federal laws for water quality as well as the demands of other water management issues like flood control within the limits and capabilities of the local government structure. As documented above, water management at the city level is driven by a number of factors, including:

- Existing state and federal water laws.
- Local water resource management issues that may be outside the total control of one unit of government (flooding, water quality, CSO issues, etc).
- Resource management issues that are impacting downstream cities and states.
- Flooding issues in a city that are impacted by actions and water management in communities upstream.
- New or anticipated regulatory requirements.

---

[1] The Stormwater Steering Committee (SSC) is a group of public and private stakeholders charged to inform, advise, and coordinate stormwater management efforts across the state. An immediate goal of the SSC is to enhance the effectiveness of existing and emerging state and local stormwater regulatory management programs, in order to build an efficient and understandable regulatory and implementation framework. (http://proteus.pca.state.mn.us/water/stormwater/steeringcommittee/index.html)

To address these issues cities and residents are commonly turning to the option of watershed-based planning, management and permitting to address the need for multi-agency coordination for water management.

11.3 The Watershed Approach

The use of the *watershed approach* presents numerous potential benefits for managing stormwater runoff volume and quality. However, the desire to use this approach and the successful implementation of a watershed-based approach has been fraught with challenges across the United States and internationally. This section will present some common organizing issues to consider, some of the common approaches used, and the critical ingredients for success.

According to Webster Dictionary, a watershed is "a region or area bounded peripherally by a water parting or draining ultimately to a particular watercourse or body of water" (Webster 1994). Watershed Management is the approach used to incorporate plans, goals, policies, objectives and activities to control or manage water and related land management issues in the watershed, often across multiple jurisdictional boundaries. Watershed activities have a wide range from direct landowner assistance for erosion and sediment control, to construction of major capital improvements for water detention and water quality treatment.

11.3.1 Scope

In the process of establishing a watershed-based program, a key task for those involved is to identify the scope of issues to be addressed by the watershed program. Many watershed programs were initially formed to address a specific issue (salmon management in the Pacific northwest, water quality of Lake Tahoe, Chesapeake Bay nutrient loading), but after initial discussion those involved often see that the key organizing issue is connected to many more (shoreline erosion, fisheries, water quality, recreation, ground water, wildlife, flooding, etc.).

Several watershed projects have had limited success due to the constrained program scope. A program focused solely on one issue will usually experience decreased support from its participants when they find that a solution conflicts with their objectives for another related issue.

11.3.2 Scale

A successful strategy for management of water resources based on a watershed must consider geographic area of interest and the appropriate political or decision-making process. Due to different political philosophies, priorities and resource issues, almost every state that has attempted to develop a watershed-based approach has used a different approach.

Watersheds naturally partition the entire country into watershed units that range in size from an area as small as a room to 1.15 million square miles for the Mississippi River basin. Decision-making at these various scales can vary from individuals, to neighborhood groups, to several city governments, to multiple states, to the federal government, and to millions of citizens. (NRC 1999).

In 1974, the USGS, in cooperation with the US Water Resources Council developed the Hydrologic Unit Classification System (HUC). The entire country was broken into accounting units and cataloging units. There are 352 accounting units in the United States and 2,150 cataloging units. The cataloging units are also referred to as the 8-digit hydrologic unit code. This scale is referred to in numerous publications as a logical scale for national or state level watershed-based management. However, catalogue units are still large, with an average size unit of 700 square miles. A nested approach of smaller watershed units within this area may be the most effective approach to watershed-based management programs for developed urban areas.

The scale of watershed management is an issue that needs close attention in the organizational phase of a watershed initiative. An effort that is too large, for example major river basin size, will be faced immediately with the difficulties of not only the coordination of numerous local governments, but also with the complexity of interstate issues. In these cases, it has been shown that varying understanding and commitment of resources can paralyze any progress for years. This result can occur where ever the geographic scale of the organization exceeds the area directly impacted by the watershed issue (NRC 2007).

While there does not appear to be any comprehensive data on watershed initiatives across the United States, most documented successful efforts appear to be in watershed basins of less than 500 square miles. Many of the documented watershed examples are in areas of less than 100 square miles (NRC 1999). Much of this success is attributable to intimate knowledge of the landscape and water resources in a smaller watershed area. It is entirely possible for a watershed board or staff to have knowledge of the issues and landscape of a watershed of 50 to 100 square miles, but it is impossible for larger areas. This integral knowledge of the smaller watershed is a great advantage when discussing issues with stakeholders or developing plans for a sub-watershed area.

In many respects, the scale of the watershed initiative will determine the topics or projects that can be addressed. For example, the large-scale Chesapeake Bay program focus only on causes of the nutrient loading to the bay. The only successful effort in the Colorado River basin has been salinity control, an effort of the Colorado River Basin Salinity Control Program. In contrast, the efforts along the Mississippi River through the Upper Mississippi River Basin Commission have been largely ineffective at addressing the numerous river channels and river bottom lands issues, without consideration to the huge watershed management issues along its many tributaries. There has not been an effective watershed initiative along the entire length of the Mississippi even with the knowledge of the Gulf of Mexico hypoxia issue and the economic effect of this nationally significant problem (NRC 2007). At these large scales it is reasonable that the organization would focus on issues of coordination among states, goals or standards to be achieved and setting priorities for actions at smaller scales by other organizations.

Watershed management efforts in the Truckee River Watershed, in California and Nevada, provide a good example of the challenges of management at a large scale. Problems and barriers identified by participants throughout the watershed included (Cobourn 1999):

- The enormity of landscape diversity and land uses.
- The huge number of stakeholders – more than 300,000 residents.
- The fear that such a large group might never reach meaningful consensus.
- The fear that the product of the effort might be a huge plan that few would read and fewer would follow.
- The fear that a consensus plan would have to be so general in order to apply to everyone that it would be useless as a working document.
- The fear that the process would take so much time and energy that many people would lose interest or not participate in the first place.
- The potential for legal issues and litigation to negate the plan.
- The problems created by crossing state boundaries.
- The needs of water rights holders.

Smaller scale watershed initiatives can focus on a wide array of important issues identified by stakeholders, communities and residents. These small scale efforts can address specific localized flooding problems caused by undersized trunk storm sewers, flood protection of homes adjacent to lakes, restoration of lakeshores for erosion control and habitat, aquatic plant control for improved fisheries production or recreational boating, implementing projects for improving water quality in an impaired water, or reducing water quantity discharge to a downstream city with a CSO issue. Small scale initiatives can realistically involve all the stakeholders to obtain agreement on a plan, program or permit coordination. This level of involvement is more difficult as the number of citizen groups, communities, counties, states and federal agencies grows in number.

11.4 Water and Watershed Management Approaches

Many major cities are often located on major water resources and need to plan for the water quantity and the water quality of flows entering and leaving the city system. How does the city manage these flows and comply with water quality standards if it has no authority to manage the planning, zoning and development standards used in adjacent (upstream) communities? Implementation of the watershed management approach to managing the cities impacts on the water resources can take two forms; developing a program using the existing political boundaries and organizations or a watershed-based approach. Both of these approaches recognize the natural watershed, but only one, the watershed-based approach, establishes a new governmental structure that is organized using the natural watershed boundaries.

11.4.1 Existing Political Boundary Approaches

Local Collaborative: Although the management of water quality and quantity is typically through a watershed approach, there are several non-watershed based approaches in use around the country. One common approach is a metro area collaborative to organize all the local governments to address a specific problem such as flood control, water quality or education. These collaboratives can take the legal form of a Memorandum of Understanding (MOU) or a Joint Powers Agreement (JPA). The MOU or JPA contain specific binding language that establishes the purpose, governing structure, funding mechanism, termination provisions and approvals for each unit of government. These mechanisms can be used to bind local, county, regional, and state agencies.

County Based Water Management: The use of the County water planning approach has been used in Minnesota through the 1985 Comprehensive Local Water Management Act. This state law required all counties in the state (outside the 7 county Twin Cities Metro Area) to prepare county-based water management plans according to basic plan content requirements of the law. The law provides that the counties should coordinate with adjacent counties to plan for drainage within watersheds. The approach holds some promise for basic water resource management programs if states enact clear plan content requirements and basic county regulatory program expectations.

City-Based Programs: Several major cities across the county have established successful local water management programs where the city has primary control over water flows into the city storm sewer system. A key ingredient to local success is a limited upstream watershed and a local commitment to water resource management. The recent federal implementation of non-point and impaired waters programs defined in the CWA provides a strong incentive for the development of effective city water management programs, collaborations or partnerships with adjacent communities.

11.4.2 Watershed-Based Approaches

Numerous watershed-based approaches have been implemented across the United States. This analysis focuses primarily on the eastern United States, where the western water rights issue does not influence organizational structure or the control of water quantity. In the process of researching this chapter, a survey was completed to attempt to inventory the watershed approaches and governance structures enabled through state legislation.[2] Numerous studies have been completed, but almost all

---

[2] Special Thanks to David Vlasin, Water Quality Monitoring and Research Intern for the Ramsey-Washington Metro Watershed District in 2007, for his research and development of Table 11.1.

previous studies analyzed specific issues as they relate to watershed governance and structure. Some of these studies focused on voluntary organizations (Lubell et al. 2002), some focused on the public participation roles of various structures (Mullen and Allison 1999), and some focused on state government approaches (NPCC 2002).

The objective for a watershed management program is to assure that the use and modification of water resources and land activities doesn't undermine the ability of our waters to sustain themselves over the long term (NPCC 2002). Accomplishing this objective within the framework of governance is the challenge. Several past publications and papers have explored and analyzed watershed governance structures and classification schemes. Some classification schemes are more complicated than others, but they can almost all be generalized into an approach presented by Michael Mullen and Bruce Allison in their analysis of water management success in the state of Alabama (Mullen and Allison 1999). In their paper Mullen and Allison presented four basic watershed management models that fit most forms of watershed management structures (these four structures are applied to the examples listed in Table 11.1):

1. *Top-Down Agency-Led Watershed Management Model*: This is likely the oldest and most common watershed approach applied in the United States. In this model, a state or federal agency or agencies design a management program, complete studies and implement a set of actions with little or no involvement of local watershed stakeholders. Local involvement typically happens when the state or federal agencies provide a cost-share program to involve local participants to implement pollution control activities identified in their basin studies. Progress toward alleviating the specific problem can be successful as long as funding is available and stakeholders see a local advantage to accept the funding to implement needed local projects.
2. *Agency-Coordinated Watershed Management Model*: This model represents the approach where an agency funds the formation of a watershed program and a staff coordinator to work with local stakeholders. The agency coordinator works with the stakeholders to study the watershed, assess problems, develop local organizational structure and assist with projects and programs. The limitation of this approach lies with the funding of the sponsoring agency. In most examples of this approach, agency funding is maintained for a short period of time within which a local stakeholder group must be established to take over the efforts organized by the agency coordinator. If stakeholder involvement is insufficient, the watershed initiative can deteriorate or fail. Another potential limitation is the ongoing funding of the local stakeholder effort. Self-sufficiency needs to be established before the withdrawal of agency funding for long-term success. Given that many watershed programs can take decades to implement, this is a critical issue.
3. *Watershed Management Authority Model*: This model represents a watershed authority that is locally initiated and provided with self-sustaining local government authorities. This option must be provided under state enabling

Table 11.1 State watershed programs

State	Code	Yr. Est.	Purpose/Goal	Program Summary	Adm. Agency(s)	Funding Source	Status/Success
Kentucky	A	1997	Coordinate existing programs and build new partnerships to improve the management of the states land and water resources	Established 5 watershed units and a state-wide planning process and schedule. Each basin has dedicated coordinator, basin team and stakeholder committee	Division of Water	Division of Water	Created Committee to address framework issues and communication and a Partner Network including governments, local businesses, and individuals.
Massachusetts	B	1993	To help agencies, non-profits, and businesses integrate on a watershed basis - identify local problems, and coordinate implementation.	Watershed teams focus on a 5 year management scheme to: Collect information; target present and future impacts; assess impacts; develop and implement activities.	Secretary of the Office of Environmental Affairs	State Watershed Round-table and partner shared funding	Have multi-discipline watershed teams in each of the 27 major watersheds; 20 full-time team leaders coordinate teams of local, state and federal groups.
Ohio	B	1990	To support monitoring and permitting programs and sample new streams/bodies of water.	Established 5 basins with District offices, hired 21 watershed coordinators, developed 5-year basin plans and reorganized to focus on basin planning for TMDLs.	Ohio Dept. of Natural Resources and Ohio EPA	319 Funds and biennial appropriations to state agencies	Revising State watershed management to focus on the TMDL program. Having difficulties coordinating priorities into the 5 year basin cycles.

11 Institutional Structures for Water Management in the Eastern US 225

Table 11.1 (continued)

State	Code	Yr. Est.	Purpose/Goal	Program Summary	Adm. Agency(s)	Funding Source	Status/Success
Florida	C	1972	Provide water management planning and implementation of projects to address water resource issues in the district.	Established 5 water management districts (with 7–11 watershed units) that cover the state. Multi-purpose water management agency tailored local conditions and issues.	Governor and Dept. of Environmental Protection	Ad Valorem tax and state funds and federal grants	All Districts are active and involved in addressing water resource issues in each District.
North Carolina	A	1991	To streamline NPDES permitting and facilitate more efficient permit re-issuance.	Developed a framework for 17 river basins using a 5 year rotating basin planning cycle. Help coordinate monitoring, assessment, federal programs & basin plans and coordinate with other agencies and stakeholder groups.	Department of Environmental and Natural Resources – Water Quality Section	State funds, 319(h) grants, loans/grants from the USDA and Clean Water Management Trust Fund	Established programs in 17 major river basins. (Example) In 1995 the Hiwassee River Watershed Coalition (HRWC) was formed to restore the Brasstown Creek Watershed.
Oregon	D	1997	To help restore salmon runs, improve water quality and help achieve healthy watersheds	Led by a 17 member board that provides support and funding for the 93 watershed councils and their projects. Provides guidance and technical assistance.	Oregon Watershed Enhancement Board (OWEB)	State and Fed funds, 15% of all State lottery and State salmon fund	Key successes are: Voluntary restorations efforts; coordination of federal, state and local tribes; monitoring; strong scientific oversight.

Table 11.1 (continued)

State	Code	Yr. Est.	Purpose/Goal	Program Summary	Adm. Agency(s)	Funding Source	Status/Success
Nebraska	C	1969	Reorganized state's 500 special purpose districts into 23 Natural Resource Districts on watershed lines. Provide comprehensive water and land planning and management.	Elected Board with authority over land and water resources, solid waste disposal, sanitary systems, drainage, pollution control, fish & wildlife mgt., parks and recreation and forest and range management.	Nebraska Office of Natural Resources Conservation Services	Local tax revenues and bonding	Second in the United States for completed projects – 2,972 acres of wetlands restored, enhanced, or created – 1.1+ million acres improved for water quality.
New Jersey	B	1998	To restore, maintain, and enhance water quality/quantity and overall ecosystem health.	Established 5 water regional bureaus with contractors and staff for Watershed Management Areas (WMAs). 20 watershed management areas nested within the 5 bureaus.	New Jersey Department of Environ-mental Protection (NJDEP)	NJDEP grants to the 20 WMAs and 319(h) grants	Each WMA has assigned staff for education and outreach programs to help involve the public. Has great education programs
Washington	D	1993 and 1998	To synchronize monitoring, inspections and permitting and to address Point and non-point source pollutions. Authorizes a local volunteer approach to watershed planning.	Focuses primarily on permitting and monitoring programs. Provides guidance and funding to establish watershed organizations and develop plans that address existing water rights and water quality related problems.	Washington State Department of Ecology. County, City and/or regional government	State Legislative appropriations	23 Water quality management areas. 62 watershed resource inventory areas (WRIA) with in the 23 WQMAs (these are local/volunteer driven)

Table 11.1 (continued)

State	Code	Yr. Est.	Purpose/Goal	Program Summary	Adm. Agency(s)	Funding Source	Status/Success
Pennsylvania	A	2002	To provide technical assistance, guidance and tools to watersheds. To provide residents interested in water resources, the ability to offer input on watershed issues.	Every 5 years the state must update its State Water Plan for each of the 6 major watersheds. Developed a Unified Watershed Assessment that helps in deciding watershed restoration priorities.	Bureau of Watershed Management, Dept. of Environ-mental Protection	Growing Greener Act (HB2) $625 million. Small Water-shed Grants Program.	6 Major Watersheds with 33 State Water Plan Sub-basins (watersheds). The Growing Greener Act (HB2) of 2005 was passed for maintenance and protection of the environment.
Michigan	A	2003	To set water quality standards, regulate public water supplies, permit the discharge of industrial/municipal wastewater, monitor quality/quantity of aquatic communities.	Reviews and approves public water supplies. Supervises wastewater collection and treatment. Provides direction to local sediment and soil erosion control programs and well pump/driller registrations	Dept. of Environ-mental Quality and Environmental Advisory Council	Clean Michigan Initiative Act, Grants, and Fee charges from members of Water-shed Alliance	Currently the watershed management approach is under review by the Environmental Advisory Council to determine if there are ways to improve the program.

Table 11.1 (continued)

State	Code	Yr. Est.	Purpose/Goal	Program Summary	Adm. Agency(s)	Funding Source	Status/Success
Minnesota	C	1952	Watershed Districts - To provide local management of water quality, flood control, drainage management, erosion control, natural resources.	Provide authority to establish a Watershed District. Boards appointed by the Counties. Formed on watershed lines. Are special purpose districts	Minnesota Board of Water and Soil Resources (BWSR) BWSRBWSR	Local ad valorem tax, grants, local cost-share funding, grants	46 Districts cover about 1/3 of the state. Budgets from $100,000 to $6 million. Actively involved in projects and TMDLs. Most have regulatory program.
		1984	County Water Planning - Requires Water Management plans for each County Outside of the Twin Cities Area	All 77 Counties have prepared County Water Plans. Plans are revised every 10 years or less.	BWSR	State cost-share grants for administrative support	County plans are completed, but most Counties give minimal support to the program.
		1982	Metropolitan Surface Water Management Act. Requires creation of Watershed Management Organizations in the Twin Cities Area.	Can be Watershed Districts or Joint Powers Boards (JPA). All are required to prepare comprehensive watershed management plans under state rules for content.	BWSR	Metro WMOs have ad valorem taxing authority for projects identified in the plans.	All WMOs are actively implementing plans and coordinating programs. Variability in size and complexity of programs. Watershed Districts are generally more active than the JPAs.

Watershed Program Approach: A (Top-down Agency Led Model); B (Agency Coordinated Model); C (Watershed Management Authority model); D (Locally led, Community-based model)

legislation. Only Minnesota and Alabama were found to provide this authority. Under this model the local watershed authority is usually formed to address a specific resource issue, but it has additional authorities to address issues identified in the watershed management planning process. The watershed also has typical local government authorities to raise funds (through taxes, fees, grants and local cost-sharing), hire staff and consultants, implement regulatory programs and construct capital improvements. Since this model reflects a process of local stakeholders requesting the formation of the watershed entity, there is an established stakeholder support group from the start. The watershed needs to implement programs, education efforts and a stakeholder involvement process to maintain the support for its programs and funding. Alabama has only one existing organization (Choctawhatchee-Pea Rivers Watershed) formed under this authority (Act 91-602). Under the Alabama example, the state provides the operational funds for the organization. Minnesota's Watershed Management Act (MS 103D–955) and the Metropolitan Surface Water Management Act (MS 103B) establishes the watershed entity as a Special Purpose unit of Government and provides watersheds with local taxing authority. There are 46 Watershed Districts and 34 Joint Powers Watershed Management Organizations in Minnesota.
4. *Locally Led, Community-Based Model*: Watershed initiatives under this model are also known as watershed groups, councils, and partnerships. There are, by some estimates, as many as 3,000 small watershed, community-based watershed groups in the United States (CTIC 2007). Watershed management under this model would be initiated, organized, led and largely funded at the local or watershed level. The organization would have representation and involvement from a wide spectrum of stakeholders. Key to this model is a strong local stakeholder involvement and local funding of the organization. Organizations under this model usually possess the following characteristics: (Kenney 1999)

- Broad and open participation.
- A focus on a specific resource issue and a defined geographic scope.
- An informal and flexible structure.
- A collaborative and consensus-based process for discussion and decision-making.
- Action orientation for dispute resolution, watershed planning, and field level projects.

These watershed organizations face significant challenges if they are to achieve significant success at achieving water resource goals. Significant among these challenges is the issue of maintaining stakeholder support and involvement over the long term, obtaining long-term financial support for administration and programs, and developing supportive collaborative relationships with the local, state, and federal agencies needed to implement actions developed by the organization.

There is some debate about whether these voluntary watershed initiatives should stay informal and voluntary or whether they should be transformed into official governmental organizations, with the authorities and responsibilities of local government. Some believe there is almost unanimous agreement that these initiatives should remain informal, voluntary associations (Kenney 1999). However, many others, particularly in the eastern United States are falling into the camp of formal governmental based organizations. This belief stems from the ultimate need for governmental action to address real water resource and water quality improvement needs, some of which are state and federal mandates. These changes ultimately become based in structural capital improvements that require major government funding or become involved in land use and development regulation to bring about changes in land development and management practices that influence water quantity and quality.

11.5 Lessons Learned

11.5.1 Implementation Difficulties/Barriers

Management of water resources on a watershed-basis faces a number of real and perceived problems. Many of these problems are practical in nature and others are political and institutional. Political and institutional issues relate mostly to the issues of collaboration and partnership. In many cities, particularly large cities, the significant internal bureaucracy presents significant issues of "turf" protection and power and control issues. Large cities often feel their issues are more important than those of the smaller adjacent communities. This is also an issue with state agency participants and their perception of the knowledge and experience of local city staff.

Before any city, state agency, or other participant commits their time and resources to a watershed initiative they need to be aware of these barriers:

- *Successful watershed collaborations take time, energy, and resources* (NPCC 2002) Participants need to recognize that organizational development and the planning and implementation process takes time. Many participants in new organizations get frustrated with the slow pace of progress. There needs to be a clear expectation at the outset of the process for all members. The danger of time is that key institutional history and understanding gets lost and key participants leave the process. Participants need to guard against the situation presented by Stuart G. Walesh in his Hydrological Cycle below. Figure 11.1 accurately depicts the common fluctuating public interest in water problems. (Walesh 1999)
- *Relevant stakeholders need to be included in the process from the beginning* (NPCC 2002) All watershed models can have some success in improving water quality and managing water resources with adequate funding, but none of the approaches will have long-term success without broad stakeholder support and community involvement. Involved stakeholders need to believe in the mission, goals, and objectives of the watershed program and allow it to grow and achieve success.

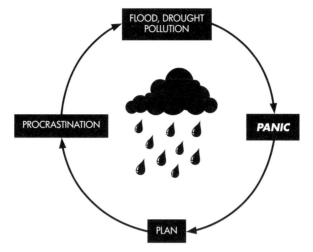

Fig. 11.1 Hydroillogical cycle

- *Resource limitations can be a major obstacle to success* (NPCC 2002) Resources refer to all the funds, personnel and equipment needed to effectively operate the business of the watershed and accomplish its tasks. This also includes the resources of leadership, advocates and volunteers. These resources don't need to be all from within the organization. Essential resource support could, and often do, come from volunteers, collaborating organizations and agencies.
- *Overlapping state and federal agency programs can cause unnecessary overlap of activities and be confusing to participants* It is common misconception that because there are numerous agencies and programs dealing with water management and regulation, that a new watershed organization or collaborative would be duplicative and wasteful. Several of the state legislatures have initiated studies to look at agency and program consolidation to improve water resource management and save tax dollars (Minnesota Planning 2002). These efforts typically document the programs, laws and rules currently in place and conclude that they serve their own purposes and are generally not duplicative. However, it is clear that numerous agencies and programs create confusion and create a significant barrier to discussions for change through the creation of yet another organization or program. The challenge for leaders seeking improvement is to be very clear about the objectives and purposes of the watershed management effort. Leaders must stay on message about the needs that are not being met with the current political structure and the need for a cross-jurisdictional approach to a coordinated action plan.
- *Risk and uncertainty are natural elements of nature and institutional settings* Organizations need to consider these factors, but not allow it to stall progress. Decision makers often ask the right question related to risk and uncertainty, but too often want all risk and all uncertainty to be resolved before taking action. Water resource management is an evolving science and practice. There are

constantly new techniques for analyzing the problem and new BMPs for addressing the issues. Some level of understanding and practice of best professional-judgment is needed. This approach leads to field testing of new approaches and the advancement of watershed science for the organization and the greater watershed community.

11.5.2 Factors Influencing Success

A number of publications provide good lists of characteristics and critical ingredients of successful watershed initiatives (NPCC 2002, Goldstein and Huber-Lee 2004). Some of these factors apply to water management planning no matter where it occurs, but is particularly acute in urban environments. These major factors need to be recognized in developing and implementing any water management program.

- *Organizational structure has important impacts on success and sustainability:* A government agency structure appears to universally work best for addressing issues related to wastewater treatment. Development of a structure that can address all flows to the system is critical. Collaborations based entirely on stormwater management and non-point pollution control can take on numerous alternative forms. As demonstrated in eastern United States examples, the more formalized structures have implemented programs that resolved significant water resource management issues. However, the overused adage is generally true; there is no "one-size-fits-all" approach. This applies to the watershed scale as well as the organizational structure. Political, social, geographic, and resource factors all influence what structure can work in a particular situation.
- *Scale is critical to development of a successful structure*: (NPCC 2002) Specific watershed problems must be addressed with distinctively different approaches. If the scale of the watershed organization is inappropriate to the scale of the issue, there will be an inherent conflict that will lead to ineffectiveness, frustration on the part of the participants, and ultimate failure.
- *Watershed management structure facilitates education and understanding of the upstream-downstream and human versus nature conflicts*: (Goldstein and Huber-Lee 2004) The watershed forum allows for the involvement of stakeholders and the mutual discussion of issues to help avoid conflict and work toward mutually beneficial solutions. This provides the opportunity for resilience and adaptability to changes in watershed issues.
- *Established regulatory authority in a watershed-based institution can result in watershed protection across political boundaries*: (Goldstein and Huber-Lee 2004) Local political issues strongly affect the development of rules and regulations for development. A watershed approach can lead to the implementation of universal standards and criteria that can directly address critical watershed issues like runoff volume control, erosion and sediment control, and wetland protection.

- *Many successful watershed-based organizations are formed to address a water management crisis*: The crisis can be an organizing factor that leads to the discussion of the need for a watershed management program or new initiative. This crisis can be a resource management issue like the impact of water quality on fisheries (salmon in the Pacific Northwest), flood management in the Red River Valley of Minnesota, or a land locked lake flood management program in the Valley Branch Watershed District in Washington County, Minnesota. The crisis could also be a lawsuit from a private citizen or another government agency under the CWA or the ESA. The crisis could also be brought on by new state or federal regulations and the threat of action for non-compliance under a NPDES permit or the need to address impaired water in the TMDL program.
- *Leadership is critical to any change*: Especially a change in governance where there are inevitably perception of winners and losers. Environmental issues often are championed by a strong advocate. Water management is not simply an issue of improving water quality or protecting people from flooding, although this may be the primary goal, the initial objective must be to create a structure for change. This objective requires a respected local leader that recognizes this need and is willing to be out front to discuss the issue and facilitate discussion and the decision making process.
- *A successful watershed-based approach requires a stable and sustainable funding method*: Successful comprehensive water and related natural resource management cannot depend upon grants or state and federal funding. A sustainable source of funding is essential to the success of the watershed initiative. Funding is needed, not only for basic organizational and administrative support, but for project and programs. Successful watershed programs will identify problems that need to be addressed. Some of these problems will require capital improvements (flood control, bank protection) and some will require program activities (public education, regulations). These efforts require a source of funds that don't require constant appeals to other units of government for contribution or being dependent on grant funds.
- *Recommendations and implementation must be based on sound science*: The credibility of the organization lies in its actions. If implementation actions are questioned and evaluated, which they should, they need to stand up to the scrutiny of sound science and engineering. Organization activities need to be based initially in comprehensive watershed planning. Resource information needs to be gathered and analyzed. Appropriate hydrologic models need to be used to characterize the water flow through watershed and identify the issues related to rate and volume. Only through this analysis can the organization begin to identify causes of problems and have rational discussion of problems and the potential solutions to those problems. This framework of information also allows for a discussion of priorities, phasing and cost. A caveat here is that "sound science" does not always mean perfect science. There is often a need to move forward with the information we have and practice adaptive management.
- *Monitoring and performance evaluation is critical to demonstrating success and motivating continued support and involvement*: Monitoring of water resources is

a primary method of determining needed actions as well as determining program success. These measures are needed to demonstrate the impact of the organization on the environment and its progress toward achieving its goals and objectives.
- *Written agreements are essential for informal watershed collaborations*: These agreements need to include state and federal participants. The time and effort to develop and approve these agreements will provide for more efficient implementation of the tasks and operation of the organization. (NPCC 2002).
- *Public education and pubic involvement is critical to the success of watershed management*: Recognizing that this is important and good, but implementing an effective program is extremely difficult. Education and public involvement needs to be recognized as critical expertise that is required in a watershed management program. Public education is not merely publishing a newsletter or brochures. Particularly in this age of non-point pollution prevention, the individual in our society plays a key role in the solution. Conveying this message requires multiple and creative techniques and delivery to multiple audiences (elected officials, city staff, city residents, school children, and the business community).

References

Cobourn, J. 1999. Integrated Watershed Management on the Truckee River in Nevada. J. Am. Water Resour. Assoc. **35**:623–632.
CTIC. 2007 Know Your Watershed Web Site, http://www2.ctic.purdue.edu/kyw/. Conservation Technology Information Center, Purdue University, Purdue University, West Lafayette, Indiana.
Goldstein, J. and A. Huber-Lee. 2004. Global Lessons for Watershed Management in the United States. Water Environment Research Foundation, Alexandria, VA. Project number 00-WSM-5.
Kenney, D.S. 1999. Historical and Sociopolitical context of the Western Watersheds Movement. JAWRA. **35**(3).
Lubell, M., M. Schneider, J. T. Scholz, and M. Mete. 2002. Watershed Partnerships and the Emergence of Collective Action Institutions. Am. J. Political Sci. **46**:148–163.
Minnesota Planning. 2002. Charting a Course for the Future: Report of the State Water Program Reorganization Project. Minnesota Planning, St. Paul.
Mullen, M. and B. Allison. 1999. Stakeholder Involvement and Social Capital: Keys to Watershed Management Success in Alabama. JAWRA. **35**(3).
NPCC. 2002. Watershed Solutions: Collaborative Problem Solving for States and Communities. National Policy Consensus Center, Portland, OR.
NRC. 2007. Mississippi River Water Quality and the Clean Water Act: Progress, Challenges, and Opportunities. Committee on the Mississippi River and the Clean Water Act, National Research Council, Washington, DC.
NRC. 1999. New Strategies for America's Watersheds.. Committee on Watershed Management, National Research Council Washington, DC.
Walesh S. G. 1999. Dad Is Out, Pop Is In. J. Am Water Resources Assoc. **35**:535–544.
Webster. 1994. Merriam Webster Collegiate Dictionary. Merriam-Webster, Inc., Springfield, MA.

Chapter 12
Adaptive Water Quantity Management: Designing for Sustainability and Resiliency in Water Scarce Regions

Jim Holway

Water quantity management in the southwestern (SW) United States is becoming increasingly complex, challenging and controversial. Key factors driving this include: Rapid population growth within multi-jurisdiction metropolitan areas, drought and an increasing awareness of climatic variability and global climate change, needs to reallocate existing supplies as well as to make water available for environmental purposes, and the magnitude of the costs and benefits at stake.

This chapter will consider water supply and demand management approaches, in particular sustainability challenges, conjunctive management and adequate supply requirements. Although we draw on the institutional framework, results, and future directions in Arizona, we briefly consider programs in other SW states. Finally, drawing on lessons learned from these efforts, we consider ways to increase the sustainability and resiliency of urban water supply systems.

12.1 Water Policy Challenges

Ensuring adequate water supplies to serve continued growth has long been a focus of municipal and state water managers in the arid SW. In keeping with the overarching themes of sustainability and resilience, this chapter opens with an overview of the key trends shaping sustainability and future water management challenges.

The sustainability of water supplies is a function of (1) population, (2) natural resources, (3) lifestyles and consumption habits, (4) technology, (5) governance, and (6) adaptation. Each of these factors will be considered in turn.

12.1.1 Population

The urban areas of the SW are among the fastest growing cities in the country, illustrated in Fig. 12.1 by population trends in Los Angeles, Phoenix, and Las Vegas.

J. Holway (✉)
Arizona State University, Tempe, AZ, USA
e-mail: jim.holway@asu.edu

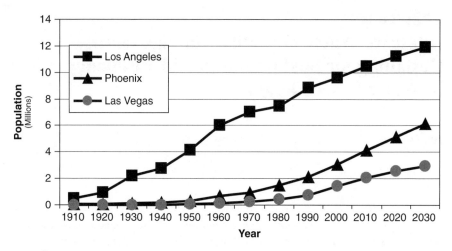

Fig. 12.1 Population trends for the central counties of three major SW cities

Rapid growth places considerable stress on limited water supplies, physical infrastructure, management institutions, and the fiscal capacity of water providers and local government.

12.1.2 Natural Resources

SW cities vary in their reliance on different sources of supply and their climate. Table 12.1 shows the mix of groundwater, surface water, and effluent being used to supply the Phoenix and Las Vegas metropolitan areas. Some of these supplies, in particular surface water, are quite responsive to annual variations in climate, others are more stable. Average rainfall over the major cities of the SW varies from 4.5 inches per year in Las Vegas, 8 inches in Phoenix, 12 inches in Los Angeles, to 15 inches in Denver. This can be contrasted with the approximately 40 inches per year received in most eastern cities. Because of the arid climate, evaporation and transpiration rates are also much higher, leading to significantly less water availability.[1] For the most part, there are no unallocated renewable water supplies in the SW, hence new water demands will be met though re-allocating existing supplies, overdrafting groundwater, reusing effluent, or desalting seawater.

Recent research on climate and river flows, as well as efforts to re-construct historic conditions, show that droughts lasting 20 to 30 years are common in the SW (Fig. 12.2). Note that the 1922 allocation of the Colorado River among the

---

[1] An illustrative example is that in Phoenix a 6-foot deep swimming pool filled in the spring would be empty by the end of the year due to evaporation; by contrast a 6-foot deep pool in South Florida that started off empty would be full two years later from the excess of precipitation over evaporation.

Table 12.1 Water supply source – metropolitan areas (percent of total supply)

Water Source	Phoenix 2003	Las Vegas 2007
Groundwater	40	10
Colorado River Water	26	86
Other Surface Water	27	0
Effluent	7	4

Note: Phoenix numbers are based on total water use in the Phoenix Active Management Area, which is comprised of multiple water providers with the City of Phoenix itself serving approximately 1/3 of the region. The Las Vegas numbers are based on deliveries by the Southern Nevada Water Authority.

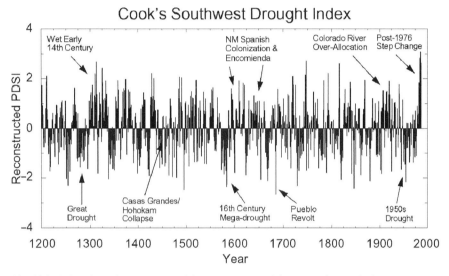

Fig. 12.2 Colorado Basin drought conditions, reconstructed from tree ring Analysis
Tree Ring reconstruction was used to estimate values for the Palmer Drought Severity Index. Values > 0 are wetter than average; values < 0 are dryer than average. Produced by J.L. Betancourt, U.S. Geological Survey, based on Cook 2000.

seven basin states occurred during a particularly wet period (Fig. 12.2), a fact that resulted in over allocating the river and created problems for future dry years. A number of major river basin allocations and much of the regional growth through the mid-1990s occurred during what now appear to have been wet periods throughout the SW.

Water managers throughout the SW will be at the forefront of efforts to deal with climate. They are beginning to grapple with the reality of the current drought and climatic variability. However, water managers must now also cope with the potential that global climate change may create both permanent reductions in available water and greater variability. Though the various climate models provide a wide range of results for future precipitation and temperatures for the SW United States, they generally agree that temperature will increase, climatic variability and the intensity of storms will also increase, and even if rainfall is not significantly impacted, runoff to the reservoir systems will be significantly reduced due to changes in seasonal runoff patterns and higher evapotranspiration. Preliminary evidence indicates that by the end of the century stream flows could be reduced by somewhere between 8 and 21% for the Colorado River and perhaps up to 30% for the Salt and Verde River system that supplies the Phoenix metropolitan area (Christensen and Lettenmaier 2006, Ellis et al. 2008, Nash and Gleick 1993, Nelson et al. 2007, Holway and Jacobs 2006, IPCC 2007).

12.1.3 Lifestyles and Consumption Patterns

Our individual consumption patterns, summarized as the gallons consumed per person per day (gpcd), together with population, determine the magnitude and nature of our demand for water. Throughout the urban SW, 25 to 40% of municipal water is used for non-residential purposes (businesses, parks, other public uses, etc.). The remaining 60 to 70% of use occurs at individual homes, and the majority of this is used outdoors (landscaping and pools primarily). Across several SW cities, total water use ranges from about 175 to 300 gpcd (Fig. 12.3). For Phoenix, 66% of total water use is residential and landscape irrigation uses more than half of water deliveries in both residential and non-residential sectors (Table 12.2).

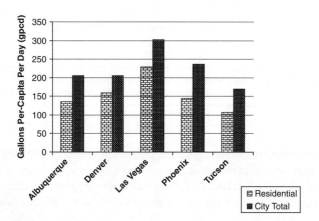

Fig. 12.3 Comparison of water demand for SW cities
Residential numbers are for use by single family households only. City totals are total water deliveries (including residential and non-residential uses), divided by population. All values are for 2001. Several of these cities have reduced their demand by 10% or more due to aggressive conservation efforts since 2001. These data also do not include certain non-potable water deliveries. For example, Denver Water customers receive approximately 32 gpcd and Phoenix residents receive approximately 30 to 35 gpcd on average in non-potable deliveries for landscape irrigation.

Table 12.2 City of Phoenix – Water deliveries by use

Sector and type of use	Percent of total	Percent of sector
Residential	66	
Landscape		51
Pools		14
Indoors		35
Non-Residential	34	
Landscape		61
Cooling		8
Other (e.g. process, sanitary)		31

Table 12.3 Statewide consumptive water use by sector (percent of total demand)

Sector	Arizona	California	Colorado	Nevada
Agriculture	74	82	86	73
Municipal	20	17	7	23
Industrial	6	1	2	4
Other			5	

Note: Figures are based on most recent statewide data available. 2006 for Arizona and Colorado and 2000 for California and Nevada. All data from state agency websites.

Agriculture is the major water user in most SW states, with municipal and industrial water use, the focus of this chapter, comprising not more than 14 to 27% of overall water demand in the individual states (Table 12.3). In Phoenix, increasing urban demand has been offset by decreasing agricultural use, with little change in overall water use (Fig. 12.4). For those regions and states with significant agricultural irrigation, converting water from agriculture to municipal use may be a principal source of supply for new urban growth, at least for a while.

The changing nature of water demands also impacts the resiliency of our water management efforts and the ability to handle variability in supply. Agricultural users deal with wet and dry years by changing their cropping mix, acres planted, and sources of supply. It is much harder to adjust demand for municipal and industrial supply. Additionally, as we institute conservation programs and reduce inefficiencies in our urban water uses, a further hardening of demand may occur that would affect our ability to obtain additional reductions during shortages. With demand pushing up against the limits of available supply, these factors are reducing our resiliency. Given the increasingly complex nature of our water supply, these changes in demand put increasing pressure on the institutions and technology used to manage water.

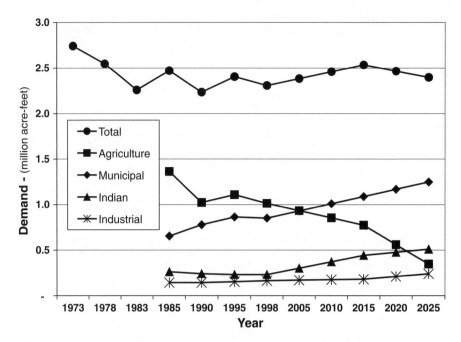

Fig. 12.4 Historic and projected sector demands for water in the Phoenix AMA.
Source: ADWR 1999, augmented with data prepared for the Governor's Water Management Commission, ADWR

12.1.4 Technology

Technology is critical to sustain an ever increasing population and demand for resources. Cities rely on technology, from water supply acquisition, storage, treatment, and distribution, through the collection, treatment and reuse or discharge of treated wastewater. Major canal systems date back over a thousand years in Phoenix and other pre-historic settlements throughout the SW. In the modern era, dams that store water over several years and reduce the damage of periodic flooding events have been critical to the settlement of the west. Most recently, canal systems stretching hundreds of miles have enabled the SW states to import water that fell as snow in far-off mountains.

In the 1940s, large-scale diesel pumps allowed extensive use of groundwater for the first time, foreshadowing the increasing importance of energy in the water sector (Chapter 4). The transportation of new water supplies over increasing distances or the additional treatment required to utilize water of lower quality will both require significantly more energy than our current supplies. In addition, environmental and public health concerns, as well as the increasing need to reuse treated water, are greatly increasing the need for higher levels of pollutant removal from wastewater, saltwater, and contaminated supplies.

12.1.5 Governance

As our cities become larger, and we approach the limits of previously secured water supplies, our institutions also grow and the rules for administering water become more complex. Governance refers to the combination of water providers, government agencies, laws, and contractual arrangements we use to manage water systems. Four particular governance issues are relevant for this chapter: (1) the multi-jurisdictional nature of metropolitan water management, (2) the need to secure new supplies or re-allocate existing supplies, (3) a consideration of the role of regulatory versus non-regulatory policies in adaptation and change, and (4) mechanisms to design our water systems to foster sustainability and resiliency.

Different SW cities exhibit very different institutional structures for water management; ranging from the highly regulated Active Management Areas of central Arizona, which rely on a large number of decentralized and fragmented water supply entities, to the centralized Southern Nevada Water Authority, which supplies raw water for all Las Vegas area water providers. The different institutional relationships and water management frameworks that exist throughout the SW provide opportunities to experiment with different approaches, to evaluate the different systems and their characteristics, and to consider which approaches are more effective at achieving the short and long-term changes necessary for the sustainability of water and our cities.

12.1.6 Adaptation and Resiliency

Each of the sustainability factors above (population, natural resources, lifestyle, technology and governance) can be evaluated based on current conditions as well as projected trends. By considering adaptation and resiliency needs, water management systems can be modified to increase their sustainability and ability to respond to changing conditions. In the following section, we examine water management frameworks and several innovative programs in Arizona and the other SW states to illustrate approaches to these water management challenges. In the final section, we will discuss the design of water management for sustainability and resilience.

12.2 Water Management: Policy Frameworks and Programs

Water management frameworks vary considerably across the SW states; and include strong state-based programs, state-local partnerships, and other approaches which are primarily local and water provider based. The nature of programs also runs the gamut from regulatory, to technical and financial assistance, to educational (Table 12.4). Some programs focus on one of these mechanisms, but most use a hybrid of different approaches.

Table 12.4 Water management framework characteristics: continuums and approaches

- Locus of control (Who is in charge):
  - State – Regional – Local – Water Provider – No One
- Geographic scope (Where does program apply):
  - Statewide – Regional – Local
- Nature of Boundaries (How management areas are determined):
  - Natural Basins – Political Jurisdictions
- Principal focus (What is managed or required):
  - Groundwater – Surface Water – Effluent
  - New Supplies – Conjunctive Mgmt – Conservation – Water Rights
- Water Rights System (Basis of water rights and priorities):
  - Prior appropriation – Hybrid/Correlative – Riparian – Allocation/Permit
- Role of Water Markets:
  - Allocations (No Market) – Hybrid – Open Market
- Program Mechanisms (What types of public policies are utilized):
  - Regulatory – Funding – Incentives – Tech and Data Assist – Education – Administration/Enforcement – None
- Regulatory Mechanisms:
  - Performance Standard versus Prescriptive Requirement
  - Direct Mandates and Controls versus Modifying the Decision Making Environment

SW water management programs have been shaped by the nature of available supplies throughout the semiarid region. As these programs evolve they will be increasingly affected by an expanding understanding of climate change and the drivers that affect water availability. Individual state and local programs have also been affected by the economics and technology of acquiring, storing, treating, and distributing those supplies; the legal framework of land ownership and water rights; the growth and pattern of water demands; and the political nature of individual areas and their major historic water users.

12.2.1 Arizona's Approach

Arizona's 1980 GMA created the Arizona Department of Water Resources (ADWR) and established a regulatory framework for managing Arizona's groundwater. The three primary goals of the GMA were (1) to control overdraft, (2) to allocate the state's limited groundwater resources, and (3) to augment Arizona's groundwater through water supply development. Politically, the driving forces for Arizona's adoption of the GWA included concerns about groundwater depletions, the threat to withhold federal funds for the Central Arizona Project (CAP)[2] and legal challenges by agriculture to the ability of mines and cities to transfer groundwater away

---

[2]The Central Arizona Project (CAP) is a major reclamation project involving a 336-mile-long canal, pumping stations, and power plants for bringing Arizona's allocation of Colorado River water into central Arizona.

12 Adaptive Water Quantity Management 243

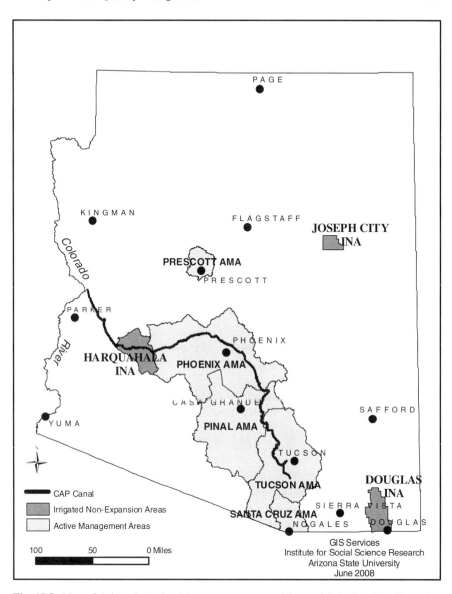

Fig. 12.5 Map of Arizona's Active Management Areas (AMA) and Irrigation Non-Expansion Areas (INA)

from the area of pumping. The 1980 act created Active Management Areas (AMAs) for regions of the state with pressing groundwater concerns, as well as the lesser regulated Irrigation Non-Expansion Areas (INA) (Fig. 12.5). The boundaries of AMAs are generally defined by groundwater basins and sub-basins rather than by the political lines of cities, towns, or counties. The goal for four of the five AMAs, as established by the GMA, is to achieve *safe yield*. "Safe yield" means establishing and maintaining a long-term balance between the annual amount of groundwater withdrawn in an AMA and the annual amount of natural and artificial recharge in the AMA. The basin wide safe-yield goal, as defined in the GMA, which does allow impacts to springs, surface flows and habitats as well as local groundwater depletions, is not the same as the concept of sustainability.

Arizona's AMA groundwater management programs utilize four major regulatory components: (1) the structure of the water rights system and its measurement and reporting requirements, (2) the requirement for adequate (or assured) water supplies for new municipal development, (3) recharge and recovery programs facilitating conjunctive management of available supplies, and (4) conservation programs for major water users.[3] In the following sections we elaborate on several of these programs and briefly consider related efforts in other SW states.

12.2.2 Adequate Water Supply Programs

The principal objective of adequate supply requirements is to ensure that new municipal development has a secure, and in some cases, renewable supply of water. Programs should be designed to influence developers who want to build, landowners who hold vacant land, and local jurisdictions that approve new subdivisions. Ideally these programs feature a strong regulatory component, with ultimate control at the state level, to prohibit local governments from permitting the subdivision of land without a secure water supply. Given the intensely political nature of local land use decisions, placing the authority for requiring adequate water supplies at a level of government higher than that responsible for approving subdivisions may be essential to withstand local political pressures to overrule hydrologic determinations. On the other hand, advocates for improving the integration of water and growth decisions want water adequacy and allocation decisions to be linked to land use decisions at the local level (Arnold 2005, Hanak 2007, Tarlock and Van de Wetering 2007). An appropriate state role may be to further encourage local authority while serving as a final stopgap, with monitoring and compliance authority, to make sure water adequacy was demonstrated prior to final subdivision approval.

Arizona's approach to water adequacy was driven by their 1980 GMA requirement to establish Assured Water Supply rules by 1995. These rules require all new subdivisions in an AMA to demonstrate a secure 100-year supply of water, primarily from renewable water supplies, before a subdivision plat can be approved by the local government.

Arizona is in a unique position relative to the linkages between water and growth. Within the AMAs, new development is subject to the strictest state-assured water supply requirements. In the non-AMA, primarily rural areas, however, Arizona has

---

[3]For greater detail on Arizona's programs see: Holway and Jacobs 2006 or Jacobs and Holway 2004. For a recent evaluation see Maguire 2007 and for a highly critical review see Hirt et al. 2008.

had one of the weakest programs in the country. In these areas, weak state rules were generally considered to prevent local governments from making stronger regulatory linkages between growth and adequate supplies. Though it is too soon to evaluate its impact, state legislation passed in 2007 grants limited water adequacy authority to local governments throughout the state, and 2005 legislation requires all water providers to prepare water plans. These requirements will, at a minimum, raise awareness of water resources issues and likely increase the call to improve available information.

Each of Arizona's neighboring states has a different approach to managing the linkages between water and growth and on requirements for adequate water supplies. California's 1983 Urban Water Plan Management Act required every water provider to prepare a long-range plan every 5 years. In 1995, California added a requirement for water supply assessments as part of the state environmental quality act reviews. Senate Bill 221 in 2001 significantly strengthened the linkages in California, by requiring all subdivisions of greater than 500 units to demonstrate a 20-year firm water supply prior to development. Local governments have the authority to establish their own standards and review adequacy for smaller subdivisions, and apparently many of them do. In Nevada, the State Engineer reviews all subdivision maps and his approval guarantees water quantity to the subdivision. In addition, the utility serving the subdivision is required by the Nevada Public Utilities Commission to acquire permanent water rights for new developments. Cities and counties have no explicit authority to set their own requirements under Nevada law. Within the Las Vegas area, the Southern Nevada Water Authority is responsible for conducting the water supply reviews.

Colorado and New Mexico have similar adequacy standards. Both states require counties to establish and enforce adequacy standards for new subdivisions within unincorporated areas. In both states many of the cities adopt and enforce their own standards. Neither state provides guidance on what constitutes adequacy (years of supply). In the Denver region, allocations are based on a 100-year standard, while one particular county, El Paso County, requires a 300-year supply. In New Mexico, most counties require adequacy for between 20 and 100 years (Hanak and Browne 2006). Colorado adopted new water adequacy legislation in 2008, Colorado House Bill 1141, which prohibits local government from approving new developments of greater than 50 lots unless the applicant has demonstrated water adequacy. However, local governments in Colorado have sole discretion to determine what constitutes an adequate water supply. Local government can also extend the requirement to smaller subdivisions. One interesting aspect of Colorado's 2008 legislation is that it explicitly requires the demonstration of adequacy to consider conservation and demand management measures as well as hydrologic variability. With the exception of Arizona's AMAs however, the other state programs do not require the water supplies to be renewable.

Adoption of stringent assured water supply requirements will, in general, be politically difficult unless alternative supplies are available to meet the requirements. Use of these types of rules to require investments in supplies for new growth is much more likely to be accepted than using such requirements to halt growth. Two mechanisms that can assist water adequacy efforts include (1) creation of an entity that can acquire supplies either for direct use or to replaced mined groundwater and (2) establishment of groundwater recharge and recovery programs to facilitate conjunctive management.[4]

12.2.3 Conjunctive Management – Recharge and Recovery Programs

Conjunctive management entails managing all types of water as a common resource in ways that allow us to maximize the utilization of each supply. It recognizes the unique characteristics of the availability and qualities of different sources of groundwater, surface water, and effluent and builds in flexibility to adjust to changing hydrologic conditions as well as institutional constraints. For example, Arizona's Salt River Project (SRP) provides an excellent example of conjunctive management of Salt and Verde River water, groundwater, and CAP water. Historically, SRP has delivered surface water when available, but switched to groundwater to preserve scarce surface water supplies during drought. More recently SRP has utilized excess CAP water to avoid draining the limited surface water in the reservoirs in the SRP system. SRP has also recently joined other Arizona interests in developing groundwater recharge projects, which can store significant quantities of surplus CAP or surface water for later use.

Conjunctive management is a key strategy for dealing with five increasingly important aspects of western water management (adapted from Bloomquist et al. 2004):

- as supplies become tight it is increasingly important to efficiently utilize all available sources of water including effluent,
- the seasonal patterns of peak demand differ from those of supply availability,
- as the region becomes increasingly urban, water supply reliability is becoming just as important as availability,
- water demands in the west will not be met without extensive storage (which can include underground aquifers) and distribution facilities,
- each of these management challenges increases the need to cooperate regionally and overcome institutional and legal constraints that limit water supply reliability, efficiency and resiliency.

---

[4]Arizona's programs provide examples of both of these mechanisms. The creation of the Central Arizona Groundwater Replenishment District, by committing to replenish groundwater used by its members, provided a mechanism for certain developments to demonstrate a renewable supply while continuing to pump groundwater. However, the Central AZ Groundwater Replenishment District also provided somewhat of a loophole in Arizona's requirement for a firm 100-year renewable supply prior to subdivision approval.

Characteristics that facilitate conjunctive management include:

- clear specification of water rights
- consistent rules for all sources of water
- ability to conserve water without loosing rights
- ability to transfer water supplies and rights
- infrastructure to move water (in particular surplus surface water)
- large aquifers with storage space
- clear standards for permitting or allowing recharge projects
- health and environmental quality regulations that facilitate effluent reuse and recharge
- clear rights to recover stored water and protect those supplies from other users
- protection of recharge zones and well sites
- ability to cooperate on financing and assessing fees, and
- regulatory incentives for conjunctive management.

In the remainder of this section we will focus on one particular aspect of conjunctive management; groundwater recharge and recovery programs. Groundwater recharge and recovery entails putting surface water and effluent back into the ground to augment existing aquifer storage and then using new or existing wells to recover the stored water.

Three principal means of conducting recharge are (1) constructing facilities such as recharge basins or ponds that allow water to soak into the ground (*direct recharge*), (2) allowing water to run down existing, often dry streambeds and passively recharge (*managed recharge*), and (3) paying a farmer to reduce groundwater pumping by accepting an alternative supply, generating "credits" to pump the saved groundwater in the future (*in lieu recharge* or groundwater savings facilities).

12.2.3.1 Contrasting of Arizona, California and Colorado Approaches

A recent study of recharge and recovery programs in Arizona, California and Colorado (Bloomquist et al. 2004) concluded that it is the differences in institutions that are responsible for differences in activities in the three states.

California: In California, conjunctive management has been practiced since the 1920s. Water management needs addressed by conjunctive management in California include: Capturing stormwater runoff that would be lost to the ocean, mitigating saltwater intrusion threats, balancing the southern California water demands with the northern California water supplies, and utilizing the underground water storage capacity that greatly exceeds the capacity of surface water reservoirs. A principal purpose for many of the recharge projects in California is to accommodate seasonal peaking and overdraft recovery. Although drought protection is another common goal, the California projects seem to focus on dealing with seasonal and other short term variability in water supplies.

Arizona: Arizona initiated groundwater recharge and recovery projects more recently than the other two states. While recharge projects in California were implemented for local water management reasons, typically without any regulatory framework, much of the recharge in Arizona is due to the state regulatory requirements for assured water supply. Arizona's water management framework facilitating recharge includes the assured water supply rules implemented in 1995 that require renewable supplies for development in AMA's, the Central Arizona Groundwater Replenishment District that will replace mined groundwater for member new developments which lack their own access to renewable supplies, and the Arizona Water Banking Authority established in 1996 to purchase and store excess CAP supplies to protect against future shortages and to facilitate Indian Water Rights settlements.

Components of Arizona's recharge and recovery program include the permitting of; storage facilities to ensure hydrologic feasibility, water storage to ensure appropriate legal rights and protect aquifer water quality, and recovery wells to protect adjacent well owners, plus the calculation of recharge credits earned and recovered. Arizona's recharge and recovery statutes provide the critical protection that an entity storing water in the aquifer can (1) retain access to that water, (2) recover the water anywhere in the same AMA, and (3) legally consider the recovered water to be from the source recharged, surface water or effluent, rather than groundwater. In addition, Arizona's recharge and recovery laws both protect ownership of recharge credits and facilitate some limited markets for transferring recharge credits.

Arizona's recharge program has resulted in the development of over 80 storage facilities with a combined capacity to store up to approximately 2 million acre-feet per year. Since the 1994 authorization of the recharge program through 2007, Arizona entities have accumulated approximately 6 million acre-feet of recharge credits

Colorado: Colorado's conjunctive management approach is very different from that of California and Arizona. The principal purpose of Colorado's initial projects was to maintain surface water flows during peak demand times while allowing continued use of tributary groundwater. Recharge projects are operated in proximity to major rivers in eastern Colorado so that recharge will naturally return to the river during low flow summer periods and ensure sufficient supplies for senior surface water right holders downstream. More recently, Colorado has expanded conjunctive management projects to meet interstate river compacts and federal environmental requirements to maintain specified flows in the South Platte and Arkansas Rivers. In addition to off-peak storage of surplus waters, basin management entities are also purchasing surface water rights that they can directly deliver to rivers during low flow periods to replace flows lost due to their members groundwater use.

12.2.4 New Supplies and Coping with Growth and Drought

Available water supplies throughout the American SW are, for the most part, fully utilized and in some cases available groundwater is being overdrafted. Therefore, water for growth will need to come principally from increasing efficiency and by re-allocating or purchasing senior surface water rights (particularly from agriculture). Effluent reuse will also be of increasing importance as a source of water for growth. Desalination of seawater and highly saline groundwater also is receiving increasing attention as a new source of significant potential. Another mechanism that could increase the reliability of municipal water supplies is long-term dry-year option agreements with agriculture, through which municipal supplies can be "backed up" by contracts with agricultural users to forgo irrigation and supply their water to cities during a drought (Holway et al. 2007).

Continued growth throughout the SW will certainly place stress on the resiliency of the water management system and the long term sustainability of water supplies. Meeting these future water demands will require significant investments in water management, anticipating and preparing for future droughts, and facilitating the ability to purchase and transfer water supplies. Areas of the SW United States without both significant water storage potential and water importation infrastructure will face significant growth limits and possibly insurmountable challenges to long term sustainability.

12.2.5 Water Demand Management Approaches – Conservation

Water conservation and demand management are also critical to water sustainability. Water made available through conservation by existing users is often the cheapest and most sustainable source of additional water. In addition to conservation, demand management techniques can focus on (1) altering the daily and seasonal patterns of demand to reduce peak loads, (2) matching different water qualities to the needs of different users, and (3) prohibiting or otherwise limiting certain water uses (typically high water using activities). Different approaches to demand management range from simply providing information to active education campaigns or establishment of voluntary conservation standards to actual investments and incentives for investments in conservation, to water rates that impact water use (Chapter 13), to directly regulating water users and uses. Arizona has taken the most regulatory approach to demand management, with water provider targets and agricultural allotments tied to the adoption of individual conservation practices.

In Arizona, mandatory conservation requirements are set for all large water users within the AMAs through a series of 10-year management plans. Agricultural groundwater rights holders with greater than 10 acres of irrigated land are given an annual allotment based on historic crops grown and an assumption of 80% irrigation efficiency. In addition, no new land can be brought into irrigation. Municipal water use is controlled through state mandated reductions in the average annual gallons per capita per day usage based on the conservation potential of each water company's service area. The per capita targets are applied not to individual water users, but to the municipal and private water companies that then have to implement programs to achieve these rates from their service areas. Alternative conservation programs, based on the use of approved best management practices, have also recently been developed for both agricultural and municipal water rights holders. In addition, certain water uses are prohibited, such as new private lakes and the use of high-water-using vegetation in public rights of way along roads, unless irrigated with effluent.

Several other SW states have achieved similar water use reductions by requiring local water providers to adopt conservation programs as a pre-condition for approving increases in water rights to serve their growing populations. Figure 12.3 illustrated the significant differences in per capita water demands across a number of SW communities. Reasons for these differences include such factors as climate, landscaping preferences, water prices, the culture of the community, the age of the housing stock and water-using infrastructure, and the types of conservation programs in place.

Per capita water use patterns are principally related to the water using behaviors and investments of individuals. One key consideration is whether regulations or educational programs are most effective in affecting individual behavior. Pricing is another key variable impacting water use: At a minimum, unit and average water prices should increase as usage levels increase. Such an *increasing block rate structure* is becoming increasingly common throughout the SW and provides some level of incentive to reduce water uses. Economic studies indicate, however, that water demand, particularly in the short run, is relatively *inelastic*, meaning that even a significant price change will result only in a limited change in water use (Western Resource Advocates 2003, Chapter 13). For example, a 20% increase in price may only result in a 2 to 3% decrease in water use.

One constraint on individual conservation efforts is the "use it or lose it" standard that applies in many prior appropriation systems. Arizona's approach to conservation, through the quantification of grandfathered groundwater rights, allowed existing groundwater users to pursue conservation opportunities aggressively without fearing that they would forfeit water rights. Landscape irrigation is the major water use in most SW cities. Although significant gains can be made through more efficient irrigation techniques, a switch to low-water-using vegetation, or reducing landscaping would be required to achieve substantial demand reductions. The potential roles of landscaping in helping cool urban areas and the implications for energy consumption have sustainability implications as well.

Another distinction of importance in the design of conservation efforts is the recognition of the difference between permanent conservation practices and practices that are needed during the short or medium-length shortages that periodically occur in arid and semiarid environments. Structural conservation practices, such as installing lower-water-using fixtures and low water use landscaping tend to be stable over time. For shorter time periods, urban water use can typically be reduced up to 20% through individual behavioral changes and sacrifices (shorter showers, not washing cars, deficit irrigation of landscapes or even allowing certain landscaping to die). However, these types of changes may not be sustainable indefinitely. Water managers worry that creating "demand hardening", by eliminating certain inefficiencies, will reduce the amount of additional conservation that could be achieved during future supply crises (Jacobs and Holway 2004).

12.2.6 Decision Making and Institutional Characteristics

Actual water use is influenced by many factors; however it is the cumulative impact of multiple individual decisions that often have the greatest role in determining the characteristics of water demand and supply. For the most part, government actions, can influence, but can not directly control these choices. Most water use occurs on private lands and it is the decisions and investments of multiple water users and providers (cities, farmers, irrigation districts, private water companies, industries, and individuals) that most strongly affect how water is used. An effective approach for state programs, particularly for demand management, is to influence the individual behaviors and investment decisions that collectively determine how water is actually used. Different types of programs, both regulatory and non-regulatory (e.g. education, incentives, and technical and financial assistance), may be effective depending on the types of decisions that need to be influenced.

Providing regulatory certainty and a clear water rights system are critical to encouraging investments in conservation and use of renewable supplies. Water rights structures and management programs should be designed to balance protection of existing users with allowing the flexibility to adapt to changing conditions.

Water management programs must be able to grow and evolve over time and, where appropriate, work with private markets to transfer water rights. One potential downside to comprehensive regulatory programs, particularly those operating at a state or federal level, is that they may become overly rigid or look for "one size fits all" solutions. Another potential constraint on flexibility is that any program with the potential for major economic impact is constrained by social and political factors. Once a program is established, stakeholders have a vested interest in preserving their benefits. Once these relationships are created they can be very difficult to alter, even if new conditions suggest changes are needed.

12.3 Adaptive Management: Designing for Sustainability with Resilience

Building sustainable urban water systems requires designing them to adapt to changing conditions. Changes in the amount and nature of water demand, as a result of population growth or increasingly consumptive lifestyles, are typically the major source of stress on urban water systems. Loss of supply, whether through historic overuse or droughts, is another key stressor. Technology, which in the past has been the key to meeting increasing demands on urban water supplies, may be unable to meet future challenges or could fail at a critical time. Additionally, our institutions must also be able to evolve as the challenges change. The ability to withstand and adapt to stress, referred to as *resilience*, is the focus of this section. The need for resiliency requires that urban water managers explore and understand each of the sustainability factors and design systems with an ability to adapt to future threats, known or unknown. The management programs and approaches discussed

Table 12.5 Designing for resiliency – key steps

- ID Major Issues and Areas of Concern
- ID Current Status of Key Sustainability Factors
- Examine Alternative Future Trends
- ID Critical Thresholds
- ID Events to which System is most Vulnerable
- Design for Adaptability
    - Anticipate specific threats
    - Increase general adaptability
        - Conduct Long Range Scenario Planning
        - Build a Diverse Portfolio
        - Secure Backup Supplies
        - Rely on a Variety of Management Approaches
        - Build Flexible Coordination and Severable Interconnections
        - Build and Test Response Plans
        - Establish Feedback Mechanisms
        - Build Adaptive Management Capacity
        - Protect Vulnerable Ecosystems
        - Anticipate and Exploit Opportunities
    - Consider Public Policy Mechanisms (Role of regulations)
- Secure Political Support

in previous sections of this chapter illustrate just a few of the technical, institutional and management mechanisms for improving resiliency that have been utilized for water management. Table 12.5 lists key steps and considerations in designing for resiliency (adapted from Leslie and Kinzig, 2009; Walker and Salt, 2006).

Identifying the major issues and areas of concern for an urban water system is the first step in any design effort. Typical areas of concern include: Adequacy of supplies for future growth, the reliability of the current and future water supplies, water quality, the cost of providing water and managing waste water, degradation of natural ecosystems, endangered species concerns, and the role of water and water use in maintaining urban and environmental amenities.

A fundamental prerequisite for effectively managing any system is good data on historic and current conditions as well as future projections. Data is required both on the key sustainability factors (population, per-capita demand, supply sources, etc.) as well as indicators for areas of concern (water quality, condition of infrastructure, water prices, etc.). Projections must include both a time-frame and a degree of certainty. This means that areas of greater uncertainty, such as the impacts of climate change and the distance into the future being evaluated, will affect both the ease and reliability of projections. More difficult projections will require more sophisticated forecasting and alternative scenario generation techniques.

A particularly critical component of resiliency is the concept of *thresholds*. Thresholds can be viewed as a sort of tipping point that, once crossed, lead to irreversible, or difficult to reverse conditions. One example would be groundwater use that grows until it exceeds the natural recharge capacity of the system. At some point, groundwater no longer provides baseflow and streams dry up for portions of the year. Overdraft can also lead to dewatering portions of an aquifer, leading to aquifer compaction and land subsidence that results in permanent loss of aquifer storage capacity and damage to the land and related urban infrastructure (roads, water lines, drainage structures, buildings, etc.). Critical thresholds for each water system can be identified based on key issues and projections of alternative futures. Understanding what constitutes a critical threshold, how to avoid crossing them and the consequences that would occur is a critical step in designing a resilient and sustainable water system. Thresholds can be a characteristic of physical systems (groundwater pumping impacts on an aquifer, waste assimilative capacity of a stream) or social and management systems (exceeding regulatory water quality standards). Typically, as a system comes closer to a critical threshold significant resources (technical, financial, management institutions) are required to maintain the system. Understanding the current trajectory of a water system (e.g. number of years of assured supplies available for current and projected population, trend in water quality measurements, changes in groundwater levels) is important to know the types of necessary management efforts.

Identifying the types of events to which a water system is most vulnerable; the ones most likely to push it over a threshold, is the next step in resiliency design. Drought cycles are a major challenge to water managers in the SW United States and are a good example of an external event over which we have little if any control, but for which we can design both adaptation (reservoirs and groundwater recharge) and mitigation (emergency conservation programs) responses to increase the resiliency and sustainability of water systems. We should also anticipate and prepare for other events more under human control, such as infrastructure failure due to accidents or inadequate maintenance.

12.3.1 Design for Adaptability

Some specific events to which urban water systems are vulnerable can be identified and anticipated, however it is never possible to anticipate and prepare for all specific potential events. Table 12.5 lists a number of the actions that can be taken to generally increase the ability to adapt and maintain the resiliency of our water systems for any number of potential threats. These general measures to increase resiliency include long-range planning, diversifying supplies, building institutional capacity and investing in technological improvements. Each of the steps described above would ideally be integrated into long-range planning efforts that would be an ongoing activity in all urban water systems. Looking long term, up to 50 or even 100 years, can be particularly important. Water resources investments sometimes take up to 50 years from conceptualization to completion and a number of the most important challenges to the sustainability of our water systems are only apparent when we look far into the future. Along with this consideration of changing conditions, a willingness to reconsider the basic assumptions used historically in designing water systems is needed (e.g. the 100 year flood standard, reservoir supply reliability standards). Future systems will need to be able to handle greater variability in demands and supplies, and in both the quantity and quality of the different supplies and water needs.

Research and the literature on resiliency point to several factors that are particularly critical in designing resiliency into any system: diversity, modularity, and feedback (Walker and Salt, 2006;121). Each of these will be elaborated upon below.

Diversity: Resiliency is obviously improved by relying on a variety of different water supplies and developing back-up supplies to cover droughts or infrastructure failures. However, resiliency theory offers further insight into the need to consider both *functional diversity* and *response diversity*. Functional diversity suggests relying upon different types of water supplies and demand reduction measures. For example, a diverse water supply portfolio might utilize surface water, groundwater, and effluent, and look to different river basins, well fields, and treatment plants and technologies for treating water. Including conservation programs and viewing conserved water as a source of supply also builds functional diversity. Response diversity looks for supply sources that will behave differently under future conditions. For example, do the different river basins used for surface water supply tend to experience droughts simultaneously or at different times? Does the groundwater system respond quickly to streamflow so that it will be stressed at the same time that surface water supplies may be limited or does it respond in an independent manner? Conservation approaches can be designed to reach different types of water users who will respond to different conditions and at different times. Back-up supplies will be different based on whether they are needed for short-term fluctuations in demand and supply (storage tanks throughout the system), seasonal variations (reservoirs), or perhaps multi-year and even multi-decadal storage (aquifer recharge). Considerations of functional and response diversity apply equally to physical infrastructure systems, management programs, and water management institutions.

Modularity: Interconnections of water systems both within urban areas and across different regions can greatly increase system resiliency, but if not designed carefully can also lead to larger system failures. This is the resiliency concept of modularity. Our electric production and distribution systems provide a good example of modularity issues. Clearly, interconnected electric power grids have made substantial improvements in the reliability, efficiency, and cost of our electricity supply. However, the August 2003 Northeast Blackout serves as an example of where a failure in one part of the power grid cascaded through the system and led to a multi-state power outage impacting 50 million people in the United States and Canada. The challenge is to design system interconnects that allow sharing of available water supplies across jurisdictions, but that can be severed, if necessary, to make sure a local emergency does not lead to a larger system failure. Modularity is also important in our governance institutions. Water, for the most part, is locally managed and regulated. This allows for a variety of approaches and mechanisms to be developed and tested throughout different state and local governments and water providers.

Feedback: Feedback mechanisms are critical for responding to changing conditions, for testing our ability to respond to potential problems, and for identifying and adapting to lessons learned. One advantage to local control of urban water systems is that it increases the likelihood that unique situations will be recognized and addressed. On the other hand, local control can also sometimes fail to identify larger regional phenomena, fail to coordinate larger regional wide responses when necessary, or be overwhelmed by a crisis. For these reasons, feedback and response mechanisms also need to be established across different levels of government, from local to regional to state to national or in some cases international.

Anticipating Opportunities and Adaptive Capacity: Improvements for the sustainability and resiliency of our urban water systems are occasionally difficult to implement. Even if water managers can agree on what is needed, change may face political opposition, financial limitations, or a general lack of public support. Understanding the types of cycles that occur in both natural and social systems can help to design appropriate actions (Walker and Salt 2006). Public support and action is often galvanized to respond to crisis or other events that capture public attention: Being able to anticipate and exploit such opportunities is often a key to implementing change. One illustration of exploiting a post-crisis opportunity would be having a new floodplain management approach developed and ready to be brought forward in the immediate aftermath of a flood. Similarly, a typical problem with drought response planning is that by the time such plans are developed it is often raining again and there is no longer public support for any difficult decisions. Typically, after a crisis, such as in the economic growth phase that follows a recession, there can be an opportunity to invest in new programs, to build new infrastructure to handle growth and to create rainy day funds, or aquifer recharge banks, to help get through future economic downturns (or droughts). Finally, most systems also go through a stable or retrenchment phase when resources and resiliency become more limited. This may be the point just before some critical threshold is crossed (groundwater overdraft leading to a significant loss of supply or aquifer subsidence, growth outstripping available supplies, beginning of a major drought cycle). This is the time to increase the focus on mitigation and response actions. This is also a point at which a focus on efficiency improvement could be counterproductive, for example reducing discretionary consumer water demand to free up water. This could be problematic if it "hardens demand" making it harder to get additional short term reductions to respond to a water shortage, particularly if water freed up through conservation is allocated to support new growth. Adaptive environmental management asks that we closely monitor conditions, use this feedback to evaluate how programs are working, and make necessary policy or technology changes. Such monitoring and adaptation may be particularly essential to recognize and protect critical ecosystems and the services they provide in terms of water quantity and quality, quality of life, and species habitat.

12.3.2 Securing Political Support

Public policies, in particular efforts to influence individual and corporate behavior though regulations, public investment, incentives, and education have a significant role in the sustainability and resilience of our water systems. All else being equal, programs that maintain flexibility and allow creative and diverse responses are preferable. Education and incentive programs tend to be better than regulation for developing flexibility. However, some efforts, such as water quality standards to protect public health or rules ensuring adequate water supplies do require regulatory approaches. When regulation is required, performance standards such as an allowable level of pollution or a target water use rate are generally preferable to prescriptive standards, such as requiring particular treatment or water use technologies. Performance standards allow greater creativity in the development of ways to meet the necessary requirements.[5]

Public support will be essential for designing resilient and sustainable urban water systems. Although investing in resiliency will have a significant long term payoff for the economy, the environment, and our quality of life; in the short-term it may cost more in terms of government intervention, money, and management complexity. Public and decision maker education about water systems and the importance of resiliency will be important to secure support for such investments. Key areas on which to focus such education include: basic understanding of water systems and our dependence on them; awareness that the past is not necessarily a good predictor of the future; and potential public policies and investments.

The effective urban water managers of the future will increasingly be called upon to participate in political processes, to collaborate across jurisdictions and to coordinate with non-water managers within their own jurisdictions. Building relationships of trust and mutual respect with key stakeholders and political officials will be key to successful efforts to build resilient and sustainable urban water systems. Open democratic decision making, efforts to engage and empower citizens, and ensuring that decision making is transparent and understandable will be critical to build this necessary support, respect and trust.

12.4 Meeting the Challenges of Today and Tomorrow

As arid regions continue to grow, water managers will need to address a number of challenges, including the need to:

- work cooperatively with water users and interested parties both locally and throughout the larger river basins and watersheds to secure the future supplies for urban growth;
- forge regional partnerships within urban areas to diversify portfolios and advance conjunctive management, including long-range aquifer management strategies that incorporate both water quantity and water quality needs;

---

[5]For an extended discussion about the tradeoffs between different policy instruments in the context of Arizona's water quantity management programs see Jacobs and Holway 2004.

- develop new institutions and mechanisms to transfer water rights and supplies to growing areas and address the impacts on donor areas;
- improve the understanding of climatic variability and future global climate change and incorporate this knowledge into long-range water management planning, almost definitely requiring more investment in storage and backup supplies to offset future shortages;
- modify the statutory, regulatory and infrastructure financing frameworks to evolve with the changing nature of water uses and supplies;
- learn from the efforts of each other as well as other regions of the country and world;
- look beyond the next 20 to 30 years and commit to longer-term objectives;
- address environmental quality, ecosystem health, and quality of life concerns as they relate to water management; and
- develop increasingly sophisticated models and other tools to facilitate comprehensive, long range, sustainable water management.

The continued rapid growth and urbanization of water short regions will require building on past water management successes and further investing in the infrastructure and water management capacity necessary to ensure long term sustainability. Numerous opportunities will exist for water managers, engineers, planners, scientists, and educators to join in these challenges.

References

Arizona Department of Water Resources, *Third Management Plan for Phoenix Active Management Area*, Arizona Department of Water Resources, Phoenix, Arizona, 1999.

Arnold, C.A., (ed.), *Wet Growth: Should Water Law Control Growth?* Environmental Law Institute, Washington, DC, 2005.

Bloomquist, W., E. Schlager, and T. Heikkaila, *Common Waters, Diverging Streams: Linking Institutions to Water Management in Arizona, California and Colorado*. Resources for the Future, Washington, DC, 2004.

Christensen, N. and D.P. Lettenmaier, "A Multimodel Ensemble Approach to Assessment of Climate Change Impacts on the Hydrology and Water Resources of the Colorado River Basin," *Hydrology & Earth Systems Sciences Discussions*, Vol. 3:3727–3770, 2006.

Cook, E.R., SW USA Drought Index Reconstruction. International Tree-Ring Data Bank. IGBP PAGES/World Data Center for Paleoclimatology Data Contribution Series #2000-053.NOAA/NGDC Paleoclimatology Program, Boulder CO, USA, 2000.

Ellis, A.W., T.W. Hawkins, R.C. Balling, and P. Gober, "Estimating future runoff levels for a semi-arid fluvial system in central Arizona," *Climate Research* Vol. 35:227–239, 2008.

Hanak, E., "Finding water for growth: New sources, new tools, new challenges," *Journal of The American Water Works Association*, Vol. 43:4:1–12, 2007.

Hanak, E. and M.K. Browne, "Linking housing growth to water supply," *Journal of the American Planning Association*, Vol. 72, No. 2:154–166, 2006.

Hirt, P., A. Gustafson, and K.L. Larson. "The mirage in the valley of the Sun," *Environmental History*, Vol. 13, No. 3:482–514, 2008.

Holway, J.M. and K.L. Jacobs, "Managing for sustainability in Arizona, USA: Linking climate, water management and growth," *Water Resources Sustainability*, (ed) Larry Mays, McGraw Hill, 2006.

Holway, J.M.., T.S. Rossi, and P. Newell, "Water and growth: Future water supplies for Central Arizona," Discussion Paper, Global Institute of Sustainability, Arizona State University, Tempe, Arizona, 2007.

Intergovernmental Panel on Climate Change (IPCC). *Climate Change 2007: Synthesis Report. Contribution of Working Groups I, II and III to the Fourth Assessment Report of the Intergovernmental Panel on Climate Change.* Core Writing Team, Pachauri, R.K and Reisinger, A. (eds.), IPCC, Geneva, Switzerland, 104 pp, 2007.

IPCC, *Climate Change 2007: Synthesis Report. Summary for Policymakers. Summary approved at IPCC Plenary XXVII. November 2007.* Available at: http://www.ipcc.ch/pdf/assessment-report/ar4/syr/ar4'syr'spm.pdf

Jacobs, K.L. and J.M. Holway, "Managing for sustainability in an arid climate: Lessons learned from 20 years of groundwater management in Arizona, USA," *Hydrogeology Journal*, Vol. 12:52–65, 2004.

Leslie, H. and A.P. Kinzig. "Resilience Science." Chapter 4 in *Ecosystem-Based Management for the Oceans*. K. McLeod and H. Leslie, (eds.), Island Press. pp. 55–73, 2009.

Maguire, R.P. "Patching The holes in the bucket: Safe yield and the future of water management in Arizona," *Arizona Law Review*, Vol. 49:361–383, 2007.

Nash, L.L. and P.H. Gleick, *The Colorado River Basin and Climatic change: The Sensitivity of Streamflow and Water Supply to Variations in Temperature and Precipitation*. U.S. Environmental Protection Agency, Washington, DC, 1993.

Nelson, B., M. Schmitt, R. Cohen, N. Ketabi, and R. Wilkinson, *In Hot Water: Water Management Strategies to Weather the Effects of Global Warming*. Natural Resources Defense Council. NewYork, July 2007.

Tarlock, A. Dan and B. Sarah, S. Van de Wetering, "Water and western growth," *Planning and Environmental Law*, Vol. 59, No. 5:3, 2007.

Walker, B. and D. Salt, *Resilience Thinking: Sustaining Ecosystems and People in a Changing World*. Island Press, Washington, DC, 2006.

Western Resource Advocates, *Smart Water: A Comparative Study of Urban Water Use Efficiency Across the 3W*, Western Resource Advocates, Boulder, CO, 2003.

Chapter 13
Demand Management, Privatization, Water Markets, and Efficient Water Allocation in Our Cities

K. William Easter

13.1 Introduction

When water is plentiful, we tend to take it for granted and overuse it. Once water becomes scarce, our neglect changes to pleas for new supplies. In other words, when water demands exceed the supply at existing low prices, shortages emerge and we appeal for an increase in supply. Yet expanding domestic water supplies has become much more difficult and expensive. This is because we have already developed the low cost sources of supply and face growing environmental constraints due, in part, to our increased awareness of the many instream services undeveloped water resources provide. The cost of developing new sources includes both the explicit financial cost of infrastructure and the opportunity cost of using water consumptively, rather than for instream or nonconsumptive services. Although nonconsumptive uses, by definition, do not involve water consumption, they can change the timing and location of water flows (hydropower), increase water temperature (cooling), or pollute (boating) the water. In addition, these instream (or nonconsumptive services) such as sewage dilution, recreation, and a healthy aquatic habitat may require increased water supplies. Balancing all these demands is a challenge and requires us to recognize that our supplies of clean, fresh water resources are quite limited. Simply increasing the supply is no longer the easy option. We must now emphasize using the water we have more wisely.

In this chapter we will explore different institutional arrangements and mechanisms that will encourage consumers to reduce their water use, particularly those mechanisms and arrangements that will facilitate "demand management". We will also consider ways to improve water management and make better use of our existing supplies through "privatization" of water utilities. We will then consider water markets and the role they might play in allocating urban water supplies. Finally, we see how water charges and markets can help reduce water pollution.

K.W. Easter (✉)
University of Minnesota, Minneapolis and Saint Paul, MN, USA
e-mail: kweaster@umn.edu

13.2 Demand Management

With the growing demand for water in urban areas and our increased concern regarding the environment, we have to develop and use institutional arrangements and mechanisms to reduce the growth in water demand. Implementing such arrangements and mechanisms has become known as demand management. More specifically, demand management primarily involves three general types of institutional arrangements: (1) water pricing, (2) water quotas-rationing, and restrictions on specific uses (watering lawns), and (3) water subsidies for technologies that use less water (low flow toilets) or legislative restrictions that only allow the purchase of these water saving technologies.

Water pricing tends to be the mechanism one thinks of when you talk about demand management. In fact, most urban water users, who are supplied water by a public utility, pay a water price (or charge) unless it is included as part of their apartment or house rent. Most urban homeowners and commercial water users have meters that measure the amount of water they consume. A few municipalities (Sacramento, California, for example) and a number of small towns do not have meters, and users are charged a fixed fee. In such cases the charge is primarily used as a way to cover the cost of supplying water to community users. It is like a land tax and sometimes it may vary by type of business or number of people living in the household. In cities with meters and where water has become scarce, water pricing has been used to help reduce water consumption by giving consumers an incentive to conserve water.

The water demand (or use) will vary by household and industrial characteristics such as income, landscape, family size, products produced and technologies used, as well as climate and season of the year. For example, households in hot, dry climates tend to use more water for garden and lawn irrigation. United States household consumption varies from about 76,000 gallons/household per year to as high as 284,000 gallons/household per year (Fig. 4.1 in Chapter 4). Outdoor water use in the United States varies from about 10 to 75% of the total household water use while indoor use averages about 70 gallons/capita per day. One key in effective demand management is to know household and industrial characteristics and use them in structuring a water management and pricing system. Reducing demand will likely require the use of more than water pricing if water supplies are low relative to the demand. However, with the appropriate pricing procedure in place, the cities' job of meeting growing demands will be much easier.

There are number of different pricing methods that can be used but each provides users different incentives to reduce water use. These methods include: (1) a fixed charge per month or quarter, (2) a constant rate per unit of water used, (3) a decreasing block rate per unit of water used, (4) an increasing block rate per unit of water used, (5) a two-price method which includes a fixed charge plus a charge per unit of water used, and (6) peak load or seasonal pricing. As discussed above, the *fixed charge* is found mostly in small towns and provides no incentive to reduce water use. It is generally set to cover the costs of operations and maintenance plus a capital charge for the infrastructure, and does not change with quantity used by a

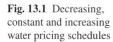

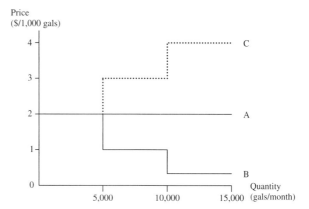

Fig. 13.1 Decreasing, constant and increasing water pricing schedules

business or household. An example would be a $25 per month charge for a household which would only change annually based on changes in the cost of supplying water.

The *constant rate* means that water users pay the same price per unit of water no matter how much they use in a given time period. For example, if the rate or price per 1,000 gals is $2 and you use 15,000 gals/month, then your monthly bill is (15 × $2) $30 (line A in Fig. 13.1). The advantage of such a pricing method is that it is easy to understand and it gives you a constant incentive to conserve water because the more you use the more you pay. Each 1,000 gals of water has the same price. Therefore, the marginal cost to the consumer is constant no matter how much is used (marginal cost is $2).

The *decreasing block rate* means that water users pay a lower rate or price, the more they use, which provides little incentive to conserve water. An example would be to pay $2/1,000 gals for the first 5,000 gals used per month but only $1/1,000 gals for the next 5,000 gals used that month and $0.40/1000 gals for all water used per month above 10,000 gals (line B in Fig. 13.1). If the user consumes 15,000 gals/month, the monthly bill would be only ($2 × 5 + $1 × 5 + $0.40 × 5) = $17 or about half of what the constant rate user would pay. Decreasing rates are generally used by towns to encourage firms to build or expand plants in their town. It favors large water consumers such as wealthy homeowners with big lawns and swimming pools, and business enterprises. Since the water charge drops as more water is used, the marginal cost to the consumer also drops as does the incentive to conserve water (marginal cost is $0.40 for water use per month over 10,000 gals). St. Paul, Minnesota has a decreasing rate that drops by $0.10 per ccf after 100,000 cf (100 cubic feet (ccf) = 748 gallons) are consumed.

The *increasing block rate* or price is just the opposite of the decreasing rate. It provides an increasing incentive to conserve water and is used by cities that are serious about reducing water consumption per capita. An example of increasing block pricing would be a price for the first 5,000 gals used per month equal to $2/1,000 gals while the price for the second 5,000 gals used per month would be

$3/1,000 gals; and for water use over 10,000 gals per month, the price would go to $4/1,000 gals (line C in Fig. 13.1). Thus, when a user consumes 15,000 gals/month, the monthly water bill would be ($2 × 5 + $3 × 5 + $4 × 5) $45. Clearly, users facing this increasing rate would have a much greater incentive to conserve water than users facing either the declining or constant rate. The price for the last block in our increasing rate example ($4/1,000 gals) is double the constant rate and ten times the declining rate example. The increasing rate should provide heavy water-using industries with a clear incentive to recycle water which can dramatically cut water use. The marginal cost of water to large water consumers increases and reaches $4 for consumption of more than 10,000 gals/month. In some cases cities have as many as five blocks, such as Irvine, California. Their block charges start out at $75 per ccf and go up to $7.28 per ccf in the fifth block.

If there is a desire to assure cost recovery and maintain an incentive to conserve water, a *two-price method* can be used. The first price is like a fixed service charge per month for anyone connected to the water system, and is thus not really a price. You pay the fixed charge whether you use water or not, and it is generally designed to cover the fixed costs of operating and maintaining the water system. The fixed charge also helps eliminate the need for very high prices to cover costs during periods of low water use and low water sales. The second price is a charge per unit of water used per month or quarter. This can be an increasing, constant, or decreasing rate and is designed to cover all the variable costs (costs that vary by the volume of water delivered). If water conservation is important, the variable charge will either be an increasing or a constant rate. To maintain efficient water allocation and conserve water, the variable rate can be set at the long-run marginal cost to the water utility of obtaining new water supplies (or the opportunity cost of water). If an increasing block rate price is used, then the highest block price used would be set at the long-run marginal cost of obtaining new water supplies.

Two other pricing concerns face many urban water managers. The first is the seasonal variability of water supplies and demand. The second is the concern that low income families will have a difficult time getting access to and paying for water. For seasonal variability, *peak load or seasonal pricing* can be used. In this case the prices or charges per unit would be raised during periods of high water demand and/or low supplies. For the United States this tends to mean high rates in the summer and low rates in the winter. For example, the rate might be only $1/1,000 gals in the winter, $4 in the summer, and $2 in the fall and spring. It might also be set lower for part of the spring season since in northern climates this is usually a period of high rainfall and snow melt. In Phoenix, Arizona, a city in the dry southwest, they charge $1.65 per ccf from December through March; $1.97 for April, May, October, and November; and $2.50 from June through September.

The concern about low income families is best handled by a system with increasing block rate pricing. The first price can be set low and the quantity cutoff point for the first or low price set so that a low income household of four or five people would receive enough water to meet their basic domestic water needs at the low price (see Fig. 13.2). This "basic minimum supply" would exclude water needed for lawns, washing cars and filling swimming pools since poor households generally

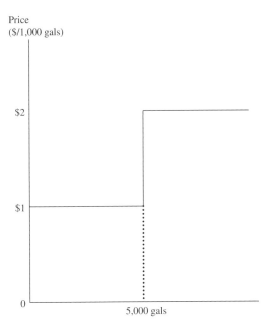

Fig. 13.2 Pricing schedule for domestic water use

do not have these water uses. The second price would be higher for all water used above the "basic minimum" per month and would be set high enough so that total collections would cover the cost of operating and maintaining the water system plus a capital depreciation charge. With this pricing system, water utilities have three instruments they can use to achieve three separate objectives: (1) recover costs from water users, (2) protect low income families from high water charges, and (3) provide an incentive for water users to conserve water. Utilities can raise or lower the two prices ($1 and $2) and they can change the "basic minimum" amount of water at which the higher price starts. For example, if they are not collecting enough money to cover costs, they can raise one or both prices by $0.25 or move the starting point for the higher price from 5,000 to 4,000 gals/month or some combination of the three. If water conservation is a big concern, they can just raise the highest price by $1, which will help achieve both the cost recovery and conservation objectives. Another advantage the two-price approach offers is flexibility in adjusting to different supply and demand conditions over time. In a sense this gives the pricing mechanism resilience. You just have to shift one of the prices, or the cutoff quantity up or down, to meet the changing water conditions. (See Hall, 2000, for an example of this type of pricing system used in the Los Angeles area).

Since the demand curve for many domestic water users (drinking and cooking) tends to be quite steep (demand is inelastic), large price changes bring about only small changes in quantity used, particularly for those families with higher incomes. This case is illustrated with steep demand curve, A, in Fig. 13.3 where higher prices

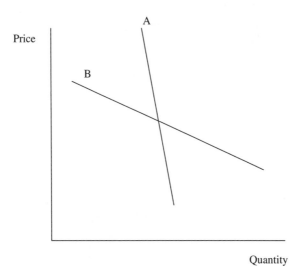

Fig. 13.3 Demand for water

have little impact on the quantity consumed. Demand curve B is not as steep and is more representative of the demand for water to wash cars and water lawns. This suggests that using only higher water prices may not achieve the water conservation needed. A more effective approach would be to use a combination of demand management strategies. An example of such a strategy would be higher water prices combined with subsidies for the adoption of selected water saving technologies, such as low flow toilets and drip irrigation for landscapes. During drought periods, restrictions on certain water uses such as car washing and lawn watering could be an additional strategy. In California's drought of the 1990s, a study of two towns in southern California showed that unless a water pricing policy is combined with other demand management mechanisms, low income families would bear a larger proportion of the reduction in water use (Renwick and Archibald, 1998). This was because the water bill for poor households was a much larger share of their income than it was for the higher income families. This was true even though the higher income families used more water per capita. Renwick and Green (2000) found that, for moderate (5–15%) reductions in demand, modest price increases combined with public information campaigns, work. However, for larger reductions in demand (greater than 15%) relatively large price increases or restrictions on water use (no lawn watering) or some combination of the two are needed.

13.3 Privatization

Another concern in the water sector that is closely related to the water scarcity concern is the low quality of service provided by some water utilities, particularly in developing countries. In such systems you find interrupted service, lack of

Table 13.1 Privatization options found in the United States

Types of agreements or concessions	Description
Acquisition – private ownership	Public utility sells the facility to private entity resulting in private ownership and operation.
Joint venture	Private entity owns facility in conjunction with public utility
Design, build, own, operate and transfer ownership to public	Private entity builds, owns, and operates the facility. At the end of the specified period, such as 30 years, the facility is transferred to a public utility.
Design, build and operate	Private entity designs, constructs, and operates the facility. The public utility retains ownership and financing risk, while the private entity assumes the performance risk for service and/or compliance.
Operate and maintain	Public utility contracts with private entity for a fee to operate and maintain the facility. The public utility owns the facility.
Designand/or build	Private entity designs and/or constructs the facilities and turns it over to the public utility to operate.
Provide specific services	Private entity contracts to provide public utility with specific services such as meter reading or billing and collection.
Management	Private entity manages and supervises the public utilities personnel.

Source: Adapted from Beecher et al., 1995

maintenance, overstaffing, leaky pipes, and large financial deficits. To help address these problems and to involve new entities in the supplying of water service, private firms have been asked to become more involved in the operation and management of such water utilities. This process has been referred to as the *privatization* of water services.

One of the key steps in this privatization has been the decoupling or unbundling of the services provided by the public utility. This decoupling allows the water utility to have private firms manage or operate different functions within the utility all the way from bill collecting and meter reading to constructing new water treatment plants. Table 13.1 lists some of the different functions private firms have performed for public water utilities in the United States. In most cases ownership stays with the public utility. Only in a few cases has there been private acquisition of water systems and this generally includes a contract for supplying water to the city. Acquisition usually occurs when a government or city is trying to raise capital to invest in the water sector. Concerns about raising capital are likely to accelerate since a growing share of U.S. water and wastewater infrastructure needs to be upgraded or replaced in the very near future (Jacobs and Hoe, 2005). Even before this recent interest in privatization, private firms have been employed to construct most of the public water infrastructure outside the old communist block of countries. Thus, a private firm providing services in the water sector is really not new. What is new is the management functions they are privatizing. In the United States most of the privatization of management has been in the smaller urban communities who want to take advantage of some of the extra services the private firms can provide, such as better laboratory facilities for water testing and more extensive technical expertise.

One of the important ideas behind privatization, besides getting more entities involved, is to introduce competitive pressures in the water industry. When contracts are offered for bid to perform different water services, the idea is that competition among private firms will lead to lower cost bids. The contracts also include performance criteria the firms have to meet. The end result, hopefully, is higher quality service provided at a lower cost than those provided by the public utility.

There are three different methods of introducing competitive pressure into the water utilities market. They are (1) product-market competition, (2) competition to supply inputs, and (3) yardstick or comparative competition. Since installing a competitive water delivery network is generally not a serious option for product-market type competition, we are limited to two other types of strategies in product-market competition. The first strategy is to allow different firms to compete to deliver water on the existing delivery network. However, opening up the water distribution network for different firms to use may create water quality problems. Water standards must be established and closely monitored to make sure that one of the competing firms does not try to cut treatment costs and introduce contaminants into the water delivery network. In actual practice, there are few cases where competing firms have been on the same network, other than in the United Kingdom (Dosi and Easter, 2003).

The other strategy is to set up competition for water concessions or agreements to manage all or part of a water utility. The water concession has become one of the most common forms of competition in the provision of urban water supplies. To make them work, it is critical that public authorities maintain regulatory pressure on the private water company during the life of the concession. However, it may be difficult to expel a poorly performing firm because of the contractual power arising from the firm's provision of essential water services. Thus, short-term contracts (5 years) are generally preferable but may not be attractive to many firms. An alternative might be a long-term contract (20 years) that must be reviewed and reapproved, based on the firm's performance, every 5 years. This would give the private firm a longer-run investment framework, as long as it maintains a satisfactory level of service.

Competition to supply inputs such as billing, revenue collection, and infrastructure maintenance is an attractive alternative that has been used in several countries. It is part of the unbundling process. This means the public utility can focus on what it does best and then contract out competitively for the other services. In Santiago, Chile, they established a "public" water company that is financially autonomous from government: It owns and manages the city water system. Since the company is financially autonomous, it has to cover all its costs with the water charges collected from its water users. As a result the water company contracted out for different services such as water billing, bill collection, and the replacement and repair of old damaged and leaking pipelines. This not only raised cost recovery levels but saved water, which was then, in turn, sold to new water users, again enhancing cost recovery. They also reduced illegal connections and broken meters.

Finally, yardstick competition can be important where there is not enough competition from private firms. In such cases the regulatory authority needs to be able to introduce a comparison of performance among firms working in different water systems and different locations. To be effective, comparable information on water charges, collection rates, and service levels must be available for a number of firms (public and/or private), including the ones being evaluated. This will only work if there are comparable firms or utilities that are providing good quality service.

Two interrelated factors seem to have stimulated United States interest in privatization. First, was the problem of financial pressure, particularly in many smaller systems and in older systems, where they faced major costs in upgrading their infrastructure. Second, the introduction of new health and environmental quality standards for domestic water supplies has raised costs. Water utilities had to meet these new stricter requirements for the removal of a range of contaminants. For example, when the new arsenic standard was introduced in 2001, about forty small communities in Minnesota failed to meet the standard the first year. Even after 2006, some of these same communities still could not regularly meet the new standard.

In 1997, a survey of 261 U.S. cities found that 40% of them had some form of private/public partnership and another 14% were considering proposals. The most common activity was the private design and construction of infrastructure, particularly water treatment plants (Callahan, 2000). Private ownership of water systems is discouraged by law, which exempts municipal debt from taxes but not private debt. There are also political concerns about private ownership partly because of the potential for local monopoly power, since the firm, usually, would be the only water supplier for a given municipality. Consequently, most water facilities in the United States are designed and constructed by private firms but are managed and owned by municipalities after completion. This model appears to be less efficient (costs more) than the design, build, operate, and transfer (ownership) model, which gives the private firm more control, particularly over how the plant is operated once it is built. Siedenstat et al. (2000) argue that the design-to-transfer model will reduce construction costs by about 25% and operating costs by 20–40% as compared to the design and build model. The Tolt River Project in Seattle, which fits the design-to-transfer model, is estimated to have had cost savings of 40% compared to the conventional model of private construction, followed by public management.

Despite the growing interest in privatization, it is likely to experience only modest future increases in the United States. It will fit some communities' needs but not others. To be effective, privatization decisions will need to be open and transparent, with public participation and periodic third-party review (Wolff and Hallstein, 2005). Clearly, the need for more funding, combined with public officials reluctant to accept the political consequences of raising taxes or fees, will maintain a continued public interest in the use of the private sector to cut costs (Jacobs and Howe, 2005).

Another area where private water development has been widespread is in the use of groundwater for irrigation as well as for commercial and domestic uses.

With the development of low-cost pumping technologies, private wells have spread around the world, particularly in Asian countries such as India, Pakistan, China, and Bangladesh as well as in the U.S. Great Plains. Although much of the private well development has been for irrigation, the irrigation wells also are used extensively as a domestic water source. In Bangladesh many rural households now have low cost pumps. Although this private well development has supplied large quantities of water, it has, in some cases, caused serious problems of overdrafting of groundwater. This is particularly true in irrigated areas with limited groundwater recharge and unclear property rights for water. Since groundwater in many cases is an open access resource (no clear water rights), users do not take into account the impact of their pumping on their neighbors. In areas with small-scale farms, farmers soon realize that if they do not use the groundwater, their neighbors will. This occurs because, in most cases, owning the land gives landowners the right to pump but does not limit the amount that can be pumped. Since water is mobile, too much pumping for irrigation will extract water from under their neighbor's land and may compete with rural towns. For example, most towns in Minnesota depend on groundwater for their water supply. In cases where overdrafting occurs, the state needs to set up limits, both on the number of wells and the pumping rates. High charges for electricity may be one method to help reduce pumping rates, although such charges have proven very hard to enforce and collect in developing countries such as India.

13.4 Water Markets

Almost the same types of concerns have been raised about water markets as have been raised about privatization: Monopoly power and access of consumers to water at reasonable prices. Currently, there have been few attempts to set up markets for municipal water, although several have been proposed (see Haddad, 2000, for proposal for a water market in Monterey, California). Of course, there is a very extensive private market for bottled water used for drinking and, in some cases, food preparation and cooking. Water markets also exist in some farming areas for irrigation water but trading is primarily among farmers (see Easter et al., 1998). For urban water use the major role water markets have had, and is likely to have in the near future, is as a mechanism for transferring water among different sectors of the economy, such as from agriculture to urban communities, or from one urban community to another (Brewer et al., 2008). So far, most governments have not used markets as a tool to transfer water, except in the western United States and some other countries such as Chile and Australia. In all of these cases markets were used in response to droughts or growing urban populations, which suggests that in the future we will likely need to use markets in this manner as urban water demands continue to grow relative to supplies.

If we want to use water markets within urban areas, the task becomes more difficult. The big problem in setting up such water markets is establishing private water rights or use rights and distributing them fairly among urban users. These rights could be in terms of volumes or shares of water available. Establishing water rights and their allocation are key to setting up water markets within cities. Who should get the water rights and how much they should have to pay for them, if anything, must be decided. The right could be based on past use, size of landholding, or family size. Since these decisions will be difficult to agree on, it is not likely that water markets will see much use within U.S. cities until water scarcity becomes much more severe. Even the water scarce Monterey Peninsula failed to introduce a proposed water market during the 1990s drought (Haddad, 2000). Yet the use of markets to reallocate more water to our growing cities does not face the same restrictions.

The advantage of using water markets for water transfers among sectors is that past users are compensated directly. It also helps improve water use efficiency by reallocating water to its higher valued uses. The problem that must be guarded against is the unintended effects or external impacts of water transfers on downstream users and the environment. When water is transferred out of a region or from agricultural to urban uses, return flows of water are reduced. In the case of downstream users of return flow, they will likely have to be compensated for any water losses caused by water trading upstream. The water trading could also be restricted to just consumptive use and the return flow would have to stay in the system. For irrigators that would mean they could only trade about half of their water rights (approximately 50% of irrigation water delivered returns to the surface flow or groundwater). The water trading issue gets a little more complicated when it comes to environmental damages. In such cases the potential environmental damage may limit the amount of water that can be transferred for urban uses and force a city to use other higher cost sources of water, which may in some coastal cities even include desalinization.

13.5 Sustainability of Water Use

The economic toolkit – including water pricing, water markets, and privatization – can play an important role in helping sustain urban water systems faced with growing water demands. Effective water pricing and selected privatization can improve the financial sustainability of water utilities by improving cost recovery (cover costs including investment costs). Of course, for water pricing to be effective in cost recovery, users must receive a level of service from the water utility they think is worth the price being charged (Easter and Lin, 2007). This will be difficult if the city faces a continued water supply constraint. Under such conditions water markets can help. Since domestic water use tends to be one of the highest valued uses of water, markets for water can be an effective way to help increase and sustain water supplies for urban areas. For example, the purchase of water rights from farmers has been an important source of reasonably priced water for the city of La Serena in northern Chile. Similarly, Phoenix, Arizona and cities in the Phoenix metropolitan area have purchased agricultural land and its associated water rights, and leased water from Arizona tribes as cost-effective ways to meet their growing water demands (Colby and Jacobs, 2007). The water market in Arizona would be more efficient if water rights were separate from land rights, as they are in other U.S. western states, such as Colorado, New Mexico, and California.

While water pricing and water markets can help assure cost recovery and water supplies, privatization can help improve the operation and maintenance of some urban water systems. If done right, privatization will complement cost recovery efforts by providing and maintaining a level of water delivery for which water users are quite willing to pay. Effective private management can also hold down costs, which will again make cost recovery easier. Finally, private firms will have a real incentive to reduce water losses and water theft because if they do, they have a larger effective supply. The large supply means more water can be sold, again, helping raise cost recovery and possibly allowing water prices per unit to be lowered. Thus, in cases where a public utility is not effectively managing its water resources, the use of selective privatization may be one answer. Through privatizing management, incentives can be changed so that they are consistent with providing a more sustainable water service.

13.6 Water Quality and Incentives

Above, we discussed how pricing, privatization, and markets can help improve water management by changing incentives. These incentives can also play an important role in managing and improving water quality. For example, selected privatization of wastewater management in urban areas can in some cases help hold down costs and improve our wastewater treatment. Much like the design and construction of water supply facilities, the private sector has played a major role in the design and construction or waste water treatment facilities. The private sector has also played an important role in installing storm drains and wastewater collection facilities.

Cities can make greater use of prices as a way to encourage firms to discharge less wastewater into the city sewage system. The price or charge would need to be based on the quantity and quality of water a firm discharges. The quantity-based charge would be similar to the ones discussed above for the purchase of city water and should be a constant, or increasing rate, per volume discharged. The quality-based charge would be a little more complicated, but should rise as the concentration of pollutants, such as phosphorous, nitrates, etc., increase. Many of the more toxic pollutants are banned. In Minnesota, the Metropolitan Council has a "strength charge" for waste water discharge based on total suspended solids, chemical oxygen demand, and volume discharged.

Markets could also help reduce the amount of pollutants discharged into our urban water ways. The United States, for example, has through regulation reduced point source pollution discharge of pollutants such as phosphorous and nitrates. Point sources are primarily factories, towns, and cities with small towns being the major point source still dumping raw sewage in our water bodies. As many as 100 small towns in Minnesota discharge raw sewage. We have had even less success in reducing the discharge of some of these same pollutants from nonpoint sources such as agriculture and urban runoff (nitrates and phosphorous).

It is possible that we can use market mechanisms, not only to reduce these nonpoint source pollutants but also to reduce the cost of pollution control in urban areas from point sources. So far, the use of markets to reallocate the right to discharge pollutants has been quite limited, but it is likely to grow in the future. The basic idea is to establish a set number of permits for key pollutants such as phosphorous, nitrates, etc., and not allow any water discharge containing those pollutants without a permit. The first step is to decide the maximum amount of a pollutant that can be discharged into a stream or river during a given period of time. Once this is established, permits can be issued and then allocated or sold to those discharging the pollutants. The maximum number of permits issued would be set to equal the maximum discharge allowance as determined by a public pollution control agency, such as the U.S. Environmental Pollution Control Agency (EPA) at the federal level, or the Minnesota Pollution Control Agency (MPCA) at the state level. The number of permits issued would be less than the amount of pollution being discharged. Consequently, those firms not obtaining enough permits to cover the amount they discharge would have to reduce their level of pollution discharge, unless they could buy more permits.

Thus, markets come in as a means to reduce the pollution load at the lowest possible cost by allowing firms to buy and sell permits (Hall and Raffini, 2005). Once the permits are made tradable, the firms with high pollution control costs will buy permits at a price below their cost of control while those with low costs of control will sell permits. Trading could also be done between point and nonpoint sources if stricter regulations were established for nonpoint sources, such as agriculture. Currently, there are only a few such point–nonpoint source trades since nonpoint sources do not have a strong incentive to participate. So far nonpoint sources have not been required to meet set discharge limits, primarily because of the large number of nonpoint source polluters and the difficulty of measuring and attributing actual discharges, which are highly dependant on rainfall events (Fang et al., 2005).

To illustrate how the market for pollution permits would work to reduce pollution control costs, let's use the example of two firms discharging one pollutant, phosphorous, into the Minnesota River. The firm with the highest marginal cost (MC) of pollution control (emission reduction) is B while firm A has a lower MC (see Fig. 13.4). Assume each firm discharges 25 units of the pollutant into the river and that the pollution control agency wants total discharges to be only 25 units in total. This means there needs to be a reduction in discharge by 25 units. The first question is how should the 25 unit reduction be allocated between the two firms? Assume the two firms are about the same size and that for equity reasons both are required to cut discharges to 12.5 units. This will be quite costly for firm B (see Fig. 13.4). In contrast, firm A has a lower MC of reducing pollution levels and can do it more cheaply. The difference in cost with each firm emitting 12.5 units is the distance x–y in Fig. 13.4. This is where a market can help reduce the cost of pollution control. If firm A sells 2.5 pollution permits to firm B at a price above A's MC of pollution control but below firm B's MC of pollution control, both firms gain. Firm

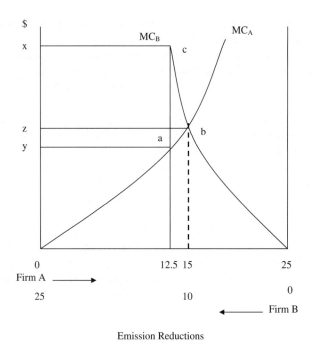

Fig. 13.4 Permit trading for pollution control

B now only has to reduce its pollution load by 10 units while firm A cuts its load by 15 units. Firm B now can discharge 15 units while firm A can only discharge 10 units of the pollutant. The only debate will be over the price of the permits. The last permit would sell at price z, where $MC_B = MC_A$ and the cost of saving from the sale of 2.5 permits is equal to the area **abc** in Fig. 13.4. These cost savings from permit trading are savings of real resources that are no longer needed to meet the water quality standard (cut emission by 25 units) because pollution levels are reduced more cost-effectively.

13.7 Summary and Conclusions

What are some of the key ideas to remember from this chapter? First, economics provides those persons managing urban water utilities with some tools that can make them more effective in serving their customers. They will also find that as water scarcity increases, they will be called upon to make good use of these economic tools in demand management including the effective use of water pricing. Yet to really reduce per capita water use, utility managers may need to combine water pricing with other demand strategies, such as subsidies for the purchase of water conserving technology.

Privatization can be an effective way to change management incentives and raise the level of services provided by water utilities. As in the past, the private sector will continue to play a major role in the design and construction of new water treatment facilities. Furthermore, the private sector is likely to play a larger role in the day-to-day management of urban water utilities, particularly in small communities that want to take advantage of the cost savings provided by private management of selected management functions. The key point being that we need to unbundle the service utilities provided and determine where the private sector can provide cost savings and/or improved service.

Although water markets are not likely to be used to allocate water within most urban communities, they can help at the next level. Water markets can be a means through which urban areas obtain water supplies from other sectors such as agriculture. In areas where agriculture is currently consuming 80–85% of a region's water, reallocating water to the urban sector may be the key to sustain urban water service. In most cases this reallocation will involve not more than 5–10% of agriculture's water. The advantage of using water markets is that they compensate agricultural water users directly and can improve the allocation of water among farmers. However, the transfer must be closely monitored to make sure that there are no negative impacts on return flows in the area selling water. If there are, those being damaged will likely have to be compensated.

Acknowledgements I would like to thank Larry Baker and Frances Homans for their very helpful comments on an earlier draft of this chapter.

References

Beecher, Janice, Richard G. Dreese and John P. Stanford. *Regulatory Implications of Water and Waste Water Utility Privatization*. Columbus, Ohio: National Regulatory Research Institute, 1995.

Brewer, Jedidiah, Robert Glennon, Alan Ker and Gary Libecap. "2006 Presidential Address, Water Markets in the West: Prices, Trading and Contractual Forms." *Economic Inquiry* 46(2)(April 2008):91–112.

Callahan, Neil V. 2000. "Once Monopolistic Water Utilities Are Becoming Competitive: Is Yours?" *Water Online* wysiwyg://211/http.www.wateronline.com

Colby, Connie C. and Katharine L. Jacobs.*Arizona Water Policy, Management Innovations in an Urbanizing Arid Region*. Washington, DC: Resources for the Future, 2007.

Dosi, Cesare and K. William Easter. "Water Scarcity: Market Failure and the Implications for Markets and Privatization." *International Journal of Public Administration* 26(3)(2003): 265–290.

Easter, K. William, Mark W. Rosegrant and Ariel Dinar, eds. *Markets for Water: Potential and Performance*. Kluwer Academic Publishers, 1998.

Easter, K. William and Yang Liu. "Who Pays for Irrigation: Cost Recovery and Water Pricing." *Water Policy* 9(2007):285–303.

Fang, Feng, P. L. Brezonik and K. William Easter. "Point-Nonpoint Source Water Quality Trading: A Case Study in the Minnesota River." *Journal of American Water Resources Association* (June 2005):645–658.

Haddad, Brent M. "Economic Incentive for Water Conservation on the Monterey Peninsula: The Market Proposal." *Journal of the American Water Resources Association* 36(1)(2000):1–15.

Hall, Darwin. "Public Choice and Water Rate Design." In *The Political Economy of Water Pricing Reforms*, ed. by Ariel Dinar. Published by Oxford University Press for World Bank, 2000, pp. 191–206.

Hall, Lynda and Eric Raffini. "Water Quality Trading: Where Do We Go from Here?"*Natural Resources and Environment* 20(summer)(2005):38–42.

Jacobs, Jeffrey W. and Charles W. Howe. "Key Issues and Experience in U.S. Water Services Privatization." *Water Resources Development* 21(1)(2005):89–98.

Renwick, Mary E. and Richard D. Green. Do Residential Water Demand Side Management Policies Measure Up? An Analysis of Eight California Water Agencies."*Journal of Environmental Economics and Management* 40(1)(2000):37–35.

Renwick, Mary E. and Sandra O. Archibald. "Demand Side Management Policies of Residential Water Use: Who Bears the Conservation Burden?" *Land Economics* 74(1998):343–359.

Siedenstat, Paul, Michael Nadal and Simon Ha Kim.*America's Water and Wastewater Industries: From Regulation to Privatization*. Dordrecht: Kluwer Academic Publishers, 2000.

Wolff, Gary and Eric Hallstein. *Beyond Privatization: Restructuring Water Systems to Improve Performance*. Pacific Institute, 2005.

Chapter 14
Principles for Managing the Urban Water Environment in the 21st Century

Lawrence A. Baker, Peter Shanahan, and Jim Holway

14.1 Introduction

14.1.1 Revisiting the Urban Water Environment

One of the key ideas of this book is the holism of the urban water environment and its importance to human well-being. As we have seen in Chapter 2, the urban water environment interconnects the built environment—the water delivery and sewage infrastructure and the impervious surfaces that alter hydrology—with the natural environment, comprising surface streams, rivers and lakes, and underground aquifers. Though conceptually simple and logical, these components are rarely considered a whole system.

We have also seen that the urban water environment includes the fabric of human society—laws and institutions. Modern cities now employ several types of water management institutions, such as regional water or sewage districts, multi-functional watershed districts, or groundwater management districts (Chapters 10, 11, and 12). Most explicitly focus on one or a few aspects of the urban water environment, and none are designed to manage all aspects of the entire urban water environment. Water markets and prices, the interplay of supply and demand, also shape how we use water and the costs and benefits of that use (Chapter 13).

Finally, water affects our psyche. As we have seen, the human relationship with water goes far beyond the physiological and sanitary functions of water; the integration of water features and water-using landscapes into the urban water environment is integral to our quality of life (Chapter 7). Managed well, water delights and sooths; managed poorly, it can be an unpleasant eye-sore.

L.A. Baker (✉)
Minnesota Water Resources Center, University of Minnesota, and WaterThink, LLC, St. Paul, Minnesota, USA
e-mail: baker127@umn.edu

14.1.2 Chapter Goal

The goal of this chapter is to synthesize what we have learned from this book. This synthesis arose from a three-day workshop convened after the core chapters had been written. We assembled to review what each of us had written, to exchange ideas, and to synthesize what we had learned. What emerged were five core principles that can help guide management of the urban water environment of post-industrial-era cities in the 21st century.

14.2 Principle 1: Influence of Urbanization

The urban environment profoundly and comprehensively affects water quantity and quality. Therefore, water managers should systematically consider all aspects of the hydrologic cycle, including the interactions between land, water and atmosphere, within both the natural and built environments.

One can lose sight of the natural environment in the city. Yet it is there, even in the most densely populated and intensively developed city centers. Basic hydrologic processes—precipitation and evapotranspiration—wield their influence on the urban setting, and they must be carefully considered in the management of the urban landscape and, particularly, the management of urban water. The complex mix of the built and natural environment that creates the city requires water managers to consider all aspects of the hydrologic cycle, including the interactions between land, water, and atmosphere, all in the context of both the natural and built environments. Such consideration requires robust tools and a systematic approach.

14.2.1 Use Water Balances to Guide Urban Water Management

We now have the technical capacity to develop hydrologic balances for cities. To consider all aspects of the hydrologic cycle, the urban water manager must necessarily operate within the construct of the water balance (Chapter 2). Certain fundamental principles emerge from this approach. First, a complete water balance for the city must include the upstream "hydroshed", which includes the local "natural" watershed as well as areas that contribute water to the city via constructed conduits (Chapter 2). For New York City, for example, the upstream hydroshed includes the Croton Reservoir watershed; for Phoenix, Las Vegas and most of Southern California, it includes the Colorado River watershed (Chapter 12).

A complete urban water balance must incorporate the hydrology of both the surface and subsurface, including ground water (Chapters 2 and 3). Not only is ground water linked to both surface waters and the human infrastructure within the city, but it also may provide all or part of the city's water supply. Further, there are many instances in which ground water interacts with and even adversely affects urban infrastructure (Chapter 3).

As explained in Chapter 2, urban infrastructure is a third major consideration for the urban water balance. While water-supply lines, sanitary sewers, and stormwater sewers and conveyances are components of the urban infrastructure with obvious implications for the water balance, other urban features are also important. These include large areas of impervious surfaces, physically altered rivers and streams, and the subsurface drainage systems at underpasses, subways, and building basements. All of these profoundly alter the movement of water through the urban environment.

Finally, the urban water balance must consider not only the upstream hydroshed and within-city watersheds but also the fate of water downstream of the city. Providing water to major cities has had implications not just for the upstream suppliers but also for downstream users and ecosystems. Urbanized watersheds generally increase the potential for downstream flooding. In desert cities, rivers may be virtually dried up as the result of water withdrawals and upstream dams. In both wet and dry climates, the water that leaves a city is often the worse for wear: It is contaminated effluent or stormwater rather than a pristine resource. Water management for the city thus necessarily entails water management both upstream and downstream of the city.

Similarly, competing water uses must also be considered. Providing water to southwestern U.S. cities like Phoenix or Denver can only be done when water is taken from other uses such as agriculture, energy or the environment; or by mining groundwater. These tradeoffs, and their social and economic effects, need to be evaluated. Water balances provide the first step in the calculation of those effects (Chapters 4 and 5). For cities in the southwest, expanding urban water supplies nearly always comes from agricultural water supply. This new urban water supply often comes from conversion of agricultural land to urban land, with transfer of water rights. In hot climates like Phoenix, the amount of water needed to support an acre of new residential development is roughly equal to the amount needed to irrigate the same acre when it was cropland. Prior agricultural land use often implies legacy contamination of surface water and particularly ground water by nitrate, salt, herbicides and pesticides. New urban land and water uses must recognize these potential compromises to the local water supply. However, urban wastewater can represent a source of water (through recycled wastewater) and soil amendments (wastewater biosolids) of value to agriculture. Nearly all "new" wastewater effluent generated in SW cities is now recycled, used for landscape irrigation, cooling towers, and other purposes. Recycled wastewater is now a significant "source" of water for urbanizing cities in the arid SW.

A carefully developed and reasonably accurate water balance is a tool for analysis of many aspects of the water environment of the city. Because it shows the total flow of water for the city, it can be used to plan for long-term water supply. On similarly long time scales, we can forecast trends in ground water levels to plan for conjunctive use of ground- and surface-water supplies and manage to prevent land subsidence and ground-water intrusion into subsurface infrastructure. If the water balance is developed with consideration of how water flows vary over time, it can be used to examine flooding of the city and downstream areas. Similarly, by looking at periods of low flow, the water balance provides the means to plan for protection of aquatic ecosystems and maintenance of both the quantity and timing or seasonality of the flow regime.

14.2.2 Manage Pollution at the Ecosystem Level

The traditional management of urban pollution has been an "end-of-pipe" approach. But, as discussed in Chapters 4 and 5, we cannot simply look at the effluent that exits the pipe from the city's wastewater treatment plant; we must also consider the redistribution and storage of pollutants within the city. The tool for this is comprehensive mass flows analysis (MFA). In Chapter 4, this type of holistic analysis for the city of Scottsdale, Arizona, led to the surprising conclusion that the city is steadily accumulating salt. This kind of mass redistribution over an extended period of time has potentially disastrous implications for the future city. Such consequences can be foreseen and averted through integrated modeling of the entire urban water system.

Holistic management of pollutants requires that mass balances be developed in conjunction with water balances. One must identify the sources of mass to the city's urban water system, how mass is transferred between components within the city system, where is it accumulated within the city system, and how it is passed from the city to the downstream and neighboring environment. Our description here is generic—we talk in terms of "mass" or "materials"— however, the approach is broadly applicable. Chapter 5 explored its application to phosphorus and lead in cities as examples of the multitude of pollutants that could be examined using this approach (see Table 5.2 in Chapter 5). A robust analysis must account for the role of all environmental media including the land and atmosphere as well as surface and subsurface water. MFA could be particularly helpful in managing nonpoint sources of pollution.

14.2.3 Practice Adaptive Management

Another tool of holistic management is adaptive management—using systems of sensors, data storage and analysis, and real-time controls to facilitate rapid assessment, managerial action, and feedback from the environment. Adaptive management has applications to many aspects of the urban water environment, including groundwater management (Chapter 3), pollution management (e.g., road salt, in Chapter 5), and stream morphology (Chapter 6).

14.2.4 Value Physical and Ecological Integrity

In addition to pollution management, urban water management must also address the physical and ecological integrity of urban rivers, lakes, and riparian areas if the city is to be provided with the desirable ecosystem services and recreational opportunities discussed in Chapters 6 and 7. We are becoming increasingly aware that impairment of ecological function of urban streams is caused as much, or more, by hydrologic alteration than by pollution. Limiting the potential adverse effects of the built environment on the natural environment is thus a multifaceted problem that requires broad integration of hydrological, biological, and chemical factors.

14.3 Principle 2: Change is Inevitable

Changes in urban water environments are inevitable but not always predictable. Water management should anticipate change and plan for it.

Hydrologic changes result from predictable variability (wet and dry cycles) and predictable trends (population growth), as well as uncertain long-term phenomena (global climate change) and the timing of "unpredictable" catastrophes (floods and droughts). Water managers need to plan for and design water systems to work under a variety of short- and long-term conditions and water demands. Building such adaptive and resilient systems first requires identifying and understanding the drivers of change and the approaches that can be taken to anticipate change.

14.3.1 Understand Drivers of Change

A first step in effective water management is being able to distinguish between three different types of changes: (1) normal variability around an average, which can be predicted with statistical techniques, (2) random events such as catastrophes, which can be anticipated and prepared for but not accurately predicted, and (3) truly uncertain events, such as global climate change, which we are just beginning to understand. Within this context, the drivers of change can be organized into five categories: biophysical, built environment, governance, behavior, and economics. Multiple and at times complex feedback mechanisms also operate within and between each of the five categories of drivers. For example, a change in climate could impact regional hydrology. The response to this change by institutions and individuals could reduce or exacerbate these impacts. Because change is ubiquitous in urban water management, understanding the driving forces of change and the relationships between these forces will be increasingly important for managers.

Biophysical factors (Chapters 2, 3, and 6) include natural resources and both hydrologic and climatic processes. Water systems have been designed to accommodate historic climatic variability. Now with global climate change, both variability as well as the long term averages may be changing. Additionally, how cities manage water resources affects the biotic environment around cities, which in turn significantly impacts water resources and the ecosystem services provided by water. The resiliency of these resources will determine how they respond to changes. Cities that are located on the outer edge of climatic or hydrologic tolerances during normal conditions, such as deserts (Chapters 4 and 12) or environments with shallow groundwater systems (Chapter 3), will likely be less resilient to change than cities located in more moderate environments.

The built environment and other land-use factors (Chapters 4, 6, 7, and 8) include everything from buildings to water management infrastructure (canals and dams), to the technology we use for water management, to our overall urban design (land use and transportation systems). Understanding the potential impacts of accidents, infrastructure failure, and shifts in climate, as well as how to respond to such challenges, is critical for water management.

Governance and institutional factors (Chapters 9, 10, 11, and 12) include our laws, policies, social customs and values. Laws and institutions affect how we respond to change, either locking us into inflexibility or providing great resilience. For example, the development of regional water management institutions (Chapters 11 and 12) can greatly improve the ability of an urban region to respond to changes. On the other hand, adhering rigidly to a water rights system based on prior appropriations (Chapter 10) could reduce resilience to long-term drought.

Human behavior (Chapters 7, 8, 12, and 13) and individual choices form the foundation for all water management activities. Our individual lifestyles and consumer choices determine the per capita demand for water. What we demand from our governments and other institutions in terms of water quality, recreational and amenity uses for water, and protection of ecosystems constitute much of the remaining water demand and the constraints on water supplies. Public support, principally through our elected officials, shapes public policy and public investments for water management. Experience with water conservation education efforts, as well as research into changing perceptions and behavior on environmental issues, provides significant information on how human behavior may respond to change.

Economic factors include competition between different water users (e.g., homeowners, farmers, and electric utilities) as well as broader economic forces (Chapter 13). Water prices alter water consumption and supply; however in many cases, water is an allocated commodity for which there is not an open market with prices determined by supply and demand. The development of water markets and additional reliance on economic signals and incentives, though controversial, could be an approach for designing adaptable water management. Finally, the ability of local, state or perhaps even the federal governments to invest in water infrastructure can, in itself or in combination with other drivers, drive change in the urban water environment.

14.3.2 Anticipate and Manage Change

The effectiveness of long-term water management depends on being able to anticipate and manage change. Collecting data, planning for the long range, managing adaptively, building redundancy and resilience into urban water systems, and building transparency are all key steps for anticipating and managing future changes that are explored in Chapter 12 and summarized below.

Collect and track data: Good data are essential to knowing what types of change are occurring, identifying areas of concern, and strategically targeting resources.

Data collection is essential to provide feedback for adaptive management (see below).

Plan for the long term: Long-range planning should include predictive scenarios that include both well-known and uncertain changes that can cause threats to the urban water environment. By identifying a range of both best and worst case scenarios, water planners can understand the potential impact of alternative future trends, identify the events to which they are most vulnerable, and prepare for responding to such events.

Incorporate adaptive management: Adaptive management should combine the knowledge gained from data tracking and scenario planning, as well as evaluating how current water management programs are working. This information can be utilized to develop program modifications before changes become critical. Building such an ongoing feedback and adjustment process for modifying water management efforts, based on data from hypothesis-driven monitoring programs under changing conditions, can be a key step in effectively building the capacity to handle future changes.

Design for resiliency: Urban infrastructure can be designed to be resilient to hydrologic change. Resilient practices include building a diverse water supply portfolio with backup supplies, building redundant infrastructure (such as cross-connections between adjoining municipal water systems), and relying on diverse management approaches to respond to crisis.

Build transparency: Transparency means making information and knowledge readily available to all interested parties. This information might include transcripts of public meetings, maps of environmental characteristics (such as groundwater tables), interactive models, etc. Transparency will be increasingly critical for anticipating and managing change. Political decision makers and water managers need to recognize the importance of transparency and design it into their water systems and institutions. For example, readily accessible data on groundwater systems would allow politicians, agencies and citizens to visualize changes that are occurring, allowing them to respond before crisis occurs. As explained in Chapter 3, "out of sight, out of mind" does not create a resilient system, particularly in the face of unanticipated change.

Some institutions have organized this knowledge in ways that can be used to communicate results in readily understandable formats. An example is Arizona State University's Decision Theater and their WaterSim project, where stake holders can readily visualize long-term changes in water supply under various climate and growth scenarios (Fig. 14.1).

14.4 Principle 3: Water and People

People affect water and water affects people. Water managers should recognize and engage the multiple parties and interests that affect and are affected by water systems.

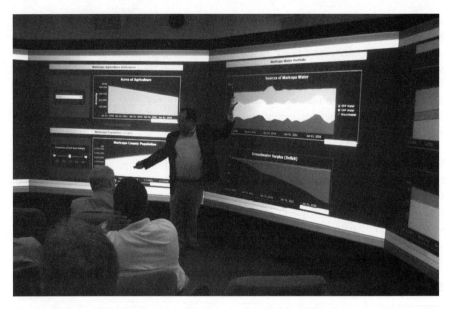

Fig. 14.1 Photo of ASU's Decision Theater. An interactive online version can be found at http://watersim.asu.edu/

A key theme of this book is the idea of the urban water environment as a core element of cities. Central to this idea is the interaction of people with their water environment, which provides not only physiological sustenance, but also opportunities for recreation, aesthetics and stress reduction (Chapter 8), broader urban resilience (Chapter 12) and even enhanced property value (Chapter 13). Achieving an optimal urban water environment therefore requires intensive interaction of water managers with citizens and organizations.

14.4.1 Engage the Public Broadly

A wide variety of people get involved in urban water issues. The list of course includes water managers, but it also includes federal, state, and municipal agencies, real-estate developers, planners, environmental activists, politicians, homeowners, and citizens. The water system is shaped by these multiple actors making independent and sometimes conflicting choices and decisions. The water manager must engage these parties and referee the public process to create effective policies.

At the base level, individuals make decisions that affect water. They help to decide the relative importance of competing decisions such as how much water is consumed, how landscape is managed, how erosion is controlled, how much impervious surface is created, and how many tax dollars are provided for water management. However, water managers are not powerless in the face of individual decisions—they have tools to influence the public's choices. For example, water managers can structure water rates to encourage efficient water use (Chapter 13) and sponsor public information programs to inform citizens of choices that are environmentally sound (Chapter 5). Thus, while individuals can make choices, managers can influence those choices. The water manager who understands the underlying individual behaviors that affect water decisions can develop "soft" policies to help the public make informed choices and ultimately achieve wise and efficient public investment and public policy.

14.4.2 Develop a Public Vision

A desire for wise and efficient policy decisions and reasonably free choice for individuals implies that effective water management is something of a balancing act. Water managers can provide direction, but ultimately individuals and communities will make their own choices. Thus, effective water management must necessarily engage individuals and communities in determining a vision for water management and the implementation of that vision in the context of comprehensive, long-term regional planning (Chapter 8). This is no mean task: Not only must public participation be sustained over long periods, but also the public needs to be educated in the arcane concepts and language of water management. Engineers and scientists must be able to communicate notions of water and mass balances with planners and stakeholders from the public at large in ways that are readily accessible and understood.

14.4.3 Visualize Change

Communicating among various stakeholders requires simple visualization tools. Simulation tools, such as the WaterSim models used in Chapter 4, enable scenario management tools for experimenting with alternative management scenarios and visualizing their results. However, more must be done to develop simulation tools that provide quantitative results to planners and the public as a foundation for planning and management. The various simulation-based computer games on the market for many years demonstrate that such tools can be made accessible to the public. Simulation tools could be implemented on the web, in decision theatres, or through other venues. Moreover, some means of interactive response from stakeholders is also essential. This could take traditional forms such as public meetings and focus groups, or new approaches such as computer gaming or web-based feedback. Regardless of the form, a shift to planning that uses analytical tools such as water and mass balances also requires a corresponding shift to more sophisticated methods to bring the resulting information to the public planning process (Chapter 8).

14.4.4 Never Forget Sanitation

Urban dwellers in the United States enjoy safe, potable water and efficient sewage disposal. Improved water and sanitation have been enormously successful in reducing mortality and increasing lifespan since its development in the late nineteenth century. Chapter 1 points to this as perhaps the most important contribution to improved human health since modern urbanization. Where water once affected urban dwellers very negatively by providing a medium for the spread of disease, modern water infrastructure now has a very positive effect, protecting the populace from sources of disease. Modern wastewater collection and treatment, however, have also transferred some pollutants from the immediate environs of the city to the peripheral environment of the city's waterways. Wastewater thus becomes a problem for downstream users, who must deal with any unaddressed residual health and environmental problems associated with treated wastewater. The potential for disease transmission via water has not disappeared. A failure in water treatment can produce disastrous outcomes, as occurred in Milwaukee in 1993, when 143,000 people became ill from *Crytosporidium*. Untreated sewage from combined and sanitary sewer overflows continues to contaminate urban waterways. We have not yet developed a mature approach for dealing with newly identified micropollutants. Finally, the use of septic systems in peri-urban areas remains a problem. These frequently fail, contributing high loads of bacteria, organic matter and nutrients to surface waters and groundwater. The failure to install sewers as population density increases in urbanizing areas can lead to serious contamination problems.

14.4.5 Value Aesthetic and Ecological Functions of Water

While protection of human health is probably the most fundamental service provided by water systems, the water environment also can provide a sense of place, recreation, wildlife, biodiversity, ecosystem, and aesthetic values. It is hard to imagine many of our cities without their defining water features: Chicago's Lake Michigan waterfront, Seattle's Pier District, or Boston's Charles River. The high value of riparian or shoreline reflects the value that society gives to these aesthetic and natural features. Similarly, as discussed in Chapter 7, most of our residents visit or recreate in waterside areas and greatly appreciate those opportunities. Water also provides important urban wildlife habitat, as long as some semblance of natural conditions area maintained (Chapter 6).

While good water quality supports all of these water-dependant values, polluted water has a deleterious effect on the urban aesthetic sense, water recreation and landscape, and even drinking water. Chapter 3 discusses the potential for urban pollution to contaminate the drinking-water aquifers that supply some cities and Chapter 5 gives a technical context to the city-wide materials fluxes that can give rise to both surface- and ground-water pollution. Cities often depend upon lakes and reservoirs for urban recreation, water landscapes, and even water supply, but those resources can be severely impaired by the nutrient fluxes created by urban and suburban living.

Urban rivers and coastlines can similarly suffer under poor environmental management. Chapters 9 and 10 explain how laws and institutions can provide protection of the water environment from pollution, and thereby also protect secondary uses of water that provide places to recreate, ecosystem services, and a sense of place.

14.5 Principle 4: Water Management Institutions

Multiple institutions affect water management. Urban water should be managed through institutions that are responsive to the hydrologic setting; capable of working across political, social, and functional boundaries; and effective at engaging all stakeholders.

Multiple entities, as described throughout the preceding chapters, shape water management. They are a diverse group of public and private interests, ranging from individuals to institutions, from farmers to major industrial users, from small communities to multi-state agencies. They include local and regional water providers, local and regional planning agencies, regulatory agencies at all levels of government, and public interest groups. Water users engage in various ways on water-related concerns. The challenge for water management institutions is to engage all of these various stakeholders and to improve the inclusiveness, efficiency and effectiveness with which we set and achieve water management objectives. Current institutions and policies too often focus on only one or two specific water management functions. For example, it is common for one public utility to manage municipal water supply and an entirely different utility (perhaps operating at a different scale) to manage wastewater (Chapter 11). Additionally, institutions operate within natural and physical (Chapters 2–6), social (Chapters 7 and 8), legal (Chapter 9), political (Chapters 10, 11, and 12), and economic (Chapter 13) contexts. These contexts create both opportunities and constraints for institutional efforts to reduce barriers across jurisdictions within a region, between different levels and functions of government, and across interrelated water management functions. As we pointed out in Principle 3 above, people are affected by water; yet in many cases, institutions mediate this relationship.

14.5.1 Develop Effective Water Institutions

We recommend the following criteria for developing effective and responsive institutions:

Address water issues at the scale of watershed and groundwater basins (see Principle 1): This will require coordination among overlapping jurisdictions. Water management is fragmented geographically across federal, tribal, state, regional, and local governmental units and private utilities with different missions, authorities and jurisdictions. A key role of regional institutions is to provide resource protection and facilitate coordination for issues that span such boundaries (Chapters 10, 11, and 12).

Manage all aspects of water in urban environments (see Principle 1): Water management is fragmented by function—water quantity versus water quality versus land use versus stormwater/flood control versus economic development—and by different rules for groundwater, surface water and effluent. Effective institutions need to recognize and avoid the types of unintended consequences that can occur when dealing with just one functional aspect of a diverse yet integrated resource such as water.

Maintain a high level of communication across agencies and disciplines: Communication is critical to overcome both the political and functional fragmentation identified above. A key part of communication is creating transparency, ensuring that all urban water management programs and institutions can acquire and understand information needed to make informed decisions.

Focus on who makes key decisions and how to impact those decisions (see also Principle 3): Key decisions in water management are made by water users and those actually building infrastructure. Most agencies work by altering the climate of public perception within which these decisions are made, through education, incentives and regulations.

Design flexible approaches for effective and efficient management: Institutions need to creatively and comprehensively integrate education, incentive programs, and regulations. Each of these approaches affects the decision-making environment in different ways, and each approach has varying strengths and weaknesses, including political acceptability. An often difficult challenge is dealing with the perception of a conflict between private property rights and projects designed to benefit the public health, safety and quality of life (Chapters 11 and 12).

Utilize sound science and adaptive management approaches: These approaches increase the flexibility, effectiveness, and legitimacy of both regulatory and non-regulatory policies (see also Principle 2). Water management institutions should take advantage of technical and managerial innovations, such as real-time monitoring, visualization of data, web-based data tools, and partnerships with universities and other research institutions to make adaptive approaches rigorous and viable.

Engage all stakeholders, including the general public, early and often: Non-governmental organizations (e.g., conservation and environmental groups) and industry associations, as well as the general public, play an increasingly important role in water policy and management. Engaging stakeholders at the very beginning and ensuring completely open and transparent public processes will help to ensure public support for difficult decisions and build the relationships that can be critical for responding to future challenges.

14.5.2 Develop the Right Type of Water Management Institutions

Institutions based on hydrologic boundaries, such as watersheds or groundwater basins, can be more effective than traditional units of government in managing regional water resources. Watershed-based organizations could, in some cases, replace previous institutions. However, new activities are also commonly designed to complement the preexisting jurisdictional based entities. Chapter 11 identifies four different models for watershed management efforts:

1. *Top-Down Agency-Led Watershed Management*: In this approach, state or federal agencies establish watershed based efforts with little local input. Notable examples would include Arizona's Active Management Areas (Chapter 12), Michigan and North Carolina.
2. *Agency-Coordinated Watershed Management*: These efforts are initiated by state agencies, but the focus is on coordinating local parties who are typically expected to pick up the leadership and perhaps the funding of programs. Massachusetts and New Jersey are examples of this approach.
3. *Watershed Management Authority*: These are independent entities that typically have taxing or regulatory authority and appointed or elected boards. Minnesota and Florida are notable examples.
4. *Locally Led Community-Based Entities*: Watershed councils and partnerships based on this model are often non-governmental organizations or informal coalitions. Such organizations exist throughout the United States.

Most states and regions employ a mix of these different models, or in some cases may even develop hybrids. Some key considerations (Chapter 11) that might lead to the formation of an effective organization include: (1) What scale of institution is needed? (2) What scope of activities is needed? (3) What level of technical sophistication is required? (4) Is a permanent taxing authority needed?

14.6 Principle 5: Interdisciplinary Framework

The urban water environment is a product of hydrology, ecology, history, land use, design, infrastructure, society, law, and the economy. Effective water management should incorporate multiple perspectives and varied expertise in an interdisciplinary framework.

As we have seen throughout this book, piecemeal management of the urban water environment has often resulted in very serious problems and, sometimes, crises. A few examples from previous chapters include the spread of cholera and typhoid following installation of sewers (Chapter 1), salt contamination from ecosystem accumulation (Chapter 4) and road salt (Chapter 5), continued lead pollution of inner cities (Chapter 5), rapid groundwater depletion, leading to subsidence, followed by rising groundwater, causing structural problems (Chapter 3), degradation of groundwater quality (Chapter 3), and increased downstream flooding and loss of ecological health caused by destruction of stream channels during urbanization (Chapter 6). These examples point toward a profound need for holism—a consideration of the entire urban water environment—to avoid these types of calamities.

14.6.1 Connect the Dots

Fortunately, we are at a point in history when we have the technological and scientific capacity to craft ecosystem-level management strategies to manage urban systems in a more sustainable fashion. What remains to be done is "connecting the dots" between parts of the ecosystem, a step that requires integration of expertise across a wide variety of disciplines. We now understand, for example, that groundwater and surface water are really one resource that should be managed conjunctively (Chapters 2 and 12), the pollutants should be managed at their source (Chapters 4 and 5), and that we can predict, within limits, the consequences of urban development on the ecological health of urban streams (Chapter 6). Understanding these connections can help us avoid unintended consequences, largely by working with natural processes. This idea of "designing with nature" is not new, but it can now be implemented more effectively than when McCarg (1969) first brought the idea into the mainstream of urban design as the result of improved science and vastly improved technology tools, from remote sensing to data storage.

14.6.2 Think Across Disciplines

This holistic vision of an urban water environment requires experts from many disciplines, working together in new ways. Traditionally, urban development was largely the purview of planners and engineers working independently of each other. This book shows that experts from many other disciplines need to be intimately involved—ecologists, hydrologists, geologists, environmental psychologists, economists, lawyers and others. To accomplish this will require breaking down the barriers that discourage multidisciplinary teamwork. In academia, these barriers include rigid disciplinary focus on tenure evaluation, lack of rewards for interdisciplinary or applied research, a reluctance to engage in "messy" politics, obsolete budget models that discourage sharing funds across departments and colleges, project funding periods that are too short to allow interdisciplinary research to flourish, and compartmentalization of research funds along traditional disciplinary lines. Fortunately, some of these barriers are being torn down. Similar problems occur throughout the "real" world. Bureaucracies are often compartmentalized along disciplinary boundaries, with narrowly prescribed mandates; communication across governmental agencies is often discouraged, or at least, not rewarded; and the funding model is often not responsive to changes in needs. An additional problem with governmental bureaucracies is that innovation is often discouraged.

Institutional changes can also promote multidisciplinary thinking. Examples include new types of planning processes (Chapter 8), the creation of bureaucracies based on watershed boundaries (Chapter 11) or groundwater regions (Chapter 12), and the development of water markets (Chapter 13). Some technical approaches also promote greater holism, for example, the development of regional water or pollutant balances (Chapters 3, 4 and 5). Finally, transparency tools can enable broad communication among agencies and stakeholders.

14.7 In Closing

The five core principles put forth in this chapter seek to guide management of the urban water environment of post-industrial-era cities in the 21st century. Each principle identifies an intrinsic aspect of the urban water system, and responds with a recommendation:

1. Urban environments profoundly affect water—thus, managers must be systematic in considering urban hydrology.
2. The urban water environment will inevitably change—thus managers must plan and prepare for change.
3. People are an integral part of the urban water environment—thus managers must engage the public in decisions about water.
4. Multiple parties play a role in deciding urban water systems—thus, managers must create effective decision-making and management institutions.
5. The urban water system is a complex product of nature, engineering, and society—thus, managers must incorporate all of these perspectives.

The onus of these recommendations falls upon you, our target audience, those who will create and manage the cities and water systems of the future.

Glossary

Acre-foot   A quantity of water that would cover 1 acre 1 foot deep; roughly 325,000 gallons.

Adaptive management   Management characterized by hypothesis-driven data collection, information feedback, information processing, and subsequent modification of management practices to improve desired outcomes, in an iterative fashion.

Alluvial channels   Stream channels that have been carved by the water into deposits of previously deposited sediment under the present hydrologic regime; stand in contrast with **non-alluvial channels**.

Aquifer   Saturated soils and rock materials that are sufficiently permeable to transmit readily useful quantities of water to wells, springs, or streams under ordinary hydrologic conditions.

Aquifer, confined.   An aquifer overlain by a confining bed which has a significantly lower hydraulic conductivity than does the aquifer.

Aquifer, unconfined   An aquifer that is not overlain by a confining bed; an unconfined aquifer has a water table.

Base flow   Water flowing in a stream channel during non-storm periods. Base flow includes m groundwater discharge to the stream plus piped flows (e.g., effluent from wastewater treatment plants or septic systems.

Best management practice   Applied to nonpoint source pollution, an acceptable practice, such as a physical structure (i.e., a detention pond), a system of inspections, a change in land use or land use practices, pollution prevention, or an educational program, used to prevent pollution from entering waterways.

Biological integrity/health   This term generally refers to the biological condition of an aquatic ecosystem relative to its "natural" biological condition, usually done using region-specific scale such as an Index of Biological Integrity (IBI).

Biosolids The sludge which accumulates in sewage treatment plants as the result of sedimentation. The sludge is mostly biological material (microorganisms), hence the term biosolids. Biosolids can be further "digested" to produce methane, applied to agricultural fields, or disposed in landfills.

Ccf One hundred cubic feet (about 750 gallons) of water; a quantity of water used mainly by American municipal systems to measure water deliveries.

Combined sewers Sewers designed to collect both urban storm water and sanitary wastes. These are no longer built, but are common in older cities.

Combined sewer overflow (CSO) Sewage from a combined sewer diverted from a wastewater treatment plant and discharged, untreated, into a river or other water body during precipitation events.

Common law Law based on a series of decisions by judges in individual cases, over the period of many years, even centuries.

Conjunctive use Combined and coordinated use of both surface water and groundwater for water supply.

Consumptive use Water withdrawn from groundwater or surface water that is evaporated, transpired by plants, incorporated into products or crops, consumed by humans or livestock, or otherwise used in a way that it is not immediately returned to the surface or groundwater environment.

Cooperative federalism A system of shared responsibility among levels of government, often in which the federal government oversees implementation of federal laws by state agencies.

Cost-effectiveness A systemic method for finding the lowest cost means of accomplishing an objective. It takes the objective as a "given" and finds the least cost means of achieving the desired outcome.

Dead zone In water quality, a portion of a river or estuary with little or no dissolved oxygen. Dead zones contain microbial life, but few if any fish. Oxygen depletion can be caused by a high external input of organic material, such as sewage, or by decaying algae.

Decreasing block rate A pricing scheme that charges less per unit of water ($/ccf) as water use increases.

Demand-side management Water management that seeks to modify water demand (generally to encourage conservation), through some combination of pricing (e.g., **increasing block rates**), use restrictions, education, and economic incentives.

Depression storage Water stored on the land surface in depressions.

Detention basin A constructed facility designed to capture stormwater and reduce the peak rate of stormwater flow released to a surface water body.

Glossary

Direct recharge Recharge of surface water or effluent accomplished through engineered recharge basins.

Disturbance In ecology, large, sudden perturbations, such as floods, fires, or human-caused catastrophes, such as chemical spills or wars.

Eco-mimicry Urban design practices intended to mimic functional aspects of the natural (non-urbanized) environment. Particularly stream flow regime.

Elasticity For water pricing, the change in quantity used in relation to price. *Inelastic demand* means that water use does not change much with price; *elastic demand* means that water use changes a lot in relation to price.

Eutrophication As generally used, extensive nutrient enrichment of a lake or estuary. Eutrophic waters are characterized by high algae abundance, often dominated by blue-green algae, and low concentrations of dissolved oxygen near the bottom. Eutrophication can be reversed by reducing nutrient inputs.

Evapotranspiration Evaporation (transformation of liquid water to water vapor) plus transpiration (release of water vapor by plants to the atmosphere).

Feedback In **adaptive management**, information obtained from a system being managed (for example, an **aquifer**) that informs future management actions.

Federal reserved rights Water rights associated with federal lands reserved for special purposes, such as Indian reservations and national parks.

Fishable and swimmable This term is a shorthand version of a specified legal goal of the 1972 Clean Water Act to attain water quality sufficient to support fish and aquatic life as well as recreation in and on the water.

Floodplain The land adjacent to a stream channel that is episodically inundated. Floodplains can be a tangible feature on the landscape, but they are often described by planners and in regulations based on the statistical recurrence interval of floods. For example, "the 10-year floodplain" is the area with a 10% chance of being flooded in any given year, whether or not there is an expressed landform corresponding to this area.

Flux In **MFA**, the movement of material across the boundaries of an ecosystem or ecosystem compartment, in terms of mass/time. (e.g., kg/yr).

Functional diversity In the context of water management, relying upon different types of water supplies and demand reduction measures to achieve resilience.

Gpcd Gallons per capita per day.

Groundwater Water under the ground within the saturated zone.

Heuristic A systematic approach, generally a series of steps, for using information to guide management or solve a problem.

Hydrograph A graph of flow rate versus time, commonly used for hydrologic description or analysis.

Hydrologic (water) cycle The natural cycle driven by the sun of the movement of water from the oceans to water vapor, to condensation to form precipitation, from precipitation to runoff and infiltration.

Hydraulic conductivity In an aquifer, the ease with which water flows. It is the amount of flow through unit cross-sectional area of an **aquifer** under unit hydraulic gradient in units of length per time.

Hydraulic gradient Change in potentiometric head (Δh) over distance (ΔL).

Hydraulic residence time (HRT) The length of time water stays in a lake or watershed, calculated as the outflow (volume/time) divided by the volume. HRT values are usually expressed in days or years.

Hydroshed The land area that is tributary to the city through either natural topography or created infrastructure.

Increasing block rate A pricing scheme that charges more per unit of water ($/ccf) as water use increases.

In-lieu recharge Recharge credit given to a farmer for using an alternative water supply. (e.g., treated effluent) rather than groundwater.

Integrated Watershed Management (IWM). An approach to encompass and coordinate all of a watershed's potential uses, services, and values in management decisions and regulatory activities, rather than managing various water resources (e.g., water supply; wastewater; aquatic ecosystems) individually.

Interception storage Precipitation captured by vegetation during a storm.

Low impact development (LID) Development intended to reduce hydrological impacts compared to normal land development practices and, ideally, to mimic natural hydrologic conditions.

Managed recharge Recharge of surface water accomplished by diverting flows into existing, often dry, and stream channels.

Marginal cost The change in total cost resulting from a one unit change in the quantity of output.

Mass Balance A mathematical accounting of the volume of water or constituents in water through some system that includes forms of water or consituents that enter, accumulate, or exit the system. The "system" may be of any size or scale, ranging from a reactor in a water treatment plant to an entire city or region.

Material Flow Analysis (MFA) Analysis of the movement of pollutants into an ecosystem and through ecosystem compartments.

Glossary

Maximum contaminant level (MCL) In the federal Safe Drinking Water Act, the maximum concentration of contaminant allowed in drinking water.

Miasmatic theory The prevalent theory of disease transmission in the early 19th century. Miasmatic theory asserted that disease was spread by "bad air" associated with filth. Miasmatic theory was supplanted by the germ theory of disease in the late 1800s.

Modularity The use of interconnections among parts of a system to increase resilience.

National Pollutant Discharge Elimination System (NPDES) Established under the 1972 Clean Water Act, the NPDES is a system of permits that stipulate effluent concentration limits, volume of discharge, and other factors for discharge of effluent to surface waters. NDPDES permits are required for any discharge of a pollutant from a point source to the navigable waters, including discharges from municipal wastewater treatment plants, industries, confined animal feeding operations, combined sewer overflows, sanitary system overflows, and urban stormwater.

Navigable waters The category of surface waters identified by the U.S. Supreme Court as subject to federal authority under the Commerce Clause of the U.S. Constitution, including all waters that have been used for any form of navigation, or which can be made suitable for navigation with improvements, and all tributaries or headwaters thereto. As used in the Clean Water Act, navigable waters define and limit water bodies subject to discharge permitting requirements, and include the territorial seas as well as wetlands that are hydrologically connected to or that otherwise have a significant nexus to navigable waters.

Non-alluvial channels Stream channels that are unable to adjust their boundaries by erosion or sedimentatoin. In urban areas, these include constructed conduits.

Nonconsumptive uses Water uses that do not "consume" water by evapotranspiration, incorporation into products, etc.; water returned to surface water or groundwater (by infiltration) after it is used.

Nonpoint source pollution Pollution from diffuse sources, such as lawns, streets, and cropland, in contrast to point sources of pollution, such as industrial or municipal sewage, which are discharged from a pipe or other discrete conveyance ("point source").

Opportunity cost of water The value of the inputs needed to make new water supplies available plus the values produced by water in its current (alternative) use.

Participatory research Research based on co-equal collaboration between community members and technical experts in all or most aspects of the research.

Permeability The capacity of a rock, soil or sediment to transmit fluid.

Place dependence The potential for a place to satisfy the needs of an individual and how that place compares in satisfaction of needs compared to another place.

Place identity A subculture of the self-identity of the person consisting of, broadly conceived, cognitions about the physical world in which the individual lives.

Point source A pipe, ditch, channel, or other discernible, confined, and discrete conveyance from which pollutants are discharged into navigable waters, subjecting the discharge to Clean Water Act permitting requirements. The statute excludes agricultural stormwater discharges and agricultural return flows.

Pore space The empty space that exists between soil and rock particles.

Porosity A measure of the extent of pore space, equal to the ratio of the volume of pore space in soil or rock to the total volume.

Potential A measure of the energy of water at a particular point in a groundwater system.

Potentiometric head A measure of potential in terms of the height of a column of water above a reference plane (usually mean sea level) that can be supported by the hydraulic pressure at a given point in a groundwater system.

Potentiometric surface A surface that represents the level to which water will rise in a tightly cased well, equal to the water table in an unconfined aquifer.

Pretreatment The practice of treating industrial or commercial on-site wastewaters, before they are discharged to sanitary sewers, to prevent discharge of toxic chemicals, oil and other contaminants that could impair the operation of municipal wastewater treatment, threaten the health and safety of treatment system operators, or result in discharge of toxic effluent or sludge.

Prior appropriations doctrine Legal doctrine used primarily in western states in which water rights are based on diversion of water for beneficial uses, and in which priority in times of shortage are based on the date of diversion ("first in time, first in right"). For contrast, *riparian doctrine*.

Publicly owned treatment work (POTW) A legal designation for a sewage treatment plant owned by a public entity.

Recharge Water infiltrating the soil that reaches the groundwater table.

Resilience The measure of an ecosystem's vulnerability to disturbance before it becomes unstable.

Response diversity In the context of water management, using supply sources that will behave differently under future conditions.

Reverse osmosis A process by which water is forced, under pressure, to flow through a semi-permeable membrane. Salt is removed from the permeate (product water) and concentrated in a brine.

Riparian zone Pertaining to the area situated on the bank of a natural body of flowing water.

Riparian rights doctrine Common law doctrine used mainly in eastern states in which water rights are owned by riparian land owners, who can withdraw water as long as they cause no harm downstream, and in which competing claims are evaluated based on the reasonableness of use among different claimants. For contrast, *prior appropriations doctrine*.

Runoff (1) That part of the precipitation, snow melt, or irrigation water that appears in uncontrolled surface streams, rivers, drains or sewers. Runoff may be classified according to speed of appearance after rainfall or melting snow as "direct runoff" or "base runoff," and according to source as "surface runoff", "storm interflow", or "base flow". (2) The total discharge described in (1), above, during a specified period of time, for example, in m^3/yr. (3) The depth to which a drainage area would be covered if all of the runoff for a given period of time were uniformly distributed over it, for example, in cm/yr.

Safe yield The maximum amount of groundwater that can be withdrawn from an aquifer without long-term declines in its water level because the withdrawal rate is equal to or lower than the rate of natural groundwater recharge.

Sanitary sewer Sewers designed solely for collection of sanitary wastes from households and commercial establishments.

Saturated zone The portion of the subsurface in which the pore space is entirely filled by water.

Sewershed The area drained by wastewater and stormwater pipes. A sewershed would often be different from the topographically-defined, natural watershed.

Specific discharge The amount of water flowing per unit time through a unit cross-sectional area of the porous medium in units of length per time.

Stakeholder A person, non-profit group, or industry which effects or can be affected by governmental action, such as regulation or planning.

Streamflow Water flowing in a surface channel, measured in units of volume/time.

Statutory law Law passed by federal, state and local legislative bodies. For contrast, see **common law**.

Stream reach A continuous part of a stream between two specified points.

Storm sewer Sewers designed solely for collection of urban runoff, in contrast to sanitary sewers and combined sewers.

Sustainable development Development that meets the needs of the present without compromising the ability of future generations to meet their own needs. This definition was developed by the Bruntland Commission (UN 1987). Many other definitions of sustainability are also in common usage.

Sustainable tourism development Tourism that "meets the needs of present tourists and host regions while protecting and enhancing opportunities for the future ... management of all resources in such a way that economic, social and aesthetic needs can be fulfilled while maintaining cultural integrity, essential ecological processes, biological diversity, and life support systems" (World Tourism Organization).

Technology-based standard A discharge standard that specifies the degree of pollutant reduction required from a NPDES-regulated waste stream for particular kinds of facilities, as determined by a designated "best" model technology.

Threshold In the context of ecosystem resilience, a "tipping point" that once crossed, leads to irreversible, or nearly irreversible, conditions.

Total Maximum Daily Load (TMDL) The maximum amount (load) of a pollutant that can be added to a water body on a daily basis without violating water quality standards. The TMDL concept is a part of the 1972 Clean Water Act.

Unit treatment process A single treatment process in a water or wastewater treatment plant, such as sedimentation, filtration, chlorination, or biodegradation.

Unsaturated zone The portion of the subsurface in which the pore space is occupied by both air and water, that is, the zone between the land surface and the water table.

Wastewater treatment plant Engineered facility that purifies domestic and industrial wastewater before discharging the treated water (effluent) to streams, rivers, lakes, oceans or groundwater. Wastewater treatment plants can be designed to remove oxygen-consuming organic matter, pathogens, toxic materials, and nutrients, depending on the need. Nearly all municipal wastewater treatment plants are designed to remove pathogens and oxygen-consuming organic materials, although many now also remove nitrogen or phosphorus.

Water budget A mass balance of components of the **water cycle**.

Water footprint For a particular region, the area of land required to provide enough water to supply one person for one year.

Water quality standards Legally-enforceable standards dictating the designated uses of particular water bodies and the ambient (instream) water quality criteria necessary to protect those uses. Water quality criteria can be stated as narrative requirements, numeric limits (minimum or maximum permissible levels of various chemicals or other measures), limits on the degree of permissible degradation from current conditions, or limits on toxicity or on changes in biological conditions.

Water quality-based treatment standards Discharge requirements in NPDES permits stricter than those specified by technology-based treatment standards, where necessary to meet ambient water quality standards.

Water rights The legal regime dictating how water resources can be used, by whom, and under what conditions. For surface waters, the most common forms of water law in the United States are the **prior appropriations doctrine** and the **riparian rights doctrine**. There are many forms of water law for groundwater. Water rights are determined by individual states.

Watershed The land area drained by a stream or lake, also called a drainage basin.

Watershed Approach (Watershed Management) Management characterized by planning and decision making on a watershed scale. This may involve integration of a variety of competing water resource priorities and goals, creation of watershed-based institutions, cooperation of multiple stakeholders and government agencies, and increased levels of public participation.

Water table The top of the saturated zone or equivalently, the elevation at which the hydraulic pressure in the aquifer is equal to atmospheric pressure.

Water treatment plant Engineered facility that takes river, lake or groundwater and purifies it into drinking (potable) water. The facility removes particles, including bacteria, protozoa and viruses, by sedimentation and filtration before adding chemical disinfectants (e.g., chlorine) to inactivate (i.e., kill) any of these pathogenic microorganisms that passed through the filters or which may enter the distribution system (pipe network) that delivers water to consumers. Other natural pollutants (e.g., arsenic, algae by-products) or human-derived pollutants (e.g., nitrate, pesticides) are removed through specialized treatment processes as necessary to comply with federal Clean Water Act requirements.

Index

A

Abiotic, 93
Accumulation, 3, 35, 60, 62, 64, 66, 74, 75, 85, 86, 88, 287
Active Management Areas (AMA), 240, 243, 244, 248
Adaptation, 151, 161, 162, 165, 166
Adaptive management, 14, 46, 70, 81–82, 86–87, 88, 233, 251, 252, 278, 280, 281, 286
Adsorption, 76
Agriculture, 2, 50, 52, 53, 73, 103
Allocation, 172, 175, 203, 204, 213, 236, 237, 242, 244, 245, 259–273
Alluvial, 94, 95, 98, 104, 112
Ammonium, 71, 76
Anthropogenic, 52, 53
Aqueduct, 10, 54
Aquifer, 2, 10, 11, 13, 18, 20, 30–45, 53, 54, 62, 63, 86, 88, 101, 177, 185, 189, 205, 246, 247, 248, 253, 254, 255, 256, 275, 284
Army Corps of Engineers (ACE), 184, 198–199, 207, 209, 211
Arsenic, 38, 44, 53, 76, 177
Assimilation, 76
Attenuation, 115

B

Bacteria, 53, 55, 57, 58, 71, 73, 141, 143, 181, 187, 284
Base flow, 17, 22, 24, 86, 96, 101
Benefits, 14, 49, 67, 84, 88, 100, 102, 107, 110, 118, 125, 126, 127, 128, 129, 137, 143, 147, 148, 149, 151, 160–163, 166–168, 177, 186, 191, 197, 200, 219, 235, 251, 275, 286
Best management practice, 11, 23, 72, 115, 249
Biological health, 105, 106, 110

Biological oxygen demand (BOD), 70, 71, 76, 81
Biosolids, 57, 70, 74, 75, 181, 212, 277
Block rate, 250, 260, 261, 262
Blood lead, 84
Boston, 1, 43, 45, 54, 143, 145, 201, 284
Boundary, 61, 74, 77, 79, 109, 112, 191, 214, 222
Buffer, 59, 113, 182, 184
Bureaucracies, 288
Bureau of Indian Affairs, 203
Bureau of Reclamation, 10, 202, 209

C

Census, 75, 78, 79, 130, 131, 151, 156
Cesspool, 5, 11, 37, 38
Chlorination, 7
Cholera, 5, 6, 11, 37, 38, 287
Clean Water Act (CWA), 7, 12, 72, 141, 165, 168, 177, 179, 180, 181, 182–189, 197, 198, 204, 211–213, 217, 222, 233
Climate, 3, 4, 14, 26, 41, 44, 46, 68, 94, 96, 106, 115, 117, 142, 143, 165, 166, 168, 203, 235, 236, 237, 242, 250, 252, 260, 262, 277, 279, 281, 286
Climate change, 3, 14, 26, 41, 46, 68, 94, 116, 117, 142, 143, 165, 235, 237, 242, 252, 257, 279
Coal, 4, 8, 9, 51, 152, 225, 287
Coastal Zone Management Act (CZMA), 165
Collaborative, 129, 185, 195, 213, 222, 229, 231
Combined sewer, 59, 72, 150, 156, 183, 184, 210
Combined sewer overflow, 59, 72, 184, 210
Common law, 44, 171, 172, 176, 179, 182, 185, 187, 190, 191
Competition, 137, 172, 204, 266, 267, 280

301

Index

Comprehensive Environmental Response, Compensation, and Liability Act (CERCL), 173, 184, 185, 188, 190, 197, 204
Confined, 30, 31, 35, 38, 53, 145
Conflict, 129, 130, 135, 136, 137, 175, 176, 178, 187, 188, 189, 190, 199, 205, 208, 219, 232, 282, 286
Conjunctive use, 46, 277
Conservation, 3, 29, 67, 73, 81, 88, 111, 113, 115, 134, 135, 145, 147, 149, 157, 161, 162, 173, 175, 178, 185, 197, 200, 209, 226, 238, 239, 242, 244, 245, 249, 250, 251, 253, 254, 255, 262, 263, 264, 280, 286
Consumptive use, 51, 57, 259, 269
Contamination, 6, 40, 42, 44, 45, 67, 73, 86, 161, 182, 183, 187, 277, 284, 287
Control volume, 17, 18, 19
Cooperative federalism, 179, 197, 211
Copper, 69, 76, 80, 142
Cost-benefit, 88
Council on Environmental Quality (CEQ), 202, 209
Cuyahoga River, 7, 129

D

Dam, 8, 9, 10, 14, 52, 102, 177, 178, 188, 200, 202, 240, 277, 280
Darcy's law, 32, 33, 34
Decay, 8, 58, 76
Decomposition, 53, 69, 76
Decreasing block rate, 260, 261
Degradation, 42, 93, 99, 105, 109, 111, 119, 133, 134, 165, 207, 252, 287
De-icing, 69, 86, 87, 142
Demand, 3, 4, 14, 35, 38, 49, 52, 53, 54, 58, 63, 67, 68, 70, 71, 73, 80, 129, 148, 155, 172, 174, 175, 176, 178, 188, 206, 208, 217, 218, 236, 238, 239, 240, 242, 245, 246, 247, 248, 249, 250, 251, 254, 255, 259, 260–265, 267, 269, 275, 279, 280
Demand management, 235, 245, 249, 251, 259–273
Denudation, 97
Denver, 9, 51, 52, 175, 236, 238, 245, 277
Depression storage, 17
Designated use, 72, 180
Detention basin, 23, 25, 86
Dewater, 38, 39, 45, 58, 178, 181, 253
Direct runoff, 96
Discharge, 1, 6, 7, 8, 17, 22, 24, 32, 33, 34, 35, 36, 38, 42, 46, 50, 52, 57, 58, 59, 62, 71, 72, 73, 95–97, 99, 101, 102, 103, 104, 107, 108, 109, 115, 116, 150, 151, 157, 161, 176, 179–184, 187, 189, 197, 198, 204, 205, 211, 212, 217, 227, 270–272
Distribution system, 22, 38, 49, 56, 198, 254
Diversity, 106, 131, 132, 145, 147, 205, 221, 254
Downcutting, 97, 103
Drinking water, 3, 7, 35, 42, 50, 51, 53, 55, 56, 61, 65, 67, 134, 177, 180, 188, 197, 204, 210, 217, 284
Drivers of change, 279
Drought, 14, 39, 41, 53, 54, 60, 68, 116, 174, 231, 235, 236, 237, 246, 247, 248, 249, 251, 253, 254, 255, 264, 269, 279, 280

E

Effective impervious area, 101
Efficiency, 53, 69, 86, 100, 160, 177, 178, 188, 206, 208, 246, 248, 249, 254, 255, 269, 285
Endangered Species Act (ESA), 141, 165, 168, 177, 188, 199, 233
Environmental Protection Agency (EPA), 35, 72, 113, 141, 142, 150, 177, 180–185, 196–198, 202–205, 209–212, 224, 271
Equitable apportionment, 175
Erosion, 24, 69, 73, 76, 81, 83, 100, 103, 107, 108, 109, 110, 113, 117, 182, 200, 219, 221, 227, 228, 232, 282
Evapotranspiration, 18, 19, 20, 25, 26, 29, 63, 115, 238, 276

F

Federal Emergency Management Agency (FEMA), 200
Federal Insecticide, Fungicide, and Rodenticide Act (FIFRA), 197
Feedback, 70, 81, 82, 87, 88, 252, 254, 255, 278, 279, 281, 283
Fertilizer, 5, 35, 53, 58, 65, 74, 78, 81, 82, 83, 156, 182
Fishable, 2, 161
Fish and Wildlife Service (FWS), 199, 201, 209
Flooding, 3, 10, 11, 12, 13, 14, 24, 44, 58, 59, 100, 107, 108, 115, 117, 143, 148, 155, 161, 163, 182, 183, 200, 207, 218, 219, 221, 233, 240, 277, 287
Floodplain, 98, 113, 114, 117, 118, 184, 188, 200, 255
Flow path, 17, 22, 96
Flux, 25, 49, 50, 62, 63, 64, 65, 66, 74, 75, 79, 80, 85, 106, 116, 284
Footprint, 9, 103, 114, 117, 157

Index 303

Fragmentation, 117, 171, 187, 189, 195, 205, 208, 210, 211, 212, 213, 286
Frame, 146–149, 151, 153, 163, 166, 252

G
Geographic, 41, 51, 52, 77, 117, 129, 144, 146, 149, 151, 155, 156, 165, 172, 189, 191, 208, 213, 214, 217, 219, 220, 229, 232, 242, 285
Geographic information system (GIS), 77, 165
Geomorphic, 93, 98, 101, 102, 106, 107, 110, 112, 116, 118
Glyphosphate, 76
Groundwater, 2, 10–15, 17–20, 22, 24, 25, 26, 29–46, 52–55, 60–68, 69, 73–76, 77, 86, 96, 100, 101, 105, 118, 151, 152, 154, 158, 159, 175, 176, 185, 186, 188, 189, 208, 212, 213, 236, 237, 240, 242–244, 246–250, 253–255, 267–269, 275, 277–279, 281, 284–288
Groundwater Management Act, 242, 244

H
Head, 8, 31, 32, 33, 35
Heritage, 130, 132–137
Heuristic, 141, 146, 152, 153, 154
Hispanic, 84, 130, 131
Holistic, 3, 12, 27, 152, 187, 189, 207, 213, 278, 288
Horton, 96, 100
Household, 5, 6, 46, 51, 57, 58, 62, 65, 67, 70, 77, 79, 80, 81, 238, 260, 261, 262, 264, 268
Hydraulic conductivity, 32, 33, 34, 36, 43, 44
Hydraulic gradient, 32, 33, 34
Hydraulic residence time, 52
Hydrograph, 23, 96, 97, 101, 109
Hydrologic model, 26, 87, 101, 102, 233
Hydrology, 2, 10, 12, 14, 29, 32, 36, 42, 44, 101, 105, 112, 145, 153, 162, 182, 203, 207, 275, 276, 279, 287, 289
Hydropower, 8, 9, 259
Hydroshed, 276, 277

I
Impervious, 2, 3, 10, 11, 21, 22, 24, 26, 36, 78, 86, 99–104, 106–108, 112, 147, 182, 200, 207, 275, 277, 282
Incentive, 148, 168, 174, 178, 200, 206, 217, 222, 242, 247, 249, 250, 251, 255, 256, 260, 261, 262, 263, 270, 271, 273, 280, 286
Increasing block rate, 250, 260, 261, 262
Inelastic, 250, 263
Infiltration, 11, 13, 24–26, 37, 42, 59, 72, 74, 96, 100, 109, 114, 117, 156, 157, 158, 159

Inflow, 17, 18, 61, 65, 71, 105, 159
Information systems, 46, 77, 165
Infrastructure, 2, 6, 12, 19, 26, 36, 37, 38, 39, 41, 42, 44, 45, 49–67, 72, 102, 104, 108, 113, 116, 117, 118, 141–146, 149–151, 153, 157, 162, 165–168, 179, 180, 182, 196, 198, 200, 206, 207, 236, 247, 249, 250, 252–255, 257, 259, 260, 265–267, 275–277, 280–281, 284, 287
Institutions, 3, 4, 11, 13, 14, 15, 111, 119, 178, 195–214, 217–234, 235, 236, 239, 241, 246, 247, 251–254, 257, 259, 260, 275, 279, 280, 281, 285–289
Interception, 17
Intermittent, 63
Irrigation, 9, 24, 25, 26, 35, 37, 44, 51, 57, 58, 60–67, 174, 175, 199, 200, 206, 207, 209, 238, 239, 243, 249, 250, 251, 260, 264, 267, 268, 269, 277

J
Joint powers, 222, 228, 229
Jurisdiction, 45, 107, 109, 111, 136, 147, 149, 150, 156, 161, 164, 167, 168, 175, 182, 187, 191, 195, 204, 205, 213, 219, 231, 241, 242, 244, 254, 256, 285, 287

L
Land use, 11, 35, 78, 85, 93, 96, 98, 104, 105, 110, 113, 116, 141, 147, 155, 160, 177, 179, 182–184, 188, 189, 199, 205–209, 212, 213, 221, 230, 244, 277, 280, 286, 287
Las Vegas, 4, 9, 54, 175, 236–238, 241, 245, 276
Lawn, 19, 69, 70, 76, 78, 82–88, 102, 141, 142, 156, 159, 178, 183, 260–264
Leach, 73, 82, 83, 85, 86
Lead, 11, 13, 19, 34, 35, 38, 40, 53, 54, 57, 58, 59, 60, 70, 72, 73, 76, 84–88, 100, 119, 136, 143, 144, 152, 155, 164, 166, 167, 178, 196, 205, 206, 224, 231, 232, 233, 236, 252, 253, 254, 255, 266, 278, 284, 287
Legacy, 69, 70, 73, 106, 149, 277
Lifestyle, 235, 238, 241, 251, 280
Load, 65, 81, 97, 99, 103, 112, 150, 260, 262, 271, 272
Low Impact Development (LID), 11, 25, 113–115, 143

M
Management, 3, 4, 8, 11, 13, 14, 15, 21, 23–27, 29, 45, 46, 60, 62, 67, 69–72, 81–88, 93, 100, 102, 106, 110–119, 126–133, 136, 137, 141, 151, 156, 161, 165, 171,

180, 184–187, 189, 195–197, 200, 201, 203, 205–210, 213, 217–234, 235–257, 259–273, 275–288
Market, 10, 80, 81, 128, 131, 132, 137, 167, 173, 182, 242, 248, 251, 259–273, 275, 280, 283, 288
Mass transfer, 75
Material flow analysis (MFA), 69, 74, 77, 79, 80, 82, 84, 85, 88, 278
Maximum contaminant levels (MCL), 177
Megacities, 41
Metals, 12, 53, 55, 56, 58, 71, 73, 76, 99, 141, 142, 151, 156, 157, 181, 182
Methemoglobinemia, 53
Metropolitan Water Reclamation District of Greater Chicago, 210
Miasmatic, 5, 7
Mobility, 76
Model, 26, 60, 61, 62, 63, 65, 66, 67, 74, 75, 76, 103, 113, 127, 136, 223, 228, 229, 267, 287, 288
Modeling, 25, 26, 50, 60–68, 84, 87, 108, 111, 112, 160, 205, 278
Modularity, 254
Monitor, 45, 158, 227, 255
Monitoring, 45, 47, 82, 87, 111, 119, 128, 142, 166, 167, 201, 205, 222, 224, 225, 226, 233, 244, 255, 281, 286
Mortality, 6, 7, 284
Multidisciplinary, 113, 118, 288

N

National Environmental Policy Act (NEPA), 176, 188, 202
National Pollution Discharge Elimination System (NPDES), 72, 180, 181, 183, 197, 198, 204, 225, 233
Natural Resources Conservation Service (NRCS), 200, 209, 226
Navigable waters, 179, 180, 181, 183, 191, 198
New York City, 9, 52, 54, 162, 174, 276
Nitrate, 53, 66, 71, 73, 75, 76, 88, 270, 271, 277
Nitrogen, 5, 35, 53, 57, 58, 71, 74, 75, 76, 134, 151
Non-alluvial, 95
Non-consumptive use, 51
Nonpoint source pollution, 23, 81, 182, 183, 212, 213

O

Outflow, 17, 18, 19, 20, 52, 61, 62, 64, 65
Output, 37, 61, 74, 75, 126, 127

Overdraft, 2, 35, 40, 236, 242, 247, 248, 253, 255, 268
Overland flow, 96, 100, 102, 114, 115

P

Participatory, 14, 131
Pathogen, 7, 55, 141, 143, 156, 157, 204
Peakshaving, 109
Percolation, 63, 66
Perennial, 96, 157
Permeability, 22, 32, 66, 101
Pervious, 2, 101, 102
Pharmaceutical, 46, 58
Phoenix, 4, 9, 51, 54, 63, 68, 74, 75, 80, 175, 235–240, 243, 262, 269, 276, 277
Phosphate, 73, 76
Phosphorus, 71–74
Place dependence, 129, 130
Place identity, 3, 129, 130
Place meaning, 125–138
Point source, 23, 69, 70–72, 81, 104, 142, 157, 179, 182, 183, 197, 211, 212, 213, 226, 270, 271, 278
Policy, 3, 10, 13, 67, 70, 80–83, 88, 112, 125, 128, 130, 143, 146, 147, 152, 171, 176, 178, 197, 198, 202, 208, 210–214, 235, 241, 252, 255, 256, 264, 280, 283, 286
Pollution, 3, 7, 8, 11, 13, 23, 37, 38, 41, 42, 46, 53, 58, 69–88, 104, 111, 115, 118, 128, 129, 134, 141, 142, 148, 156, 167, 179, 180, 182–190, 196–198
Population, 3, 4, 8, 12, 37, 41, 42, 46, 52, 60, 68, 72, 75, 78, 79, 84, 88, 99, 102, 104, 105, 111, 130, 131, 134, 137, 151, 152, 190, 217, 235, 236, 238, 240, 241, 250, 251, 252, 253, 268, 279, 284
Porosity, 33, 34
Post-consumption, 73
Potable, 19, 25, 26, 50, 51, 53, 54, 57, 58, 60, 61, 63, 207, 238, 284
Potentiometric, 31, 32, 33, 35
PowerSim, 62, 65, 68
Precipitation, 2, 9, 11, 17–20, 25, 26, 29, 36, 37, 60, 61, 63, 64, 71, 76, 84, 87, 96, 97, 112, 115, 117, 133, 182, 183, 188, 205, 236, 237, 276
Pre-consumption, 73
Pricing, 15, 68, 80, 178, 188, 250, 260, 261, 262, 263, 264, 269, 270, 272
Prior appropriation, 172, 174, 176, 188, 203, 204, 242, 250, 280
Private property, 184, 190, 191, 205, 207, 286
Privatization, 15, 49, 259–273

Index

Publicly Owned Treatment Works (POTW), 209, 212
Public trust, 190, 191
Pumping, 32, 35, 38–42, 49, 53, 54, 57, 61, 64, 65, 67, 176, 243, 247, 253, 268

R

Rainfall, 9, 13, 41, 59, 61–64, 67, 96, 101, 109, 112, 142, 157, 236, 238, 262, 271
Reasonable use, 173, 176
Recharge, 13, 17, 22, 24, 25, 26, 29, 35–37, 41, 54, 61, 62–66, 100, 101, 105, 244, 246–248, 252–255, 268
Reclamation, 10, 80, 202, 209, 210
Recreation, 2, 13, 14, 53, 58, 60, 93, 99, 119, 125–138, 151, 166, 168, 172, 180, 186, 196, 198, 201, 204, 207, 219, 221, 226, 259, 278, 280, 282, 284
Recycle, 63, 82, 85, 262, 277
Regulation, 23, 35, 53, 67, 72, 81, 104, 136, 137, 147, 150, 151, 155, 168, 171, 172, 173, 176–190, 196, 197, 198, 202, 211, 230–233, 247, 250, 252, 255, 256, 270, 271, 286
Renewal, 8, 13, 99, 115, 126, 186
Reservoir, 2, 9, 10, 18, 29, 49, 50, 52, 53, 55, 56, 173, 206, 210, 238, 246, 247, 253, 254, 276, 284
Resilience, 14, 39, 45, 54, 67, 112, 232, 235, 241, 251, 256, 263, 280, 282
Resource, 1, 3, 5, 12, 26, 29, 34, 35, 41, 44, 45, 46, 49, 67, 73, 99, 101, 105, 113, 125–138, 151, 152, 161, 162, 165, 167, 171–176, 178, 179, 182, 184–186, 189, 190, 191, 195, 197, 199, 200–204, 208–213, 217–257, 259, 268, 270, 275, 277, 279, 280, 284–288
Resource Conservation and Recovery Act (RCRA), 185, 188, 197, 204
Responsible parties, 185, 186
Restoration, 14, 40, 106, 107, 112, 113, 116–119, 133, 149, 162, 186, 204, 221, 225, 227
Return period, 22, 23
Reuse, 26, 46, 58, 61, 62, 181, 182, 185, 240, 247, 249
Reverse osmosis, 58, 59, 62, 65, 71
Riparian, 2, 22, 24, 81, 86, 93, 103, 105, 107, 110, 111, 113, 116, 117, 118, 172–174, 176, 182, 184, 186, 188, 191, 200, 203, 207, 242, 278, 284
Riparian rights, 172–174, 176, 188
Runoff, 2, 8, 9, 17–25, 42, 53, 61, 62–66, 69, 70, 73, 82–84, 88, 94, 96, 101–109, 114–116, 141–143, 148, 150, 151, 155, 157, 158, 160, 163–167, 172, 175, 182, 183, 205, 212, 219, 232, 238, 247, 271

S

Safe Drinking Water Act (SDWA), 177, 185, 188, 197, 204, 205
Safe yield, 54, 244
Salt, 3, 12, 33, 35, 39, 41, 49, 51, 53, 55, 57–67, 69, 70, 72, 73, 79, 80, 86, 87, 142, 161, 171, 175, 195, 210, 236, 238, 240, 246, 247, 254, 255, 277, 278, 287
Salt Lake City, 175, 210
Sanitary sewer, 5, 6, 42, 43, 57, 59, 71, 72, 77, 150, 183, 277
Sanitation, 5, 7, 11, 284
Scale, 26, 44, 45, 49, 60, 64, 66, 77, 79, 94, 105, 111, 115–118, 130, 144, 148, 150, 156, 160, 166, 191, 218, 219, 220, 221, 232, 240, 268, 277, 285, 287
Scenario, 25, 26, 60, 66, 106, 252, 280, 281, 283
Scottsdale, 51, 60, 61, 63, 65, 67, 68, 278
Sediment, 12, 18, 53, 85, 86, 93, 94, 95, 97, 98, 101, 103, 104, 106, 107, 109, 110, 112, 116, 156, 157, 159, 160, 161, 219, 227, 232
Sedimentation, 7, 55, 56, 57, 70, 71, 74, 76
Septic system, 19, 22, 24, 37, 284
Sewage, 2, 5, 6, 7, 12, 14, 42, 49, 50, 51, 57–59, 65, 69, 70–77, 81, 104, 133, 143, 150, 153, 157, 159, 172, 180–184, 187, 188, 198, 204, 205, 210, 212, 217, 259, 270, 275, 284
Sewer, 3, 5, 6, 26, 36, 38, 43, 44, 49, 59, 72, 141, 150, 154, 156, 181, 183, 184, 206, 207, 210, 212, 218, 222
Sewershed, 21, 78
Site-based, 148, 149, 155
Snow, John, 5, 37, 43
Social marketing, 81
Soft policies, 81, 283
Standards, 5, 7, 63, 67, 71, 81, 86, 93, 104, 108, 109, 134, 142, 143, 147, 148, 149, 151, 155, 166, 168, 177, 178, 179, 180, 181, 185, 187, 188, 197, 204, 211–213, 221, 227, 232, 242, 245, 247, 249, 250, 253, 254, 256, 266, 267, 272
Statutory law, 187
Storage, 10, 17, 18, 19, 20, 37, 42, 44, 54, 56, 61, 62, 100, 112, 114, 115, 154, 156, 157, 173, 188, 198, 199, 200, 206, 207, 240, 246, 247, 248, 249, 253, 254, 257, 278, 288
Storm sewer, 2, 3, 5, 6, 10, 22, 26, 38, 72, 77, 86, 87, 100, 107, 183, 198, 218, 221, 222

Storm water, 72, 86, 171, 172, 183, 184, 196, 204, 205, 207
Stream, 13, 22, 24, 35, 71, 86, 93–118, 149, 150, 151, 157, 158, 159, 160, 165, 172, 173, 174, 180, 182, 183, 201, 204, 238, 253, 271, 278, 287
Stream channel, 93, 96, 97, 100, 102, 107, 108, 110
Streamflow, 18, 19, 20, 22, 26, 62, 96, 97, 117, 254
Subsidence, 34, 38, 44, 54, 253, 255, 277, 287
Subsidy, 2, 10, 34, 38, 39, 44, 54, 72, 73, 80, 81, 178, 202, 206, 253, 255, 260, 264, 272, 277, 287
Subsurface, 7, 17, 18, 22, 30, 32, 33, 34, 35, 36, 40, 41, 42, 43, 44, 45, 54, 63, 96, 100, 112, 115, 276, 277, 278
Superfund, 185, 186, 197
Supply-demand, 80
Suspended solids, 71, 270
Sustainable, 14, 25, 26, 34, 37, 42, 66, 67, 73, 87, 111, 117, 118, 132, 133, 137, 143, 148, 171, 178, 186, 187, 189, 190, 195, 208, 233, 249, 250, 251, 253, 256, 257, 270, 288
System, 1, 3, 6, 14, 17, 19, 21, 22, 25, 26, 35, 36, 38, 43–46, 49, 50, 54–59, 62, 65, 66, 67, 72–75, 78–82, 98, 102, 112, 114, 115, 135, 136, 142–161, 165, 167, 172, 177, 179, 180, 181, 183, 197, 199, 201, 203–212, 218, 220–222, 232, 238, 242, 244, 246, 249, 251, 252–254, 260, 262–263, 266, 269, 270, 278, 282, 288

T
Target, 81, 84, 155, 156, 224, 256, 289
Targeting, 280
Technology-based standard, 180
Theory of Planned Behavior, 80–81
Threshold, 109, 112, 118, 147, 252, 253, 255
Top-down, 81, 223, 228, 287
Total dissolved solids (TDS), 58, 65
Total impervious area, 101, 147
Tourism, 126, 130, 132, 133, 137, 161, 168
Toxic, 42, 46, 58, 66, 71, 72, 142, 181, 182, 197, 212, 270
Toxic Substances Control Act (TSCA), 197
Toxins, 53, 58
Transdisciplinary, 14, 88
Transparency, 280, 281, 286, 288
Transportation, 1, 8, 10, 125, 132, 142, 143, 152, 155, 162, 163, 165, 168, 185, 280
Twin Cities, 15, 73, 79, 80, 134, 135, 222, 228

Two-price method, 260, 262
Typhoid, 6, 7, 11, 187, 287

U
Unconfined, 30
Unit treatment, 55
Urbanization, 2, 3, 4, 6, 9–11, 17, 21, 22, 24, 26, 93–119, 142, 182, 188, 189, 201, 257, 284, 287
Urban penalty, 6
U.S. Geological Survey (USGS), 19, 20, 36, 45, 50, 201, 220, 237
Usufructory, 174

W
Wastewater, 1, 3, 7, 8, 11, 13, 19–22, 26, 35, 37, 38, 40, 41, 46, 49, 50, 51, 54, 57–67, 70, 75, 80, 153, 171, 172, 179, 180, 196, 200, 205, 207, 209, 210, 227, 232, 240, 265, 270, 277, 278, 284–285
Wastewater treatment, 1, 3, 8, 26, 38, 50, 51, 54, 57, 58, 59, 62, 66, 70, 71, 72, 74, 75, 153, 180, 200, 207, 209, 232, 270, 278
Water table, 22, 24, 29, 31, 34, 35, 38, 39, 40, 42, 43, 44, 62, 113, 153, 154, 158, 159, 281
Water budget, 14, 17–27
Water closet, 5, 11
Water cycle, 17, 21, 22, 25, 26, 172, 210
Water demand, 4, 38, 52, 54, 63, 67, 175, 178, 236, 238, 239, 242, 246, 247, 249, 250, 251, 255, 259, 260, 262, 268, 269, 279, 280
Waterfront, 1, 8, 9, 133, 152, 284
Water right, 10, 172–176, 188, 189, 203, 204, 206, 221, 222, 226, 242, 244, 245, 247, 248, 249, 250, 251, 257, 268, 269, 270, 277, 280
Water scarcity, 13, 264, 269, 272
Watershed, 2, 9, 10, 13, 14, 15, 18, 19, 20, 22, 23, 26, 53, 73, 77, 78, 79, 84, 86, 93–119, 142, 146, 147, 149, 152, 155, 156, 160, 173, 174, 185, 188, 196, 201, 210, 213, 214, 218–234, 256, 275, 276, 277, 285, 286, 288
Watershed district, 15, 228, 229, 233, 275
WaterSim, 281, 282, 283
Water supply, 1, 2, 4, 6, 9, 11, 12, 13, 14, 19, 24, 25, 26, 32, 33, 34, 36–39, 43, 46, 50, 52, 53, 62, 65, 67, 68, 86, 119, 133, 161, 171–178, 187, 189, 195, 196, 200, 202, 205–210, 213, 214, 217, 218, 235, 237, 239, 240, 241, 242, 244–246, 248, 254, 268, 269, 270, 276, 277, 281, 284, 285

Water treatment, 1, 7, 8, 13, 16, 26, 38, 50, 51, 53, 54, 55, 56, 57, 58, 59, 60, 62, 66, 70, 71, 72, 74, 75, 153, 171, 177, 180, 188, 196, 200, 207, 209, 210, 232, 265, 267, 270, 273, 278, 284
Water use regime, 25
Wetland, 2, 10, 11, 72, 76, 114, 145, 154, 156, 159, 160, 161, 162, 182, 184, 200, 210, 211, 226, 232

Y

Yield, 3, 20, 34, 54, 81, 97, 103, 104, 111, 116, 244

Z

Zinc, 69, 76, 142, 156
Zoning, 81, 117, 127, 184, 186, 188, 189, 205, 206, 207, 208, 209, 221

Printed in the United States of America